Sustainable Language Support Practices in Science Education:

Technologies and Solutions

Felicia Zhang
University of Canberra, Australia

Brett Andrew Lidbury
Australian National University, Australia

Alice Marion Richardson
University of Canberra, Australia

Brian Francis Yates
University of Tasmania, Australia

Michael Guy Gardiner
University of Tasmania, Australia

Adam James Bridgeman
University of Sydney, Australia

Jurgen Schulte
University of Technology, Sydney, Australia

John Cameron Rodger
University of Newcastle, Australia

Karen Elizabeth Mate
University of Newcastle, Australia

Information Science
REFERENCE

Senior Editorial Director:	Kristin Klinger
Director of Book Publications:	Julia Mosemann
Editorial Director:	Lindsay Johnston
Acquisitions Editor:	Erika Carter
Development Editor:	Hannah Abelbeck
Production Editor:	Sean Woznicki
Typesetters:	Mike Brehm, Christen Croley
Print Coordinator:	Jamie Snavely
Cover Design:	Nick Newcomer

Published in the United States of America by
Information Science Reference (an imprint of IGI Global)
701 E. Chocolate Avenue
Hershey PA 17033
Tel: 717-533-8845
Fax: 717-533-8661
E-mail: cust@igi-global.com
Web site: http://www.igi-global.com

Library of Congress Cataloging-in-Publication Data

Sustainable language support practices in science education: technologies and solutions / Felicia Zhang, editor.
p. cm.
Includes bibliographical references and index.
Summary: "This book presents cases on the results of a study done in Australia on first-year university students and the impact of new techniques of language acquisition on science education"--Provided by publisher.
ISBN 978-1-61350-062-0 (hardcover) -- ISBN 978-1-61350-063-7 (ebook) -- ISBN 978-1-61350-064-4 (print & perpetual access) 1. Science--Study and teaching
(Higher) 2. Science--Terminology. 3. Language and education. 4. Science--Study and teaching (Higher)--Australia--Case studies. I. Zhang, Felicia.
Q181.S86 2011
507.1--dc23
2011024238

British Cataloguing in Publication Data
A Cataloguing in Publication record for this book is available from the British Library.

Editorial Advisory Board

Table of Contents

Foreword x

Preface xii

Acknowledgment xix

Chapter 1
Language Problems of First Year Science Students 1
Felicia Zhang, University of Canberra, Australia

Chapter 2
Language Support Strategies in First Year Chemistry 18
Adam J. Bridgeman, University of Sydney, Australia

Chapter 3
First Year Chemistry Language Support at the University of Tasmania 58
Brian F. Yates, University of Tasmania, Australia
Michael G. Gardiner, University of Tasmania, Australia

Chapter 4
Language Focus for Genetics and Molecular Biology Students 98
Brett A. Lidbury, Australian National University, Australia

Chapter 5
Evaluating a Genetics Concept Inventory 116
Felicia Zhang, University of Canberra, Australia
Brett A. Lidbury, Australian National University, Australia

Chapter 6
Language Support for First Year Human Physiology and Biology 129
Karen Mate, University of Newcastle, Australia
John Rodger, University of Newcastle, Australia
Brett Lidbury, Australian National University, Australia

Chapter 7
The Benefits of Teaching Students the Language of Physics 160
University of Technology, Sydney, Australia

Chapter 8
Clickers in a First Statistics Course.......... 195
Alice Richardson, University of Canberra, Australia

Chapter 9
A Successful Collaboration Between Language Arts and Science.......... 226
Felicia Zhang, University of Canberra, Australia

Compilation of References 230

About the Contributors 240

Index 243

Detailed Table of Contents

Foreword x

Preface xii

Acknowledgment xix

Chapter 1
Language Problems of First Year Science Students 1
Felicia Zhang, University of Canberra, Australia

The chapter discusses the background of a project which aimed at addressing the language needs of a diverse student body (both domestic and international student body) by embedding strategic approaches to learning and teaching in first year sciences in tertiary education. These strategies consisted of active learning skills which are widely used in language learning. The disciplines covered by the project were Biology, Chemistry, and Physics and involved the University of Canberra, University of Sydney, University of Tasmania, University of Technology, Sydney, and University of Newcastle in Australia. This project was funded by the Australian Learning and Teaching Council (ALTC).

Chapter 2
Language Support Strategies in First Year Chemistry 18
Adam J. Bridgeman, University of Sydney, Australia

In-class and online active learning strategies which have been implemented at The University of Sydney are described to response to the challenge of teaching chemistry at a first year level, to students with a wide range of abilities and levels of motivations. It was found that online resources are useful for helping students become familiar with chemical language and symbolism and to provide them with a means of practicing their use. Online quizzes are an invaluable means of students self-assessing their progress and of providing meaningful assessment of their level of mastery. In-class activities involving student response systems and student-centred, inquiry based approaches are built around active learning and on-going formative assessment.

Chapter 3
First Year Chemistry Language Support at the University of Tasmania..58
Brian F. Yates, University of Tasmania, Australia
Michael G. Gardiner, University of Tasmania, Australia

A number of in-class and post-lecture learning strategies are described that have recently been incorporated into the first year chemistry program at the University of Tasmania, Hobart. These methodologies are centered on supporting the various language based challenges that students face in achieving a firm introduction to the discipline. Activities included mobile phone-based technologies for in-class audience surveys, word matching exercises, and chemical songs. The post-lecture approaches included (i) online based quizzes (ii) developing a broader range of weekly assignment questions to focus on language issues; (iii) chemical Sudoku puzzles, and (iv) crossword activities. Evaluations of these approaches are presented.

Chapter 4
Language Focus for Genetics and Molecular Biology Students..98
Brett A. Lidbury, Australian National University, Australia

This chapter examines the role of scientific language comprehension and confidence for senior undergraduate students in Genetics and Molecular Biology, and the impact of language-centred learning strategies to assessment outcomes. A number of online and tutorial language exercises and strategies are described that were designed to promote scientific language competence and subsequent genetics learning. The effect of these interventions was analyzed through grade and assessment performance comparisons with earlier traditionally taught Genetics cohorts. Language was found to be most influential for middle range performing students. Despite at least a year of tertiary education, students still had difficulty interpreting some everyday words in a science context.

Chapter 5
Evaluating a Genetics Concept Inventory...116
Felicia Zhang, University of Canberra, Australia
Brett A. Lidbury, Australian National University, Australia

This chapter examines the reliability and validity of a subset of the Genetics concept inventory (S. Elrod, 2007) to discriminate good students from poor performing students in undergraduate units called Genetics and Molecular Biology at the University of Canberra, Australia. These two units went through a series of reforms since 2005. These reforms included the implementation of a number of online and tutorial language exercises and strategies designed to promote scientific language competence and subsequent genetics learning. The effect of these interventions was analyzed through grade and assessment performance comparisons with earlier traditionally taught Genetics cohorts as well using the genetic concept inventory. The genetic concept inventory questions used at the University of Canberra have been found to be reliable and valid according to a number of statistical tests.

Chapter 6
Language Support for First Year Human Physiology and Biology..129
Karen Mate, University of Newcastle, Australia
John Rodger, University of Newcastle, Australia
Brett Lidbury, Australian National University, Australia

This chapter examines student performance in Human Physiology and Biology in response to a number of interactive language-focused and career-oriented interventions that were integrated into both lecture and tutorial formats. The study took place at the University of Newcastle, Australia and involved around 1000 students on two campuses. The major impact of the language and other interventions was in the level of student motivation and engagement with the course topic.

Chapter 7
The Benefits of Teaching Students the Language of Physics 160
University of Technology, Sydney, Australia

First year physics courses for non-physics majors are among the most difficult course to cope with for science and engineering students alike. Not only are students confronted with physics specific concepts and mathematics applied in unfamiliar environment, students also have to learn the specific lingo of physics and keep it separate from common language and language used in their own field of major. The author investigated the link between language of physics used in our first year physics lectures, understanding of related physics concept, and student performance after a language focused intervention. The author found that a language conscious approach in first year physics not only improved the performance of students of non-English background but also benefited the performance of students of native English speaking background.

Chapter 8
Clickers in a First Statistics Course...195
Alice Richardson, University of Canberra, Australia

This chapter reports on a three-year project to study to improve student outcomes in an introductory statistics course. Collaboration with a language expert led to the introduction of a variety of student-centred activities during the course. Some were online such as matching terms to definitions and multiple choice quizzes. Some were face-to-face such as the introduction of an student response system (commonly known as clickers). Results show a small but significant increase in grades following the introduction of these two interventions. A meta-analysis of research into the use of clickers was also conducted. This shows a small but significant increase in exam marks for students who used clickers in class.

Chapter 9
A Successful Collaboration Between Language Arts and Science...226
Felicia Zhang, University of Canberra, Australia

This chapter summarizes the findings of the preceding chapters in order to summarize the findings for the project addressing language difficulties in science education in large classes.

Compilation of References 230

About the Contributors 240

Index 243

Foreword

The effective communication of science through language, including reading, writing, listening, speaking, and visual representation, is an essential part of scientific learning, understanding, and practice. Language is the medium by which scientific reasoning occurs, whether be it formal language or symbolic representations of scientific phenomena. Language permits one to reason critically and analogically, to be creative, to formulate explanations, to integrate knowledge and to transfer and apply knowledge in a sound manner to the solving of problems.

Language also allows one to perform the process of science, including constructing hypotheses, formulating research questions, selecting appropriate methods, analyzing and reasoning deductively and inductively about data, identifying limitations and exceptions, making generalizations, and relating new finding to existing knowledge. Besides verbal language, scientists rely heavily on the use of symbolic language, including scientific and mathematical symbols and representations (e.g. pictures, images, diagrams, graphs, tables, charts, animations, models), to illustrate and interpret scientific phenomena and results. As does verbal language, representations all have limitations that can make it difficult for students to understand and communicate science.

Thus scientific language and literacy is indispensable to scientific thinking and reasoning and needs to be developed in parallel with the development of scientific knowledge and understanding. But whereas most instructors expect higher-level scientific reasoning and expression from their students, few have the tools and techniques necessary for promoting and supporting students' use of scientific language. This book in emphasizing this problem has as its major focus how language support might be introduced and sustained in science education.

In addressing this issue the authors focus on a wide range of practical approaches, strategies, and solutions imported from language studies that could be readily implemented by practitioners to improve the scientific literacy of students across multiple scientific disciplines, particularly those from diverse language and educational backgrounds. For example, Zhang (Chapter 1) advocates the use of active learning strategies that are widely used in language learning, to enhance deep conceptual understanding in science by students from diverse backgrounds. Bridgeman (Chapter 2) in contrast reports on the use of online tutorials to enhance students' symbolic language used in chemistry. Yates and Gardiner (Chapter 3) on the other hand report on activities that help avoid the problems of limited access to technology at some institutions, by introducing activities that develop verbal and symbolic language through the use of mobile-phone based technologies, word-matching exercises, crossword and Sudoku puzzles, and songs.

Lidbury (Chapter 4) advocates the use of various paper-based tutorials and online language exercises and strategies designed to promote language competence, motivation, and scientific reasoning in genetics and molecular biology with a specific focus on the drawing and symbolic representation of genetic concepts. Zhang and Lidbury (Chapter 5) report on the use concept inventories to diagnose genetics learning while Mate and Rodger (Chapter 6) report on various lecture and tutorial based interactive language focused interventions in the context of human physiology and biology. Schulte (Chapter 7) focuses on physics literacy, Richardson (Chapter 8) on the use of clickers in enhancing statistical literacy, and finally Zhang (Chapter 9) summarizes the findings in this book by proposing a model for addressing language difficulties in science education, based on strategies transferred from the language arts.

Over the years there has been a dearth of projects involving cross-disciplinary teams addressing the crucial issue of the interaction between language and conceptual understanding and learning in science. This book will go a long way towards filling this gap in our knowledge and at the very least it will raise awareness among science educators of the different types of verbal and symbolic language used to communicate science, the demands they make on the students and the difficulties they could encounter that might necessitate remediation.

Trevor R. Anderson
University of KwaZulu-Natal, South Africa

Trevor Anderson *is a member of the IUBMB Educational Committee and the editorial board of "Biochemistry & Molecular Biology Education".." In 2007 he was a recipient of the Monash University, Biomedical Sciences "Scholar in Residence in Education" award. He has extensive experience in science education research focusing mainly on conceptual understanding, visual literacy, scientific reasoning, and curriculum change. He is particularly interested in symbolic language and improving the design and use of representations for the development of conceptual knowledge and visualization literacy. He has published widely and presented numerous keynote talks and faculty development workshops. He is the author of the Bridging-the-Gap series in "Biochemistry and Molecular Biology Education" that promotes the application of educational research to teaching practice. Central to his faculty development activities has been the design of the conceptual and reasoning difficulties (CARD) e-resource (http://www.card.unp.ac.za) aimed at developing teacher competence at tertiary and secondary levels.*

Preface

This case book is a result of a project in science education funded by the Australian Learning and Teaching Council. The project, "A cross-disciplinary approach to language support for first year students in the physical sciences," commenced in October 2007. It addressed the language needs of a diverse student body by investigating and testing language oriented strategic approaches to learning and teaching in first year sciences. This project was concerned with the acquisition of language specific to science in addition to the implicit teaching of meta-cognitive skills required in doing science. The disciplines covered by the project were biology, chemistry, and physics. Around 3400 students were involved in the project from the University of Canberra, the University of Technology, Sydney, the University of Sydney, the University of Tasmania, and the University of Newcastle in Australia.

BACKGROUND

Student retention and progression rates are a matter of concern for most institutions in the higher education sector (Burton & Dowling, 2005; Simpson, 2006; Tinto & Pusser, 2006). There is also substantial literature concentrating on the first year experience at university (Kerri-Lee Krause, Hartley, James, McInnis, & Centre for the Study of Higher Education, University of Melbourne, 2005).

Currently, there are two broad approaches to providing extra academic (rather than language) support to help students succeed during their first semester at university: targeting all students who wish to participate in extra learning opportunities or targeting only those students deemed to be at risk.

For example, the peer assisted study support schemes at the University of Wollongong, University of Queensland (Miller, Gregg, & Kelly, 2000), and now at the University of Technology, Sydney and a number of other universities, offered academic support to all interested students. Students usually self-select to participate in these schemes. While there are considerable resource implications associated with such broad-based schemes, they are widely reported to be effective (O'Byrne, Britton, George, Franklin, & Frey, 2009).

However, the problem with both of the approaches above is that students either have to self-select or be selected for such extra academic support. This works on the assumption that students who are not selected are all coping with their first year science study. This project questions this assumption and offers proof that as far as language in science is concerned, all students need support. Thus, we aimed to offer language support to all students who attended lectures and tutorials thereby developing an approach of academic support that supports *all* students.

PROJECT AIMS

Specifically this project aimed to:

1. Target the issue of language in science and suggest ways of solving some of the language issues by importing techniques and strategies frequently used in the teaching of foreign languages;
2. Create innovative online teaching modules that directly address the language difficulties in the targeted disciplines in science;
3. Expand and reshape the current teaching approach to include a language focus in the teaching of science in the face to face mode;
4. Increase student awareness of the language used in the targeted disciplines by presenting student and staff insights of the particular types of language used in that context;
5. Rigorously evaluate the implementation of these learning strategies on student learning to enable their transportability to other teaching contexts in higher education in Australia.

SCIENCE BACKGROUND OF STUDENTS

Students undertaking tertiary studies in the physical and biological sciences are a highly diverse group, and that diversity is increasing (Harris, et al., 2007). For instance, at the University of Sydney, Australia there are usually around 2000 students from various faculties in the first year chemistry cohort. Some of these students had little or no Higher School Certificate studies in chemistry (especially students in Sports Sciences). However, others have very high Universities Admissions Index chemistry scores (>98 for Veterinary Science students). With such a diverse group there is also, naturally, a wide range of interest in and aptitude for the subject. Such diversity is typical for classes in biology, chemistry, and physics at a number of Australian universities (see Table 1).

THE PROJECT METHODOLOGY

The project protocols were developed through collaboration among all stakeholders. These included a specialist in language learning and science subject specialists who were also e-learning developers and students. Designs built on the team members' knowledge of research into online learning (Schulte, 2006), computer-assisted language learning (Zhang & Barber, 2008), linguistics (Zhang, 2006), empirical research in science (Ellem & McLaughlin, 2005), and pedagogical practice. A student-centered approach was emphasized in the design process and in the design itself.

The subjects in this project were taught by lecturers who hold broadly constructivist views of learning as described by Bruner (1986). In this view of learning, learners are considered to bring different conceptualizations, intentions, styles, and approaches to the learning situation (Kolb, 1984; Marton, Hounsell, & Entwistle, 1984; W. G. Perry, 1988). Students' active engagement in learning activities was also an essential ingredient.

Furthermore, these activities should be based on direct experience as far as possible (Boud, 1993) and reflection was seen as important in building understanding (Schon, 1987). Finally, the project was also informed by Lave and Wenger's ideas of situated learning (Lave & Wenger, 1991). Students and

Table 1. Summary of discipline areas studied through a language focus, host universities for each discipline and features of large student cohorts and learning environments

Name of University	**Discipline**	**Average UAI**	**No. of students in subject/unit**	**Web supp-ort**	**Prerequisites**
University of Technology, Sydney	Physics (1st year)	77	410	Yes	None
University of Sydney	Chemistry (1st year)	83	2000; 200 per class	Yes	High School Certificate (HSC) in Chemistry and no HSC Chemistry.
University of Newcastle	Biology (1st year)	N/A	450	Yes	None
University of Tasmania	Chemistry (1st year)	82.5	250	Yes	Year 12 Chemistry
University of Canberra	Genetics (2nd year)*	82.9	85 - 90	Yes	1st year Biology or Chemistry

*Second year Genetics at the University of Canberra was the target unit for the first language-based study in university science education performed by Drs. Zhang and Lidbury in 2006

staff were participating in academic communities of practice. Therefore, the classroom was no longer a site for the transmission of knowledge but rather a site for social practice. To be included in such a social practice environment, the language of that environment must be learnt.

In this process, the roles of teaching and lecturers were changing too. Science lecturers worked alongside an educationalist and contributed to educational research and scholarship. Just as experiencing change in how they learned took place over two years for the students involved in this study, academics teaching also experienced changes. The science academics involved in this project were extremely accomplished and knowledgeable individuals in their own disciplines. By participating in this project, they were positively recognizing the possible contribution education theories and practices could make to their teaching. The involvement of the educationalist was a way of establishing a mutually beneficial learning relationship so that science academics and the educationalist could gain new knowledge from each other. The educationalist involved in the project had very little scientific background or knowledge in the targeted disciplines. She, in a sense, was like a student who chooses to do science without the necessary pre-requisites.

In this model, changes in teaching approaches were explored through a co-teaching or peer coaching approach (Ladyshewsky, 2006; Roth, 1998; Roth, Tobin, Zimmermann, Bryant, & Davis, 2002) in which the education/language expert shared with the science academic techniques and strategies used in teaching in a constructivist model. The science academics taught the education expert the content and pedagogy used in a particular science discipline. This coaching practice before lectures and tutorials in private between the educationalist and the lecturers was an essential element in successfully implementing the change in science academics' lecturing styles in the face to face context. During the coaching practice in private, the educationalist and the lecturers worked together to anticipate areas that students might not understand. This preparedness enhanced the delivery of the content using the new face to face protocol.

The staff project participants were instrumental in ensuring the sustainability of the project processes and findings. At each participating university, the staff project participants involved were instrumental in disseminating the outcome and findings of this project in higher education to colleagues within their own disciplines in their own universities and across the sector. Teaching project staff participants consulted with staff in units for the promotion of teaching and scholarship in higher education such as the Teaching and Learning Centre at the University of Canberra. Teaching staff project participants also conducted workshops to train lecturers in their disciplines in using the online and face to face protocols. Project participants were also involved in a peer-mentoring program organized at each individual institution. During the peer mentoring program, project participants mentored colleagues who intended to adapt and implement the strategies tested in this project, and they also conducted workshops and seminars to showcase the processes and outcomes of this project at their own institutions.

In the project, we undertook to do the following:

- Conduct an online language difficulty survey to ascertain the problems students might have with scientific language;
- Implement the following two protocols in teaching in all five universities

We also implemented the following protocols:

1. During each lecture, the lecturer built into the lecture materials short survey questions made available on VotApedia <HU*http://www.urvoting.com*UH> or audience response devices such as clickers <HU*www.keepadinteractive.com*UH> to offer feedback on lecture content;
2. During tutorials, interactive activities were introduced. Such interactive activities could include small group discussions involving the linking of concepts learned.

Ethics approval for surveying and communicating with participants was given and monitored by the University of Canberra Ethics Committee; approval number: 01-119. Each participating institution also obtained ethics approval for their participation in this project.

TARGET AUDIENCE

This casebook will be of interests to educators, science educators, both education and science lecturers themselves, as well as staff in teaching and development units. Researchers in higher education might also be interested as this cross-disciplinary approach to science education can be a sustainable model for the professional development of the staff.

DESCRIPTION OF EACH CHAPTER

The chapter submissions in this volume include nine chapters detailing successful collaboration in science education.

Chapter 1 documents the results of language difficulty surveys distributed in the participating institutions to ascertain the extent of the language difficulties encountered by first year science students.

Chapter 2 documents the successful implementation of various Chemistry language strategies in the first year Chemistry curriculum at the University of Sydney, Australia.

Chapter 3 is another instance of successful implementation in Chemistry at the University of Tasmania, in Hobart, Australia.

Chapter 4 documents the longitudinal implementation of language strategies in the curricula of Genetics and Molecular Biology at the University of Canberra, Canberra, Australia.

Chapter 5 details the verification of the Genetic Concept Inventory (GenCI) (Elrod, 2007) which was used extensively at the University of Canberra in the Genetics unit.

Chapter 6 of this book documents a successful transfer of the strategies to the subject of Human Physiology at the University of Newcastle, Australia.

Chapter 7 contains a description of the implementation of language strategies in the first year Physics curriculum at the University of Technology, Sydney, Australia.

Chapter 8 describes another example of successful transfer of knowledge to the discipline of first year Statistics teaching.

Chapter 9 is a concluding chapter which summarizes findings from the project for science education starting with the issue of language difficulties.

CONCLUSION

One of the key achievements of this project was that it succeeded in changing the teaching practices of science lecturers, not by imposing a set of "best practices" onto them, but by directly involving them in designing, testing, and implementing practices that they would use in their own teaching. With the idea of disseminating sustainable strategies in mind, the guiding principle of selecting suitable strategies was that the strategy must be easy to use and flexible enough to be modified to suit different institutional contexts. As a result of the cross-disciplinary collaboration between the science lecturers and the educationalist (the project leader, Dr. Felicia Zhang), some twenty-five language oriented exercises created using the freeware Hot Potato software (http://hotpot.uvic.ca/), over forty critical thinking activities, and over forty five multiple choice questions and a large number of VotApedia questions (http://www.urvoting.com) in the disciplines of chemistry, biology, and physics have been created. These teaching materials can be used either online or in face-to-face contexts, in lectures or tutorials, and are not constrained by institutional computer infrastructure such as the Learning Management System (LMS). This case book contains many of the teaching materials used in different project sites and is intended to be a one stop shop for science language activities used in the project.

Finally, this book contains a wealth of ideas, practical exercise sheets, and technological advice targeted at a number of specific topics in the Physics, Chemistry and Biology curricular. In some disciplines such as Chemistry, materials concerning all the usual topics have been included. Though coverage in other disciplinary areas has been isolated to a few topics, the model of creating language oriented exercises, creating exercises that integrate the use of technology can be readily transferable to other areas of science and other disciplines such as the discipline of business and health. What distinguishes the materials in this book from other curriculum advice is that our learning materials are not like those in any science textbooks, are produced by the disciplinary science academics themselves rather than by the educationalist, can be easily implemented with or without technology, and most importantly, have proven to produce significant improvement in first year science students' achievement.

Felicia Zhang
University of Canberra, Australia

Brett Andrew Lidbury
Australian National University, Australia

Alice Marion Richardson
University of Canberra, Australia

Brian Francis Yates
University of Tasmania, Australia

Michael Guy Gardiner|
University of Tasmania, Australia

Adam James Bridgeman
University of Sydney, Australia

Jurgen Schulte
University of Technology, Australia

John Cameron Rodger
University of Newcastle, Australia

Karen Elizabeth Mate
University of Newcastle, Australia

REFERENCES

Boud, D. (1993). Experience as the base for learning. *Higher Education Research & Development*, *12*(1), 33–44. doi:10.1080/0729436930120104

Bruner, J. S. (1986). *Actual minds, possible worlds*. Cambridge, MA: Harvard University Press.

Burton, L. J., & Dowling, D. (2005). *In search of the key factors that influence student success at university*. Paper presented at the HERDSA 2005 International Conference: Higher Education in a Changing World. Retrieved from http://eprints.usq.edu.au/

Ellem, G. K., & Mclaughlin, E. A. (2005). *Tales from the coalface: From tragedy to triumph in a blended learning approach to the teaching of 1st year biology*. Paper presented at the 2005 National UniServe Science Symposium: Blended Learning in Science Teaching and Learning. Retrieved from http://sydney.edu.au/

Elrod, S. (2007). *Genetics concept inventory*.

Harris, K.-L., Krause, K., Gleeson, D., Peat, M., Taylor, C., & Garnett, R. (2007). *Enhancing assessment in the biological sciences: Ideas and resources for university educators*. Retrieved 1st April, 2007, from http://www.bioassess.edu.au

Kolb, D. A. (1984). *Experiential learning: Experience as the source of learning and development*. Englewood Cliffs, NJ: Prentice-Hall.

Krause, K.-L., Hartley, R., James, R., Mcinnis, C., & Centre for the Study of Higher Education, University of Melbourne (2005). *The first year experience in Australian universities: Findings from a decade of national studies*. Retrieved from http://www.dest.gov.au/

Lave, J., & Wenger, E. (1991). *Situated learning: Legitimate peripheral participation*. Cambridge, UK & New York, NY: Cambridge University Press.

Marton, F., Hounsell, D., & Entwistle, N. J. (Eds.). (1984). *The experience of learning*. Edinburgh, UK: Scottish Academic Press.

Miller, V., Gregg, G., & Kelly, B. (2000). *Peer Assisted Study Sessions (PASS) in a first year biology teaching program-strategies developed by undergraduate PASS leaders*. Paper presented at the Effective Teaching and Learning at University Conference. Retrieved from http://www.tedi.uq.edu.au/conferences/

O'Byrne, J. W., Britton, S., George, A., Franklin, S., & Frey, A. (2009). Using academic predictors to identify first year science students at risk of failing. *CAL-laborate International, 17*(1), 15-25. Retrieved from http://sydney.edu.au/

Perry, W. G. (1988). Different worlds in the same classroom . In Ramsden, P. (Ed.), *Improving learning: New perspectives*. London, UK: Kogan Page.

Schon, D. A. (1987). *Educating the reflective practitioner: Toward a new design for teaching and learning in the professions*. San Francisco, CA: Jossey-Bass.

Schulte, J. (2006). *Delivering first year physics assignments with limited resources - An Australian three-centre study*. Paper presented at the 2006 National UniServe Science Symposium: Assessment in Science Teaching and Learning. Retrieved from http://sydney.edu.au/

Simpson, O. (2006). Predicting student success in open and distance learning. *Open Learning: The Journal of Open and Distance Learning, 21*(2), 125–138.

Tinto, V., & Pusser, B. (2006). *Moving from theory to action: Building a model of institutional action for student success*. Paper presented at the National Symposium on Postsecondary Student Success. Retrieved from http://nces.ed.gov/npec/papers.asp

Zhang, F., & Barber, B. (Eds.). (2008). *Handbook of research on computer-enhanced language acquisition and learning*. Hershey, PA: Information Science Reference. doi:10.4018/978-1-59904-895-6

Zhang, F. Z. (2006). *The teaching of Mandarin prosody: A somatically-enhanced approach for second language learners*. Unpublished PhD thesis, University of Canberra, Australia.

Acknowledgment

The authors wish to thank the Australian Learning and Teaching Council (ALTC) for awarding an ALTC Competitive Grant for this project and casebook. Without their support, this important project and book would not be possible.

The authors also wish to thank participating and non-participating staff in Physics, Chemistry and Biology in the five institutions for their support in the lead up to the development of the activities for this project and during their implementation. It has been rewarding that a number of the activities developed have continued to be used and developed further. Thanks are also due to the students that provided useful feedback to us and helped to improve the material that we developed.

Particular thanks are due to Mr. Bryce Lockhart-Gillett who did a fantastic job in writing many of the Concept Tests and VotApedia questions for the whole first year program in Chemistry at the University of Tasmania, Hobart. Bryce's familiarity with the unit material through private student tutoring and laboratory demonstrating was very helpful for his efficient completion of this large task that had seemingly endless rolling deadlines throughout 2008. But, more than anything, his deep understanding of the material and the goals of the Australian Learning and Teaching Council project, and good communication with the topic lecturers and project investigators led to an excellent set of materials having been developed. Similarly, thanks are due to the tutors who participated enthusiastically in implementing the language based strategies at the University of Sydney. Without their hard work and support, the project could not have been completed satisfactorily. Team members of the ALIUS Australian Learning and Teaching Council project are also to be thanked in connect with some of the strategies from that project that meshed very well with the language support focus of this project.

We authors pay great appreciation to Dr Felicia Zhang for the numerous visits that she made to Hobart, Sydney, and Newcastle that really helped to open our eyes to some of the language-based issues facing students, developing some starter materials that got us on track, proofing many of our lecture materials for language based vagueness that we had been completely blind too and keeping on top of the grant requirements.

Special thanks go to the publishing team at IGI Global, whose contributions throughout the whole process, from inception of the initial idea through to final publication, were invaluable. In particular, we would like to thank the managing development editor Hannah Abelbeck, who helped keep the book on schedule and who gave valuable and timely assistance with detailed advice.

The authors also need to acknowledge with gratitude the encouragement, patience and understanding of our partners and family members for making sacrifices in order to leave us with the time and energy to complete the book.

In closing, we trust that our readers will find themselves on an exciting journey of learning as they make their way through this casebook.

Felicia Zhang
University of Canberra, Australia

Brett Andrew Lidbury
Australian National University, Australia

Alice Marion Richardson
University of Canberra, Australia

Brian Francis Yates
University of Tasmania, Australia

Michael Guy Gardiner
University of Tasmania, Australia

Adam James Bridgeman
University of Sydney, Australia

Jurgen Schulte
University of Technology, Australia

John Cameron Rodger
University of Newcastle, Australia

Karen Elizabeth Mate
University of Newcastle, Australia

Chapter 1
Language Problems of First Year Science Students

Felicia Zhang
University of Canberra, Australia

ABSTRACT

The chapter discusses the background of a project which aimed at addressing the language needs of a diverse student body (both domestic and international student body) by embedding strategic approaches to learning and teaching in first year sciences in tertiary education. These strategies consisted of active learning skills which are widely used in language learning. The disciplines covered by the project were Biology, Chemistry and Physics and involved the University of Canberra, University of Sydney, University of Tasmania, University of Technology, Sydney, and University of Newcastle in Australia. This project was funded by the Australian Learning and Teaching Council (ALTC).

INTRODUCTION

Specialist terminology in Biology, Chemistry and Physics has proved difficult for most students (Wellington & Osborne, 2001). Students have difficulty recognizing where a concept begins and ends and therefore cannot differentiate concepts. Zhang and Lidbury (2006) identified difficulties with language as contributing significantly to problems students experience in studying science (specifically Genetics). In this study, we seek to implement language oriented strategies (Table 1) developed by Zhang and Lidbury (2006) for First Year Biology, Chemistry and Physics lectures and tutorials with the aim of evaluating the benefits of those methods.

One distinctive feature of this project is that the language expert worked collaboratively with lecturers in the disciplines closely as a novice first year student with no previous HSC experience (but an expert in language learning and teaching). In this project, the language expert 'learned' the

DOI: 10.4018/978-1-61350-062-0.ch001

Table 1. Proposed language oriented techniques to be implemented in the project

1. Small group work in tutorials using guided questions.	6. Attaching sound files to vocabulary
2. Students are provided with a list of terms and, through the process of group work, place these terms in relation.	7. Breaking down long words to aid memory by identifying prefixes and suffixes, and exploring the roots and origin of words.
3 Giving students opportunities to put forward their points of view in groups.	8. Using warm up activities such as matching scientific terms to definitions for revision purposes.
4. Using online language exercises such as crosswords, gap-fill (Cloze) exercises.	
5. Providing stimulus quizzes for lecture and tutorial materials on WebCT thus encouraging students to prepare before the lecture.	10. Role playing: students practise conveying complex scientific discoveries to the public.

content of the disciplines by direct experience and in this process, attempts to (a) make lectures more accessible; (b) look for opportunities for small group participation; (c) look for opportunities to develop students' communication skills in communicating scientific knowledge to lay communities; (d) create opportunities for providing formative feedback both face-to-face and through online activities by negotiating and collaborating with the disciplinary lecturers. Some of the activities focused on language issues, others focused on active learning strategies.

Context and Characteristics of the Students

Students undertaking tertiary studies in science are a highly diverse group. For instance, at the University of Sydney, Australia in 2008, there were 969 students from various faculties in the first year Chemistry cohort. Three-hundred and eighty eight students had little or no High School Certificate Chemistry while 116 students have very high University Admission Index Chemistry scores (greater than 98 for Veterinary Science students). With such a diverse group there was, naturally, a wide range of interest in and aptitude for the subject. Such diversity is typical in cohorts in Biology, Chemistry and Physics at the participating Australian universities.

Administration of the Language Difficulty Questionnaires

In this project, we adopted a model which varied slightly from that of Jacobs (1989). We distributed one questionnaire through the Learning Management System (LMS) of the participating institutions. In this questionnaire, we tested comprehension of ten common terms. The examined terms are; 'research', 'power', 'concentration', 'equilibrium', 'graph', 'system', equation', 'experiment', 'model' and 'significant'. The criteria for selection of the words were that they have to be:

- Words used as basic currency in physics, chemistry and biology lectures and for which definitions would be assumed unnecessary; and
- Words which in lay contexts acquire more flexible and approximate meanings.

For example, in the question related to 'research' for the Chemistry and Physics disciplines, the following was provided:

"We carry out research to find out the answers to scientific questions"

What is meant by the word 'research' in the sentence above? (There may be more than one correct answer; see Table 2). They are then provided with five answers for this question:

Table 2. Answers to the question "What is research?"

	Student response	Correct answer	Value
A	observing the results of a series of chemical reactions	yes	33.33%
B	Answering an exam question.	No	-10%
C	Looking for information in the library or on-line	Yes	33.33%
D	Copying information out of a textbook	No	-10%
E	Testing a hypothesis	yes	33.33%

Table 3. Confidence question

	Student response	Value
A	Yes, I understand the meaning of this word	100%
B	No, I do not understand the meaning of the word	-100%
C	I have some idea of the meaning of this word.	0%

Each correct answer is allocated 5 marks each. If a student chose the three correct answers, she/he should score 15/15. This indicates a *complete* understanding of the term in different contexts. Any score less than 15 suggest an incomplete understanding of the term. All 10 terms followed a similar format. Then in the second part following each question was a related confidence question which assessed how confident a student was of the choice(s) he/she made (See Table 3):

The University of Sydney and University of Tasmania were using the *Blackboard* LMS (formerly *WebCT*) which allowed printable statistics for a cohort of students to be obtained for each questionnaire. From the printable statistics, the percentage of students who demonstrated a complete understanding of a term in different contexts and how confident they were of their understanding was obtained. If we divide the 'Yes' confidence level (Table 4, column 2) by the percentage of students demonstrating a 100% understanding (Table 4, column 3), we are able to establish a 'delusion index'. A high value in the 'delusion index' column suggests that students are highly delusional about their understanding of a particular term.

For instance, for the term 'power' in Table 4, a delusion index of 2.14 suggests that students are more realistic about their understanding of this term when compared to their understanding of the term 'significant' which has delusion index of 15.13. This is demonstrated by only 3.2% of the student body understood this term completely but 48.40% of students thought they understood the term completely (indicated by the value in the corresponding 'Yes' column). A ranking of the difficulty of the terms (from the easiest=1 to the hardest=10) is also provided by examining the percentage of students who achieved 100% understanding of any terms (column 3).

ANALYSIS OF QUESTIONNAIRE DATA

Chemistry Discipline

Table 5 is a composite table of the rankings obtained from the three cohorts from the University of Sydney. The data in Table 5 suggests that terms such as 'model', 'significant', 'research', 'power' and 'system' are the most difficult for science students. Note that 'research' was one of the most

Table 4. Chemistry 1A (mixed cohort) at the University of Tasmania

Terms	Yes	100% understanding	Delusion index	Ranking (easiest=1,hardest=10)
Significant	48.40	3.20	15.13	10
Research	73.50	9.10	8.08	9
Model	48.40	12.90	3.75	8
System	58.10	19.40	2.99	7
Power	48.40	22.60	2.14	6
Concentration	87.10	25.80	3.38	5
Equilibrium	76.70	32.30	2.37	4
Equation	77.40	35.50	2.18	3
Graph	87.10	41.90	2.08	2
Experiment	83.90	71.00	1.18	1

Above is a table showing the results of this questionnaire from the University of Tasmania obtained in 2008.

Table 5. Chem1001/1101/1901 at the University of Sydney, 2008

Terms	Ranking (easiest=1, hardest=10)
Significant	10
Model	9
Power	8
System	7
Equilibrium	6
Research	5
Concentration	4
Graph	3
Equation	2
Experiment	1

Key: Chem1001 (Fundamental students with no previous HSC chemistry) at the University of Sydney

Chem1101 (Students with HSC Chemistry) at the University of Sydney

Chem1901 (students with HSC chemistry and high UAI) at the University of Sydney

problematic terms. From the example which described the question on 'research' on page 3 of this paper, we can see that in order to achieve a 100% complete understanding of this term, students had to tick all the correct answers (a, c, and e). The three answers referred to the understanding of the term in scientific and non-scientific contexts.

Table 6. Physical modeling students at the University of Technology, Sydney, 2008

Terms	100% understanding	Ranking (easiest = 1, hardest = 5)
Research	15	5
Equilibrium	17.5	4
Power	25	3
density	27.5	2
Graph	42.5	1

The more popular choice ticked was 'e=Testing a hypothesis' which was the scientific definition with which they might be most familiar. This could be interpreted as symptomatic of students' ability to transfer knowledge gained in science to other realms of knowledge.

Physics Discipline

The questionnaires were constructed slightly differently at the University of Technology, Sydney. Only 5 terms were used and no confidence questions were posed (Table 6) due to technical difficulties with administration of the questionnaires. Out of the five terms, 'concentration', which was used in the Chemistry questionnaires at the University of

Table 7. University of Newcastle, Biology 1002, 2008

Terms	Yes	100% understanding	Delusion index	Ranking (easiest=1,hardest=10)
Significant	44.44	33.33	1.33	9
Equation	72.22	38.89	0.54	8
Research	77.78	44.44	0.57	7
Equilibrium	66.67	50.00	0.75	6
Concentration	77.78	61.10	0.79	5
Power	44.44	66.67	1.50	4
Graph	88.89	77.78	0.88	3
Experiment	66.67	83.30	1.25	2
System	72.22	88.89	1.23	1

Table 8. University of Canberra, 2nd year Genetics 6531, 2008

Terms	Yes	100% understanding	Delusion Index	Ranking (easiest=1, hardest=10)
Model	9.50	4.00	2.38	10
Equation	21.90	15.00	1.46	9
Power	25.70	15.00	1.71	8
Equilibrium	25.00	18.00	1.39	7
Research	66.80	33.00	2.02	6
System	65.00	36.00	1.81	5
Significant	57.10	40.00	1.43	4
Experiment	64.40	43.00	1.50	3
Concentration	72.00	46.00	1.57	2
Graph	69.40	48.00	1.45	1

Sydney and the University of Tasmania, had been changed to 'density' as this was more suited to the physics context. However, it was not possible to derive a 'delusion index' because the confidence questions were not deployed at the University of Technology, Sydney. Again 'research' seemed to be the hardest term for students.

Biology Discipline

The questionnaires were also distributed to first year students at the University of Newcastle and second year students at the University of Canberra in 2008 (Table 7 and Table 8, respectively). Since both these universities were involved in the discipline of Biology, the same language difficulty questionnaire appropriate to Biology was distributed (See Table 7). At the University of Newcastle, one item (significant) was repeated (which was due to human error) and the item on 'model' was not included. Therefore the table shows the data for nine items rather than ten.

In conclusion, the results reported above show that the students from the five institutions had problems with the ten terms tested. This signals an urgent need for language focused training in the teaching of first year science.

IMPLICATIONS

These results would suggest that, in the teaching of physics at university level, insufficient attention is being paid, by lecturers and students alike, to the language used to express and explain scientific concepts. In the field of physics, for example, since the vocabulary used sounds familiar to students, they (and their lecturers) assume understanding where in fact little or none exists.

It would seem impractical, however, to suggest that lecturers take on the additional task, during lectures, of redefining and explaining basic vocabulary. Responsibility for accurate understanding should rest, at university level, with the students. However, they do need to be alerted to which words might be problematic and be guided in overcoming them.

Jacobs (1989) suggested the administration of a more comprehensive set of tests on many more words within the first half of the first-year physics course. As she put it "this alone could shock or inspire the students sufficiently to recognize physics discourse as a specialized subset of ordinary language, requiring constant alertness to precision and the possibility of idiosyncratic meaning"(p.398).

Since this project involved the three disciplines of Biology, Chemistry and Physics, it was not feasible to administer a more comprehensive set of tests. The language difficulty surveys were used as a diagnostic tool only. However, the strategies developed for the project concentrated on the language of science.

LITERATURE REVIEW: THE ROLE OF LANGUAGE IN SCIENCE

Specialist terminology in biology, chemistry and physics has proved difficult for most students (Wellington & Osborne, 2001). Previous research in the language of science (Gardner, 1977; Gardner & Australian Science Education Project., 1972; Pickersgill & Lock, 1991; Wandersee, 1988) further suggests that students have problems with both technical and non-technical vocabulary. This applies especially with logical connectives such as 'and', 'or', 'but', and 'although'. In other words, students have difficulty recognizing where a concept begins and ends and therefore cannot put different concepts into relation. This is despite the often obvious meaning of ordinary conjunctions such as those listed.

Research into the problem of enabling students to better acquire scientific vocabulary suggests that 'technical' words make up only a small percentage of vocabulary in scientific texts. They therefore pose fewer difficulties than vocabulary which is used in everyday English as well as in a science context. Wellington (cited in Marais & Jordaan, 2000a) suggested that scientific words can be classified according to their level of abstraction. The higher this level, the more difficult they are to understand. In the opinion of Wellington, mathematical words (known as symbolic language) represent the highest level of abstraction. The next level of abstraction includes concept words which can be theoretical constructs such as 'energy' and other sensory concepts which might be in common everyday use. A lower level of abstraction includes process words and less difficult naming words.

With words that are peculiar to science, certain concepts seem to be easily derived from everyday experience. Other words that have both a scientific and everyday meaning such as 'work', 'energy' or 'power' can cause confusion for learners. As students bring the everyday association of the words with them, these same words, when introduced in a physics class, can cause confusion. Itz-Ortiz, Rebello and Zollman (2003) suggested that a weak version of the Sapir and Whorf hypothesis might be dictated by the language habits of our community predisposing certain choices of interpretation of words. This would cause confusion in students with scientific terms. Researchers have studied semantics in physics (Touger, 1991; Williams, 1999).

Scientists became interested in how to teach science as far back as the 1980s. Initial attempts had been made to describe and understand what it is that we would like students to learn. This is evident from titles such as 'Scientific approaches to science education' or 'Learning to think like a physicist: A review of research-based instructional strategies' (Yore, Florence, Pearson, & Weaver, 2006). However, as Jay L Lemke pointed out (1998) such research constitutes a cognitive model of science education which cannot tell us enough to become better teachers of science as 'it lacks the necessary vocabulary to tell us just what we must lead students to do in order to learn to reason and act scientifically' (p. 1).

In order to transfer to students what they must do in learning to reason and act scientifically, the teacher is the role model 'who can model for students how scientists talk and write and diagram and calculate, how scientists plan and observe and record, how we represent and analyze data, how we formulate hypotheses and conclusions, how we connect theories, models, and data, how we relate our work and result to those of other researchers.' (Lemke, 1998, p. 1). In addition, teachers do not model these abilities by simply telling students a list of concepts they need to 'understand'. They participate in social semiotic dialogues (Lemke, 1998) in which teachers and students become engaged together in scientific sense-making and scientific doing. These dialogues are carried out in languages of science which integrate multiple languages of representation such as visual representation, mathematical symbolism and languages of experimental operations. We would argue that these should be seen to include the meta-language of learning such as checking, problem solving, reflecting and so on, to learn how to synthesize, monitor, critique and reflect on the information gathered in the process of learning. Furthermore, acquiring scientific terms is, of course, a basic part of scientific literacy education which would include the ability to read and write a science text which is, in turn, a dominant genre used by practicing scientists (Yore, Hand, & Prain, 2002).

In order not to fall into the trap of 'activitymania' (Yore & Treagust, 2006), we place teachers in the centre of learning by asking them to explicitly model (a verb) how to test reality 'by checking, monitoring, coordinating, and controlling deliberate attempts to execute learning activity (Koch, 2001). This is why, in the lectures and tutorials involved in this project, interventions were designed to increase student to student and student to staff discussions in activities which promote sense-making rather than just completing calculations.

From a foreign language teacher's point of view (this is the project leader's area of expertise), it is heartening to see that the three-language problem (home language, instructional language, science language) that exists for most science language learners is acknowledged as parallel to that of English language learning. In her opinion, this also applies to foreign language learning, in which learners need to move across many discourse communities of their family, school, science and foreign language cultures. However, as pointed out by Yore & Treagust (2006, p. 297): "There has been little consideration between science education and language arts reforms...because it is difficult to convince science majors in teacher education programs that language is an essential part of doing science." This project thus represents one of the first attempts to link practices frequently used in language teaching to the teaching of science by providing examples of how and what to include in science curricula. These examples have been chosen for their simplicity and for the fact that they can be easily constructed with or without the help of technology. In other words, this project is an exemplification that even if one is not a linguist, one can still incorporate language-focused practices in one's teaching.

CHANGED TEACHING PRACTICES REGARDING THE USE OF LANGUAGE

This project also documents the change process the lecturers went through in order to build language-focused and inquiry-based practices into their day to day teaching.

Gunstone (1994) argued that metacognition is central to constructivist perspectives of learning. This idea is reaffirmed and embraced by Yore & Treagust (2006). Metacognition was first used by Flavell in 1976 (Flavell, 1976). He describes it in these words:

"Metacognition refers to one's knowledge concerning one's own cognitive processes or anything related to them, e.g., the learning-relevant properties of information or data. For example, I am engaging in metacognition if I notice that I am having more trouble learning A than B; if it strikes me that I should double check C before accepting it as fact." (J. H. Flavell 1976, p. 232).

A number of strategies were deployed. These are outlined in the following (Table 9) *which includes* the research that supported the use of these strategies.

Table 9. Strategies implemented in the project and the research that supported them

Strategies	Experimental sites	Meta-cognitive skills practiced	Research supporting use.
1. Small group work in tutorials using guided questions	All institutions	Learning to use oral language to express and explain scientific ideas	(Kempa & Ayob, 1995; Ritchie, 2001)
2. Students are provided with a list of terms and, through the process of group work, place these terms in relation	University of Newcastle and University of Canberra	Physical concept mapping, exploring relationships between each term	(Roth & Roychoudhury, 1994; Wellington & Osborne, 2001)
3 Giving students opportunities to put forward their points of view in groups	All institutions	Creating a supportive atmosphere for idea exploration and debate	(Chin & Brown, 2002)
4. Using online language exercises such as crosswords, gap-fill (cloze) exercises and simplified scientific readings	University of Canberra	Explicit instruction, practice, applying concepts	No research discovered in science education
5. Providing stimulus questions for lecture and tutorial materials on WebCT* thus encouraging students to prepare before the lecture	University of Tasmania, University of Newcastle	Preparation or reflection, just in time learning, online feedback	(Zhang & Lidbury, 2006)
6. Breaking down long words to aid memory by identifying prefixes and suffixes, and exploring the roots and origin of words	University of Newcastle	Explicit instruction on how to acquire new vocabulary and how to see patterns in the roots and origins of words	(Wellington & Osborne, 2001)
7. Using warm up activities such as matching scientific terms to definitions for revision purposes	University of Newcastle	Categorizing and systematizing terms	(Richardson & Zhang, 2008; Richardson, Zhang, & Lidbury, 2008)
8. Using of flashcards for vocabulary revision, creating a glossary	University of Newcastle, University of Canberra	Meta-linguistic maintenance	(Zhang & Lidbury, 2006)

* Now called Blackboard LMS

BUILDING ON EXISTING KNOWLEDGE OF PEDAGOGICAL ADVANCES

Research into why students did so poorly in tertiary level physics courses has been going on since 1985. The 'common sense theory' of the physical world, which the students brought with them into the classroom, had long been blamed as one of the major causes for instructional failure in introductory physics (Halloun & Hestenes, 1985; Hestenes & Wells, 1992; Hestenes, Wells, & Swackhamer, 1992; Jacobs, 1989; Touger, 1991). Work has been carried out on two fronts: (1) to establish diagnostic instruments to diagnose students misconceptions such as concept inventories; and (2) to explore different ways of organizing learning activities in order for students to learn.

Since the publication of the 'Mechanical baseline Test' (Hestenes & Wells, 1992) and the 'Force Concept Inventory' (FCI) (Hestenes et al., 1992) which appeared in 1992 in the Journal *Physics*, a large number of concept inventories have been developed and validated in different areas of science. These are the Brief Electricity Magnetism Assessment (BEMA) (Pollock, 2008) and the Conceptual Survey of Electricity and Magnetism (CSEM) (Maloney, O'Kuma, Hieggelke, & Van Heuvelen, 2001), the Chemistry Concept Inventory (CCI) (Krause, Birk, Bauer, Jenkins, & Pavelich, 2004), Genetics Concept Inventory (Elrod, 2007) and Biology Concept Inventory (BCI) (Klymkowsky & Garvin-Doxas, 2008). While these inventories are useful, they have essentially been established as diagnostic instruments to diagnose students' misconceptions or conceptual understanding. Data from such diagnostic tools, while informing educators what the problem areas are, do not offer advice on how these problem areas can be addressed in teaching.

In the meantime, many researchers have also invested a great deal of energy into investigating 'how' problem areas identified by the diagnostic tools cited above can be taught better or how learning activities can be organized better to enable better learner outcomes in science education.

For example, in chemistry, radical suggestions for reform have been about whether to teach introductory level chemistry from the macro and tangible, then the sub-micro atomic and molecular and then the representational use of symbols and mathematics (Johnstone, 2000). However, judging from the popular chemistry textbooks published by leading publishers, this debate is clearly having little impact. The textbooks used in the two chemistry cohorts involved in this project are Chemistry by Blackman, Bottle, Schmid and Mocerino; and Chemistry: The Molecular Science by Stanitski and Jurs. Both follow the traditional sequence in curricular design by teaching the sub-micro atomic and molecular first in conjunction with the representational use of symbols and mathematics. They very rarely teach the macro level of chemistry first. Some researchers argued for a complete departure from the traditional sequencing of chemistry content in introductory courses (Dreyfus, Jungwirth, & Eliovitch, 1990), However, other researchers utilized planned cognitive conflicts by confronting students with a phenomenon that cannot be explained with their prior knowledge (Dreyfus et al., 1990; Nieswandt, 2001).

In physics, many studies in designing instructional sequences in teaching concepts such as force, motion and Newton's Third Law have been carried out (Alonzo & Steedle, 2008; Halloun, 1998; Savinainen, Scott, & Viiri, 2005). Analogies have also been used extensively to remediate misconceptions in physics since the early 1990s. (Brown, 1992; Dupin & Johsua, 1989).

Interactive Engagement (IE) strategies have been shown to enhance the gain in learning. In an IE class, the lecturer sets up a demonstration and asks the students to predict and write down, with consultation among themselves, their prediction of how the demonstration will work out. Then the demonstration is carried out, the results discussed and the relevant theory presented. The central idea is to offer students opportunities to engage with each other, commit to a position, confront this with reality and then use the event to trigger reflection and foster understanding. Extensive research has been done using IE strategies and the finding is that the 'use of Interactive Engagement (IE) strategies can increase the effectiveness of conceptually difficult courses well beyond that obtained with traditional methods (Hake, 2007). Professor Hake's claim is based on the achievement scores obtained using standardized tests such as the FCI.

Harvard's physics Professor Eric Mazur implemented 'Peer Instruction' (PI) which also uses interactive-engagement strategies. Data obtained in classes using PI in a variety of disciplines show that learning gains nearly triple. Most importantly, students not only perform better on a variety of conceptual assessments, but also improve their traditional problem-solving skills (Crouch & Mazur, 2001).

In recent years, IE and PI, in various scientific disciplines, have been enhanced by the use of computer technology such as concept mapping tools, visualization tools (Wu, Krajcik, & Soloway, 2001) audience response devices such as clickers and other web-enhanced strategies (Martyn, 2007; McDaniel, Lister, Hanna, & Roy, 2007; Odom & Kelly, 2001).

In the late 1990s, studio classes replaced straight lecturing in a number of universities in the US. Studio teaching consisted of a mixture of student exercises, instructor coaching, and sometimes laboratory experiments. It drew inspiration from the idea of interactive learning and generally taking advantage of modern technology to deliver instructional materials (Cummings, Marx, Thornton, & Kuhl, 1999; Pipes & Wilson, 1996; Roy, 2003).

CONTRIBUTION OF PREVIOUS RESEARCH TO THE PROJECT

The project leader/educationalist visited four institutions in semester one and semester two (from July to November), 2008 to observe lectures and tutorials. These were the University of Sydney, the University of Technology, Sydney, the University of Tasmania and the University of Newcastle. Several inhibiting factors emerged which prevented the teaching of first year sciences in the form of studio classes or IE classes from the observation sessions:

- Physical space: lecture theatres restricted movement. Perhaps due to the impersonal nature of such physical space, the student body tended to be less engaged and lecturers tended to deliver lectures in a more transmissive mode.
- The large number of students in the classes (which ranged from less than 100 to 500 students per class) affected not only course delivery but also assessment. In order to manage such large numbers, assignments and tests were mainly conducted online and involved multiple-choice questions so that the LMS system could automatically calculate and assign grades to each student.
- Difficulty of the class tests is another factor. For instance, in the mid-semester test for Chemistry 1A at the University of Tasmania, students were required to answer some 50 questions in fifty minutes. Class tests generally involve short answer questions which require calculation and transformation of items such as from a chemical formula to a diagram and so on. Therefore, in order to pass the test, students really need to know the answers automatically without thinking. The difficulty of the

test was confirmed by 26% of the student cohort (total n=214) failing the test.

- Coverage and the service teaching nature of the units: the coverage of the content is enormous. In a typical Physics Modeling lecture at the University of Technology, Sydney, 20 slides were covered, with about three concepts on each slide in two hours. The amount of coverage needed in each science unit is out of the control of the unit convenors because these first year units are service units. Service units such as Chemistry 1001 or Physical Modeling serve a large variety of disciplines such as mechanical engineering, electrical engineering and pharmacy. Even though reducing content to incorporate new learning strategies might be necessary to improve learning outcomes, academics in the feeder disciplines might not necessarily embrace it. Furthermore, some students in these units might only be doing Chemistry 1001 or Physical Modeling for one semester because this particular unit is a pre-requisite for other units in their degree.
- Lack of communication skills: from the class observations, students in all three institutions did not demonstrate skills in speaking and writing about science and they were not practiced at transferring or communicating what they learned to other people.

Due to these factors, these units were delivered in the traditional transmissive mode at each institution prior to 2009. Despite these inhibiting factors, the project still intended to change the teaching of these subjects in the participating institutions to a more interactive mode of teaching by incorporating a number of interactive strategies implemented by the lecturers themselves. Studio teaching and radical changes in curriculum design were simply either too expensive to implement or too disruptive to other lecturers who co-teach with the participating lecturers. Methods of intervention are briefly described below.

METHODS OF ADDRESSING DIFFICULTIES IN THE LANGUAGE OF SCIENCE

Having identified in year 1 of the project that there was a gap in students' understanding of common scientific words, we intended to enhance student awareness of the language used in each disciplines by *inspiring* them 'to recognize that scientific discourses are a specialized subset of ordinary language, requiring constant alertness to precision and the possibility of idiosyncratic meaning' (Jacobs, 1989). In order to 'inspire' students, we implemented a range of language specific education techniques to alert students that (1) precision of these terms in science is important and (2) how to check that they are being precise when using these terms. Zhang and Lidbury (2006) implemented a range of language specific educational techniques to encourage such thinking. Results of Zhang and Lidbury (2006)'s study suggested that students favoured the use of small group work to acquire difficult and abstract genetic concepts. We decided to implement (1) a face to face (FTF) learner-centred, interactive lecturing protocol and (2) online content and language support for learners (ONLINE) in the experimental stage of the project.

The FTF protocol consists of the following phases:

1. During each lecture, the lecturer build into the lecture materials, short survey questions made available on Votapedia (http://www.urvoting.com) or audience response devices such as clickers (www.keepadinteractive.com) to offer feedback on lecture content
2. During tutorials, interactive activities are introduced. Such interactive activities can include small group discussions involving

the linking of concepts learned (Techniques 2 in Table 9) and activities related to technique 7, or 8 or 10 in Table 9.

The VotApedia tool (www.urvoting.com) was used at the University of Tasmania in first year Chemistry in semester one, 2009 and 2010 and at the University of Newcastle in Biology. According to the main page of the website 'VotApedia is an audience response system that does not require issuing clickers or need specialist infrastructure. Known users can create surveys and edit the surveys on the site. Once signed on, students can participate in surveys either through mobile phones, online or through SMS. Both online through mobile phones are free to the students but there will be a charge through SMS' (http://www.urvoting.com/index.php?title=Main_Page).

The project was carried out in two years. During the control part of the project in 2008, no intervention took place except the implementation of language difficulty questionnaires. Lecturers taught the subject matter as how they would normally deliver the material and students were assessed in the normal fashion. Assessment data collected during this phase constituted baseline data to which data from the experimental phase of the project in 2009 were compared. 2008 student cohorts constituted the Control Groups of students. Students, in 2009, in the participating universities constituted the Experimental Groups of students.

CONCLUSION

In the following chapters, the implementation process of the language oriented strategies at each project site is described in detailed. In some chapters, the description included the full scale transfer of successful strategies from this project to the whole first year program in some disciplines and other subjects outside the original disciplines involved in the project. It is the sincere hope of the authors that this book and all it contains will provide inspiration for educators around the world to start the journey of addressing issues in science education.

REFERENCES

Alonzo, A. C., & Steedle, J. T. (2008). Developing and assessing a force and motion learning progression. *Science Education, 93*(3), 389-421. doi: 10.1002/ sce.2030310.1002/ (SICI)1098-237X(199907)83:4<397::AID-SCE1>3.0.CO;2-#

Brown, D. E. (1992). Using examples and analogies to remediate misconceptions in physics: Factors influencing conceptual change. *Journal of Research in Science Teaching*, *29*(1), 17–34. .doi:10.1002/tea.3660290104

Burton, L. J., & Dowling, D. (2005). *In search of the key factors that influence student success at university*. Paper presented at the HERDSA 2005 International Conference: Higher Education in a Changing World, Sydney, Australia. Retrieved from http:// eprints.usq.edu.au/

Chin, C., & Brown, D. E. (2002). Student-generated questions: A meaningful aspect of learning in science. *International Journal of Science Education*, *24*(5), 521–549. .doi:10.1080/09500690110095249

Crouch, C. H., & Mazur, E. (2001). Peer instruction: Ten years of experience and results. *American Journal of Physics*, *69*(9), 970–977. .doi:10.1119/1.1374249

Cummings, K., Marx, J., Thornton, R., & Kuhl, D. (1999). Evaluating innovation in studio physics. *American Journal of Physics*, *67*(S1), S38–S44. .doi:10.1119/1.19078

Dreyfus, A., Jungwirth, E., & Eliovitch, R. (1990). Applying the cognitive conflict strategy for conceptual change - Some implications, difficulties, and problems. *Science Education*, *74*(5), 555–569. .doi:10.1002/sce.3730740506

Dupin, J. J., & Johsua, S. (1989). Analogies and "modeling analogies" in teaching: Some examples in basic electricity. *Science Education, 73*(2), 207–224. .doi:10.1002/sce.3730730207

Elrod, S. (2007). *Genetics concept inventory*. Retrieved from http:// bioliteracy.colorado.edu/ Readings/ papersSubmittedPDF/ Elrod.pdf

Flavell, J. H. (1976). Metacognitive aspects of problem solving . In Resnick, L. B. (Ed.), *The nature of intelligence* (pp. 231–236). Hillsdale, NJ & New York, NY: Lawrence Erlbaum Associates.

Gardner, P. L., & Australian Science Education Project. (1972). *Words in science: An investigation of non-technical vocabulary difficulties amongst form I, II, III & IV science students in Victoria, a report*. Melbourne, Australia: Australian Science Education Project.

Gardner, P. L. (1977). Logical connectives in science: A summary of the findings. *Research in Science Education, 7*(1), 9–24. .doi:10.1007/ BF02643108

Gunstone, R. F. (1994). The importance of specific science content in the enhancement of metacognition . In Fensham, P. J., Gunstone, R. F., & White, R. T. (Eds.), *The content of science: A constructivist approach to its teaching and learning*. London, UK: Routledge.

Hake, R. (2007). Six lessons from the physics education reform effort. *Latin-American Journal of Physics Education, 1*(1), 24-31. Retrieved from http://www.journal.lapen.org.mx/

Halloun, I. (1998). Schematic concepts for schematic models of the real world: The Newtonian concept of force. *Science Education, 82*(2), 239-63. doi: 10.1002/(SICI)1098-237X(199804)82:2<239::AID-SCE7>3.0.CO;2-F

Halloun, I. A., & Hestenes, D. (1985). The initial knowledge state of college physics students. *American Journal of Physics, 53*(11), 1043–1055. .doi:10.1119/1.14030

Hestenes, D., & Wells, M. (1992). A mechanics baseline test. *The Physics Teacher, 30*(3), 159–166. .doi:10.1119/1.2343498

Hestenes, D., Wells, M., & Swackhamer, G. (1992). Force concept inventory. *The Physics Teacher, 30*(3), 141–158. .doi:10.1119/1.2343497

Itza-Ortiz, S. F., Rebello, N. S., Zollman, D. A., & Rodriguez-Achach, M. (2003). The vocabulary of introductory physics and its implications for learning physics. *The Physics Teacher, 41*(6), 330–336. .doi:10.1119/1.1607802

Jacobs, G. (1989). Word usage misconceptions among first-year university physics students. *International Journal of Science Education, 11*(4), 395–399. .doi:10.1080/0950069890110404

Johnstone, A. H. (2000). Teaching of chemistry - logical or psychological? *Chemistry Education Research and Practice in Europe, 1*(1), 9-15. Retrieved from http:// www.uoi.gr/ cerp/

Kempa, R. F., & Ayob, A. (1995). Learning from group work in science. *International Journal of Science Education, 17*(6), 743–754. .doi:10.1080/0950069950170606

Klymkowsky, M. W., & Garvin-Doxas, K. (2008). Recognizing student misconceptions through ed's tools and the biology concept inventory. *PLoS Biology, 6*(1), e3. .doi:10.1371/journal.pbio.0060003

Koch, A. (2001). Training in metacognition and comprehension of physics texts. *Science Education, 85*(6), 758–768. .doi:10.1002/sce.1037

Krause, S., Birk, J., Bauer, R., Jenkins, B., & Pavelich, M. J. (2004). *Development, testing, and application of a chemistry concept inventory*. Paper presented at the 34th ASEE/IEEE Frontiers in Education Conference, Savannah, GA. Retrieved from http:// fie-conference.org/ fie2004/

Krause, K.-L., Hartley, R., James, R., McInnis, C., & Centre for the Study of Higher Education. University of Melbourne. (2005). *The first year experience in Australian universities: Findings from a decade of national studies*. Retrieved from http:// www.dest.gov.au/ sectors/ higher_education/ publications_resources/

Ladyshewsky, R. K. (2006). Peer coaching: A constructivist methodology for enhancing critical thinking in postgraduate business education. *Higher Education Research & Development*, *25*(1), 67–84. .doi:10.1080/13600800500453196

Lemke, J. L. (1998). *Teaching all the languages of science: Words, symbols, images, and actions*. Paper presented at the International Conference on Ideas for a Scientific Culture. Barcelona. Retrieved from http://www-personal.umich.edu/~jaylemke/ sci-ed.htm

Maloney, D. P., O'Kuma, T. L., Hieggelke, C. J., & Van Heuvelen, A. (2001). Surveying students' conceptual knowledge of electricity and magnetism. *American Journal of Physics*, *69*(S1), S12–S23. .doi:10.1119/1.1371296

Marais, P., & Jordaan, F. (2000). Are we taking symbolic language for granted? *Journal of Chemical Education*, *77*(10), 1355. .doi:10.1021/ed077p1355

Marton, F., Hounsell, D. J., & Entwistle, N. J. (Eds.). (1984). *The experience of learning*. Edinburgh, UK: Scottish Academic Press.

Martyn, M. (2007). Clickers in the classroom: An active learning approach. *EDUCAUSE Quarterly*, *30*(2). Retrieved from http:// www.educause.edu/.

McDaniel, C. N., Lister, B. C., Hanna, M. H., & Roy, H. (2007). Increased learning observed in redesigned introductory biology course that employed Web-enhanced, interactive pedagogy. *CBE Life Sciences Education*, *6*(3), 243–249. .doi:10.1187/cbe.07-01-0003

Miller, V., Gregg, G., & Kelly, B. (2000). *Peer assisted study sessions (pass) in a first year biology teaching program-strategies developed by undergraduate pass leaders*. Paper presented at the Effective Teaching and Learning at University Conference, Duchesne College, The University of Queensland. Retrieved from http:// www.tedi.uq.edu.au/ conferences/

Nieswandt, M. (2001). Problems and possibilities for learning in an introductory chemistry course from a conceptual change perspective. *Science Education, 85*(2), 158-79. doi: 10.1002/1098-237X(200103)85:2<158::AID-SCE40>3.0.CO;2-3

O'Byrne, J. W., Britton, S., George, A., Franklin, S., & Frey, A. (2009). Using academic predictors to identify first year science students at risk of failing. *CAL-laborate International, 17*(1), 15-25. Retrieved from http:// sydney.edu.au/

Odom, A. L., & Kelly, P. V. (2001). Integrating concept mapping and the learning cycle to teach diffusion and osmosis concepts to high school biology students. *Science Education*, *85*(6), 615–635. .doi:10.1002/sce.1029

Perry, W. G. (1988). Different worlds in the same classroom . In Ramsden, P. (Ed.), *Improving learning: New perspectives*. London, UK: Kogan Page.

Pickersgill, S., & Lock, R. (1991). Student understanding of selected non-technical words in science. *Research in Science & Technological Education*, *9*(1), 71–79. doi:10.1080/0263514910090107

Pipes, R. B., & Wilson, J. M. (1996). A multimedia model for undergraduate education. *Technology in Society, 18*(3), 387-401. doi: 10.1016/0160-791X(96)00020-6

Pollock, S. J. (2008). *Comparing student learning with multiple research-based conceptual surveys: CSEM and BEMA*. Paper presented at the Physics Education Research Conference 2008, Edmonton, Alberta, Canada. Retrieved from http:// www.colorado.edu/

Richardson, A., & Zhang, F. (2008). *Language support for statistics learners*. Paper presented at the OZCOTS 2008 - 6th Australian Conference on Teaching Statistics, Melbourne. Retrieved from http:// sky.scitech.qut.edu.au/ ~macgilli/ ozcots2008/

Richardson, A., Zhang, F., & Lidbury, B. A. (2008). *Activating multiple senses in learning statistics*. Paper presented at the Assessment in Science Teaching and Learning Symposium, The University of Sydney, Sydney. Retrieved from http:// sydney.edu.au/

Ritchie, S. M. (2001). Actions and discourses for transformative understanding in a middle school science class. *International Journal of Science Education*, *23*(3), 283–299. .doi:10.1080/095006901750066529

Roth, W.-M. (1998). Science teaching as knowledgability: A case study of knowing and learning during coteaching. *Science Education, 82*(3), 357-77. doi: 10.1002/(SICI)1098-237X(199806)82:3<357::AID-SCE4>3.0.CO;2-B

Roth, W. M., & Roychoudhury, A. (1994). Science discourse through collaborative concept mapping: New perspectives for the teacher. *International Journal of Science Education*, *16*(4), 437–455. .doi:10.1080/0950069940160405

Roth, W. M., Tobin, K., Zimmermann, A., Bryant, N., & Davis, C. (2002). Lessons on and from the dihybrid cross: An activity-theoretical study of learning in coteaching. *Journal of Research in Science Teaching*, *39*(3), 253–282. .doi:10.1002/tea.10018

Roy, H. (2003). Studio vs interactive lecture demonstration–Effects on student learning. *Bioscene: Journal of College Biology Teaching, 29*(1), 3-6. Retrieved from http:// acube.org/ pdf/ bioscene/ v29-1.pdf

Savinainen, A., Scott, P., & Viiri, J. (2005). Using a bridging representation and social interactions to foster conceptual change: Designing and evaluating an instructional sequence for Newton's third law. *Science Education*, *89*(2), 175–195. .doi:10.1002/sce.20037

Schon, D. A. (1987). *Educating the reflective practitioner: Toward a new design for teaching and learning in the professions*. San Francisco, CA: Jossey-Bass.

Schulte, J. (2006). *Delivering first year physics assignments with limited resources - An Australian three-centre study*. Paper presented at the 2006 National UniServe Conference, Symposium of Science Teaching and Learning Research, The University of Sydney, Sydney. Retrieved from http:// sydney.edu.au/

Touger, J. S. (1991). When words fail us. *The Physics Teacher*, *29*(2), 90–95. .doi:10.1119/1.2343227

Wandersee, J. H. (1988). The terminology problem in biology education: A reconnaissance. *The American Biology Teacher*, *50*(2), 97–100. Retrieved from http:// www.jstor.org/.

Wellington, J. J., & Osborne, J. (2001). *Language and literacy in science education*. Buckingham, UK & Philadelphia, PA: Open University.

Williams, H. T. (1999). Semantics in teaching introductory physics. *American Journal of Physics*, *67*(8), 670–680. .doi:10.1119/1.19351

Wu, H.-K., Krajcik, J. S., & Soloway, E. (2001). Promoting understanding of chemical representations: Students' use of a visualization tool in the classroom. *Journal of Research in Science Teaching*, *38*(7), 821–842. .doi:10.1002/tea.1033

Yore, L. D., Florence, M. K., Pearson, T. W., & Weaver, A. J. (2006). Written discourse in scientific communities: A conversation with two scientists about their views of science, use of language, role of writing in doing science, and compatibility between their epistemic views and language. *International Journal of Science Education*, *28*(2-3), 109–141. .doi:10.1080/09500690500336601

Yore, L. D., Hand, B. M., & Prain, V. (2002). Scientists as writers. *Science Education*, *86*(5), 672–692. .doi:10.1002/sce.10042

Yore, L. D., & Treagust, D. F. (2006). Current realities and future possibilities: Language and science literacy-empowering research and informing instruction. *International Journal of Science Education*, *28*(2 - 3), 291–314. .doi:10.1080/09500690500336973

Zhang, F., & Barber, B. (Eds.). (2008). *Handbook of research on computer-enhanced language acquisition and learning*. Hershey, PA: Information Science Reference. doi:10.4018/978-1-59904-895-6

Zhang, F., & Lidbury, B. A. (2006). *It's all foreign to me: Learning through the language of genetics and molecular biology*. Paper presented at the 2006 National UniServe Science Symposium: Assessment in Science Teaching and Learning Research, The University of Sydney, Sydney. Retrieved from http:// sydney.edu.au/

Bruner, J. S. (1986). *Actual minds, possible worlds*. Cambridge, Mass.: Harvard University Press.

Ellem, G. K., & McLaughlin, E. A. (2005). *Tales from the coalface: From tragedy to triumph in a blended learning approach to the teaching of 1st year biology*. Paper presented at the Symposium of UniServe Science at the University of Sydney: Blended Learning: Design and Implementation, Sydney, NSW. Retrieved from http:// sydney.edu.au/

Harris, K.-L., Krause, K., Gleeson, D., Peat, M., Taylor, C., & Garnett, R. (2007). *Enhancing assessment in the biological sciences: Ideas and resources for university educators*. Retrieved from http://www.bioassess.edu.au

Kolb, D. A. (1984). *Experiential learning: Experience as the source of learning and development*. Englewood Cliffs, N.J.: Prentice-Hall.

Lave, J., & Wenger, E. (1991). *Situated learning: Legitimate peripheral participation. Cambridge* [England]. New York: Cambridge University Press.

Simpson, O. (2006). Predicting student success in open and distance learning. *Open Learning: The Journal of Open and Distance Learning*, *21*(2), 125–138. doi:.doi:10.1080/02680510600713110

Supovitz, J. A., & Turner, H. M. (2000). The effects of professional development on science teaching practices and classroom culture. *Journal of Research in Science Teaching*, *37*(9), 963–980. 10.1002/1098-2736(200011)37:9<963::AID-TEA6>3.0.CO;2-0.doi:10.1002/1098-2736(200011)37:9<963::AID-TEA6>3.0.CO;2-0

Tinto, V., & Pusser, B. (2006). *Moving from theory to action: Building a model of institutional action for student success*. Paper presented at the National Symposium on Postsecondary Student Success, Washington, DC. Retrieved from http:// nces.ed.gov/ npec/ papers.asp

Walker, J., & Halliday, D. (2008). *Fundamentals of physics* (8th ed.). Hoboken, NJ: Wiley.

Yeo, S., & Zadnik, M. (2001). Introductory thermal concept evaluation: Assessing students' understanding. *The Physics Teacher*, *39*(8), 496–504. .doi:10.1119/1.1424603

Zhang, F. Z. (2006). The teaching of Mandarin prosody: A somatically-enhanced approach for second language learners. (PhD thesis, University of Canberra, Canberra).

KEY TERMS AND DEFINITIONS

Cognitive Conflicts: Activities which have embedded in them a surprise or a misconception to the normal predictable way of thinking.

Interactive Engagement (IE): Interactive Engagement (IE) strategies have been shown to enhance the gain in learning. In an IE class, the lecturer sets up a demonstration and asks the students to predict and write down, with consultation among themselves, their prediction of how the demonstration will work out. Then the demonstration is carried out, the results discussed and the relevant theory presented.

Peer Instruction (PI): A technique Professor Eric Mazur to help make lectures more interactive and to get students intellectually engaged with what is going on. In this method the instructor presents students with a qualitative (usually multiple choice) question that is carefully constructed to engage student difficulties with fundamental concepts. Next, the students consider the problem on their own and contribute their answers in a way that the fraction of the class giving each answer can be determined and reported. Then, students then discuss the issue with their neighbors for two minutes and vote again. Finally, the issues are resolved with a class discussion and clarifications.

Studio Reaching: Consisted of a mixture of student exercises, instructor coaching, and sometimes laboratory experiments in a studio environment.

VotApedia Tool: A free audience response system that does not require issuing clickers or need specialist infrastructure. Known users can create surveys and edit the surveys on the site. Once signed on, students can participate in surveys either through mobile phones, online or through SMS. Both online through mobile phones are free to the students but there will be a charge through SMS.(http://www.urvoting.com/index.php?title=Main_Page).

Chapter 2
Language Support Strategies in First Year Chemistry

Adam J. Bridgeman
University of Sydney, Australia

ABSTRACT

In-class and online active learning strategies, which have been implemented at The University of Sydney, are described to respond to the challenge of teaching chemistry at a first year level, to students with a wide range of abilities and levels of motivations. Core to the design of these activities is the belief that students learn chemical concepts most effectively when they are actively engaged in doing and talking, rather than rote learning and listening. The strategies described have been developed within the context of large classes and limited resources. They are fully adaptable to other topics beyond first chemistry and to other sciences. They are also necessarily designed to be scalable to large or small classes and to be sustainable. Online resources are useful for helping students become familiar with chemical language and symbolism and to provide them with a means of practicing their use. Online quizzes are an invaluable means of students self-assessing their progress and of providing meaningful assessment of their level of mastery. In-class activities involving student response systems and student-centred, inquiry based approaches are built around active learning and on-going formative assessment. To develop language skills, social interaction via peer instruction and group discussions can be utilized to ensure that classes are vibrant and engaging.

DOI: 10.4018/978-1-61350-062-0.ch002

INTRODUCTION

Chemistry is the central science. An understanding of chemistry is important in many scientific disciplines and for an informed appreciation of technological, environmental and medicinal topics. First year chemistry is thus mandatory or at least highly recommended for most science students and is also chosen by students on many other degrees desiring knowledge of the scientific method. For example, around 20000 students undertake chemistry units across 35 Australian universities (Barrie, Buntine, Jamie, & Kable, 2001).

Teaching chemistry has been compared to teaching a foreign language (Markow, 1988). It is necessarily an abstract subject, dealing with how processes and events at the microscopic level have an effect at the macroscopic level and in everyday life (Ben-Zvi, Eylon, & Silberstein, 1988; Maskill & Cachapuz, 1989; Zoller, 1990) (Nakhleh, 1992). Many choose to study chemistry because they want, for example, to design new medicines and materials or want to understand biological and environmental systems. Before a student can understand these applications of chemical science and the way chemist interpret chemical phenomena, a good basis in its language, grammar and symbolism is required.

As noted by (Markow, 1988), chemists often use common words to mean something either much more specific or even completely different to the everyday usage. Examples include 'weak', 'strong', 'significant', 'concentration' and 'favorable'. To a chemist, these terms have a precise, conceptual connotation which is unlikely to be the same as that of the lay reader or a new student. Scientists have appropriated everyday adjectives like 'weak' and 'strong' and, at the same time, many scientific terms such as 'gas' and 'quantum' have entered the general vocabulary and had their original meaning changed or blurred. Perhaps most unfortunately, even the words 'chemical' and 'chemist' are often used in a completely different way in the media to how a practitioner would define them.

Probably the easiest language issues for an educator to deal with are the new words that a student will encounter. Sensitivity to the unfamiliarity and the pronunciation of terms like 'nucleophile', 'electrophile' and 'stoichiometry' is important. Such words can form a barrier to comprehension and a glossary can be an important addition to a course or text. Success in chemistry, however, requires a deeper mental representation and conceptual understanding of chemical terms than rote learning of definitions. For example, as well as being able to recite "mass per unit volume" and perform a calculation, a successful chemist will have a mental image of what density *is* (Beek & Louters, 1991).

As well as dealing with the different uses of words in everyday language and in chemistry, the educator must also be aware of misunderstanding and misconceptions in the definition of chemistry terms that students bring with them to university. The acquisition and development of chemical understanding and language is not built upon a blank slate (Beall, 1994; Bradley & Brand, 1985; Granville, 1985; Mason, 2006; Mayer, 2010; Nakhleh, 1992; Sanger & Greenbowe, 1999). Prior knowledge from high school can be particularly hard to alter, requiring awareness of the issue from both the educator and the student. In first year chemistry courses, students may not be open to such limitations in their knowledge and to their own misunderstandings. Preconceptions about the meaning of words are not easily changed (Cary, 1984).

Introductory courses also rapidly introduce and then rely completely on symbolic notation. According to Wellington and Osborne (2001), symbolic language have the highest level of abstraction of scientific words. As discussed by (Marais & Jordaan, 2000a), three types of symbolic notation are used in chemistry:

1. Letter symbols are used to represent the elements, units and physical quantities. Thus, the letter "K" is used to represent the element potassium, the unit of absolute temperature and (in italic form) the equilibrium constant.

The 'fluent' chemist can instantly infer which of these is meant from the context.

2. Cons such as arrows, braces (curly and square) and mathematic symbols (+ and -) are used to represent key chemical terms which can also depend on their context. Their uses are logical but the interpretation is subtly different to that used in mathematics.
3. Combinations of icons and letters are used to modify or narrow their meaning. For example, the symbol Δ is used to indicate change in a quantity so that ΔH is read as the 'change in the enthalpy (H)'.

A symbol such as Na is used to represent both a macroscopic lump and a microscopic quantity of sodium. The fluent chemist will apply meaning to the representation based on its context. The plus sign in the symbol Na^+ is used to indicate an excess of positive charge due to the *loss* of an electron whilst the negative sign in the symbol Cl^- is used to indicate an excess of negative charge due to the *addition* of an electron. A plus sign in a chemical equation is used to indicate 'reacts with' rather than 'added to'.

Reading and comprehension of a chemical equation requires a student to correctly translate and interpret each of the three symbol types based on their context and the order in which they appear. Even an introductory examination typically requires a student to use a large number of cognitive steps just to comprehend the questions. After the chemical equations are translated and interpreted, the student must solve the problem at hand and then report their answer in the same symbolic language. A successful chemist is, of course, fluent enough that the thought processes are performed directly in the symbolic language and these steps become unnecessary. A chemist will think in the symbolic language of chemistry just someone who is fluent in a foreign language will think in that language when reading

This chapter examines the role of language in the learning of chemistry amongst first year chemistry students at the University of Sydney, Australia. It describes a number of interventions that have been implemented to improve student learning and engagement. First year chemistry at the University of Sydney has one of the largest intakes in Australia. Typically, around 3800 students take first year chemistry units each year. These represent students from every faculty within the university. The teaching for this cohort is divided up with units tailored for those with and without high school chemistry as well as specialist units for students on professional degrees such as Pharmacy and Veterinary Science. Although the average admissions index is reasonably high, the student body represents a reasonably wide variety of ability levels. This covers 'Special Studies' and Veterinary Science students, who possess some of the highest possible high school chemistry grades, and others who enter with no post-16 chemistry knowledge. These students are taught in separate classes so that the content and its assessment are at an appropriate level and are based on assumptions about prior knowledge.

Just as importantly, the interest and motivation of the students varies hugely with only about a tenth planning to pursue chemistry beyond their first year. Motivation is a key element of success in science (Oliver & Simpson, 1988). The abstract nature of many introductory chemistry topics and the need to develop language skills can contribute to de-motivation of students with the 'pay-off' from the required learning often belated. It is just these aspects which give chemistry its reputation as being 'hard to learn' (Millar, 1991).

The engagement, retention and progression of students from a wider range of backgrounds and nationalities are now a core issue for all faculties. Traditional approaches to teaching designed to underpin the 'elitist' system of the past are unlikely to support the 'mass education' system now needed (Eilks & Byers, 2010). Although a consideration of language difficulties should lead to an improvement in learning for all students (Zhang, et al., 2010), language is a key accessibility issue for ensuring wider participation in the sciences.

Table 1. Answers to the question of "What is research?"

	Student response	Correct answer	Value
A	observing the results of a series of chemical reactions	Yes	33.33%
B	Answering an exam question.	No	-10%
C	Looking for information in the library or on-line	Yes	33.33%
D	Copying information out of a textbook	No	-10%
E	Testing a hypothesis	Yes	33.33%

LANGUAGE DIFFICULTY SURVEYS AND CLASS OBSERVATIONS

Students in three units were tested on comprehension of ten common terms used in first year chemistry: 'research', 'power', 'concentration', 'equilibrium', 'graph', 'system', equation', 'experiment', 'model' and 'significant' (see Appendix C, test 1). These words are used as basic currency in physics, chemistry and biology lectures and a typical educator might assume that no definitions would be necessary. They have precise meaning in chemistry and are also used in lay contexts with more flexible and approximate meanings. The three units, CHEM1001, CHEM1101 and CHEM1901, correspond to 'fundamentals' (no high school chemistry), mainstream (university average performance in high school) and advanced (high achieving school chemists) respectively were chosen to provide a good overview of the whole cohort. In the first part of the survey, students were provided with a chemical use of each term and were then asked to define its meaning by picking one or more definitions from a list.

For example, the following was provided as a use of the word 'research':

"We carry out research to find out the answers to scientific questions."

The students were asked:

'What is meant by the word "research" in the sentence above'

(There may be more than one correct answer)

They were provided with the five answers below in Table 1. The correct answers and the value ascribed to each are also indicated.

Each correct answer was allocated five marks each so that a student choosing all three would score 15/15, indicating a complete understanding of the term in different contexts. A score less than 15 suggests an incomplete understanding of the term. All 10 terms followed a similar format.

In the second part of the survey (see Appendix C, test 2), the students were asked to relate their confidence in their answer to these questions and these were marked according to Table 2.

The surveys were delivered through the learning management system (LMS) and were completed 60, 95 and 19 students in the three units corresponding to 15.5%, 20.5% and 16.4% for CHEM1001, CHEM1101 and CHEM1901 re-

Table 2. Student understanding

	Student response	Value
A	Yes, I understand the meaning of this word	100%
B	No, I do not understand the meaning of the word	-100%
C	I have some idea of the meaning of this word.	0%

Table 3. Language survey results for CHEM1001 (students with no high school chemistry) at The University of Sydney

Terms	Confidence level (%)	Understanding (%)	Delusion index	Ranking (easiest=1, hardest=10)
Significant	56.90	1.90	29.95	10.00
Model	64.70	15.40	4.20	9.00
Power	31.70	24.60	1.29	8.00
System	45.10	26.90	1.68	7.00
Equilibrium	59.30	32.10	1.85	6.00
Research	64.10	36.90	1.74	5.00
Concentration	80.40	43.90	1.83	4.00
Graph	84.90	49.10	1.73	3.00
Equation	82.70	53.80	1.54	2.00
Experiment	78.80	75.00	1.05	1.00

Table 4. Language survey results for CHEM1101 (students with average high school chemistry) at the University of Sydney

Terms	Confidence level (%)	Understanding (%)	Delusion index	Ranking (easiest=1, hardest=10)
Significant	51.80	4.80	10.79	10.00
Model	67.50	18.10	3.73	9.00
Power	41.60	19.30	2.16	8.00
Research	62.40	20.00	3.12	7.00
System	53.00	28.60	1.85	6.00
Equilibrium	83.50	50.60	1.65	5.00
Concentration	80.20	51.20	1.57	4.00
Graph	83.30	53.60	1.55	3.00
Equation	78.30	57.80	1.35	2.00
Experiment	67.10	68.70	0.98	1.00

spectively. The results for the three sets of students are listed in Table 3, Table 4, and Table 5 below.

As noted above, preconceptions and misconceptions about scientific terms are particularly difficult to shift. Based on the data in the first two columns of the survey results, a crude 'delusion index' can be defined as the ratio of confidence to understanding by simply dividing the confidence value by the corresponding understanding score. An item for which the cohort's confidence is matched by their understanding has an index close to 1. An item for which the confidence is high but the understanding is low corresponds to a high index and decidedly delusional students. The delusion index is included as the third column in Tables 3 – 5. Lastly, the difficulty of the terms is ranked in the fourth column corresponding to the percentage of students who achieved 100% understanding of any terms.

The results suggest that terms such as 'model', 'significant', 'research', 'power' and 'system' are the most difficult for science students. In order to

Table 5. Language survey results for CHEM1901 (students with very good high school chemistry) at The University of Sydney

Terms	Confidence level (%)	Understanding (%)	Delusion index	Ranking (easiest=1, hardest=10)
Significant	64.70	5.60	11.55	10.00
Research	60.00	28.60	2.10	9.00
Power	26.30	28.60	0.92	8.00
System	50.00	31.60	1.58	7.00
Model	64.70	50.00	1.29	6.00
Graph	94.40	52.60	1.79	5.00
Equation	94.10	55.60	1.69	4.00
Equilibrium	88.90	57.90	1.54	3.00
Concentration	84.20	60.00	1.40	2.00
Experiment	82.40	77.80	1.06	1.00

display 100% understanding of the term 'research', students had to tick all three correct answers (A, C and E in the scheme above), covering uses of the term in scientific and non-scientific contexts. The more popular choice ticked was 'E = Testing a hypothesis' which is presumably the scientific definition with which they are most familiar.

Alongside this attempt to provide a numerical analysis of the language difficulties in first year chemistry at Sydney, a number of lectures and tutorials were observed by an expert in foreign language teaching (Zhang, Lidbury, Schulte, Bridgeman, & Yates, 2008). Several factors emerged as possible issues in the implementation and use of language orientated strategies in chemistry classes:

- The physical space and layout of traditional, steeply tiered lecture theatres encourages a didactic style of teaching and a passive learning style for the students.
- The large class sizes makes the traditional lecture the most cost efficient means of 'knowledge transfer' and leads to online and multiple choice questions being the most economical means of assessment.
- The amount of content covered encourages surface and rote learning.
- Students' ability to communicate, both in writing and orally, are often not distinct learning outcomes and are rarely directly assessed.

LANGUAGE METHODS EMPLOYED TO SUPPORT LEARNING IN FIRST YEAR CHEMISTRY

The interventions that have been developed and trialed in first year chemistry courses at The University of Sydney, Australia are described. A number have also been extended for use in intermediate and senior units. The activities are described in two sections covering those designed to be delivered online and those most deliberately constructed to promote active learning and language skills in the lecture theatre and tutorial. Despite this convenient division, it should be noted however that both there is some blurring of this distinction in their pragmatic use.

The language surveys and class observations outlined above are consistent with the literature discussed in the introduction: language difficulties are a significant issue in students learning, engaging and succeeding in chemistry. A number of strategies were considered and implemented in first year chemistry class at The University of Sydney in response to this research. These interventions coincided with a tightening of budgetary restrictions, focusing the need for the changes to be sustainable and adaptable enough to be adopted by a variety of educators in a research intensive school.

The interventions were designed to (a) make lectures more accessible, (b) increase opportunities for small group participation; (c) increase opportunities to develop students' communication skills and (d) create opportunities for providing formative feedback through both face-to-face and online activities.

ONLINE ACTIVITIES

Use of Flashcards and Games

Familiarity with chemical vocabulary is an essential starting point for a student to develop conceptual understanding. This 'phrase book' includes the names of the elements of the periodic table and common chemicals. Although there are now systematic rules for naming chemical compounds, students are also faced with (at least a thousand years of) historically-derived and very entrenched names. It is also vital that students rapidly become able to 'translate' between these names and their chemical symbols. A certain amount of 'rote learning' of these terms and symbols is thus a useful and necessary first step in developing key concepts.

We have developed software (A.J Bridgeman, 2010) to make online flashcards to allow students to rehearse important chemical words and to practice pairing of terms and symbols. This software requires the educator to input terms and definitions. The 'definitions' can range from symbols and formulae to images and animations. The software generates two-sided flashcards which are available for students to play back in random or controlled order in their own time and at their own time. The author has also used the flashcards to display on the 'big screen' in the lecture theatre whilst students are arriving in class. The database of terms and definitions is also used to generate an online glossary, a simple mix and match memory game and, where appropriate, randomly populated crosswords using the definitions as the clues and the terms as the answers.

Figure 1 shows a screenshot of the two sides of flashcards produced for (a) common cation and anion names and their chemical symbols and (b) matching an acid with its conjugate base. The first of these illustrates the use of flashcards in rehearsing basic vocabulary and chemical symbolism. The second shows how they can be used for a higher level activity involving making the connections between two chemical types as well as for linking chemical formulae and structures.

Although primarily designed to provide online resources, we have also utilized printed versions of the flashcards and crosswords in tutorial classes to provide ice breakers or revision tools. The generation of flashcards by students also makes a useful tutorial exercise in developing effective learning styles.

Where appropriate, sound files are attached to the terms so that they play back in sync with the flashcard. Chemical names and terms may be derived from a wide number of languages and can be longer than most other words that students are likely to have encountered. The pronunciation of chemical terms is thus not always obvious (even to chemists) and many educators will have experienced student embarrassment that can result from this. By simultaneously exploiting the visual and aural senses and by using 'the human voice behind the words' (Sutton, 1992), the activity becomes more effective for students with English as a first and second language.

Flashcards, crosswords, su doku, mix and match, simple ordering and drag and drop games (A.J. Bridgeman, 2005; A.J Bridgeman, 2010) are a useful alternative to the question types available in learning management systems. These online and face-to-face strategies provide repetition, forma-

Figure 1. Chemical flashcards used to relate (a) chemical names and formula for common cations and anions and (b) the formulae and structures of conjugate acid / base pairs

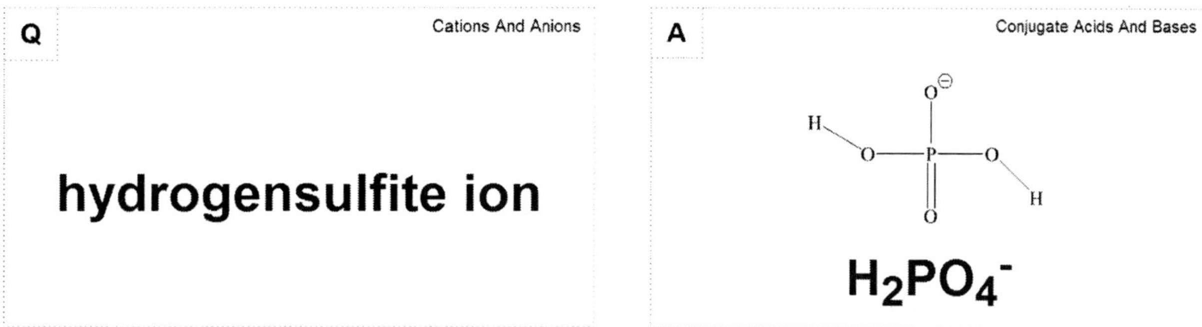

(a)

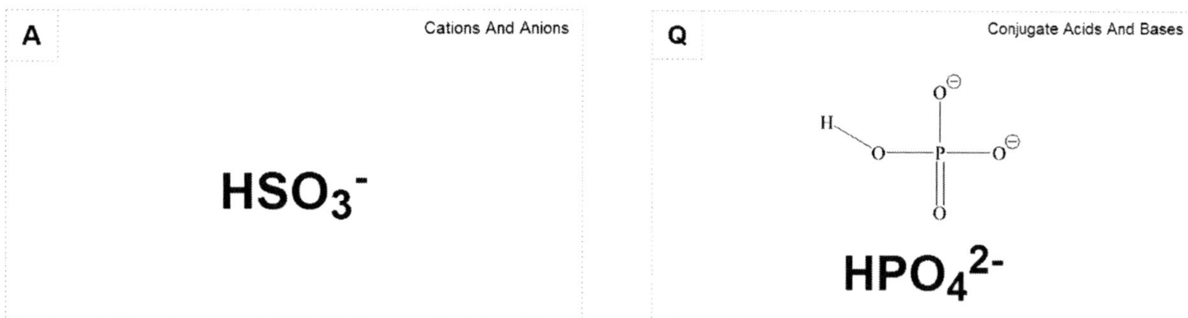

(b)

tive assessment, with instant feedback, and help encode information into the long term memory (Svinicki, 2004).

DESIGN OF ONLINE QUIZZES

Online, pre-laboratory quizzes are used in first year chemistry to encourage students to come prepared to class. Traditionally, these have involved multiple choice questions based on reading the laboratory handbook. One of the main problems with these questions is that encourage students to passively skim read the notes searching for the particular topics. We have also tended to overuse calculation questions, possibly because it is easy to quickly write alternative versions to reduce plagiarism. These have now been augmented by cloze (fill in the blanks or gap-fill) exercises, matching and jumbled sentence activities. These require students to more actively read the text in order to pick out words and phrases and to be able to reshape them and construct the meaning.

A particularly suitable activity for pre-laboratory preparation, and one which requires comprehension, is sentence ordering. Typically, laboratory instructions at first year level involve a list of instructions and hence may resemble a 'recipe' for the students to follow. We have used two types of sentence ordering quizzes. One involves simple ordering of the instructions which are repeated verbatim from the laboratory handbook. The second involves ordering instructions which are condensed and rewritten versions of those in the handbook. As well as requiring comprehension and engagement from the student, both activities build towards the higher level skills of students designing their own experimental instructions and flowcharts.

With large cohorts of students, these types of questions allow meaningful activities to be constructed, delivered and assessed. Active learning is encouraged and students are provided with instant feedback on their understanding.

We have extended these activities to include mini-research tasks on chemical safety and on using chemical databases. Students have to access technical websites, identical to those used by research workers, and are given guidance on selecting and evaluating the information embedded on these sites. Materials Safety Data Sheets (MSDS), for example, are used in many industrial and business contexts as well as in chemical laboratories. An ability to understand the technical language on these online sources is an important graduate skill that a technician should possess. Equally, a comprehension of the *meaning* and limitations of the warnings is an important attribute that a professional scientist should acquire. Thus, alongside questions asking students to match safety definitions and terms, we also include cloze activities which involve students constructing their own MSDS forms for the chemicals they have met in the laboratory and to interpret whether a procedure is safe or hazardous.

For the 'Fundamentals of Chemistry A' unit (CHEM1001), we introduced weekly, post-lecture quizzes to encourage students to review and reflect on their understanding of each week's material. As the students in this class have no high school background, it is extremely important that the students keep up a steady work rate and gets to grips with the basic language of chemistry in the early weeks of the course. In particular, it is important that students quickly become familiar with the grammar of chemical names and the symbolism used for representing compounds, cations and anions (Marais & Jordaan, 2000b). The quizzes were built using the quiz template available in Flash, modified by the author to read the questions from a database. The quizzes used short answer, drag and drop and matching formats as well as multiple choice questions.

The quizzes were based around a short online review of the concepts and were constructed using randomly selected questions from a large database. To encourage them to reflect and to make effective use of the feedback they received, students were allowed multiple attempts at each quiz within a short assessment period, linked to the appropriate part of the lecture timetable, with their higher mark being recorded. Figure 2 shows screenshots from three of these quizzes.

Figure 1(a) shows a quiz develop to develop understanding of atomic symbols. These use combinations of letters and icons which students are expected to be able to read and extract information from. As noted by (Marais & Jordaan, 2000b), we often take an understanding of such symbols for granted and it is important that we address this early (Markow, 1988) and that students instead actively rehearse their understanding.

The second two quizzes shown in Figure 2 are designed to develop comprehension of the nomenclature of inorganic compounds. The first of these, Figure 2(a), is a multiple choice quiz which enables the common incorrect answers to be used as distracters. These include the small errors of basic 'grammar', such as incorrect use of spaces and capitalization, which students often dismiss as unimportant but which ultimately make learning the language so difficult. Students receive appropriate feedback when such an answer is chosen.

The second nomenclature quiz, Figure 2(b), utilizes the same database but requires the student to type in the correct name of the compound. This is requires a much better understanding of the topic than simply choosing from a list and is just the writing skill that they will need when writing reports or exam answers. Although it is much more difficult to construct personalized feedback for short response questions, the database

Figure 2. Examples of online, weekly post-lecture quizzes on (a) extracting information from atomic symbols, (b) and (c) the grammar of chemical compound nomenclature

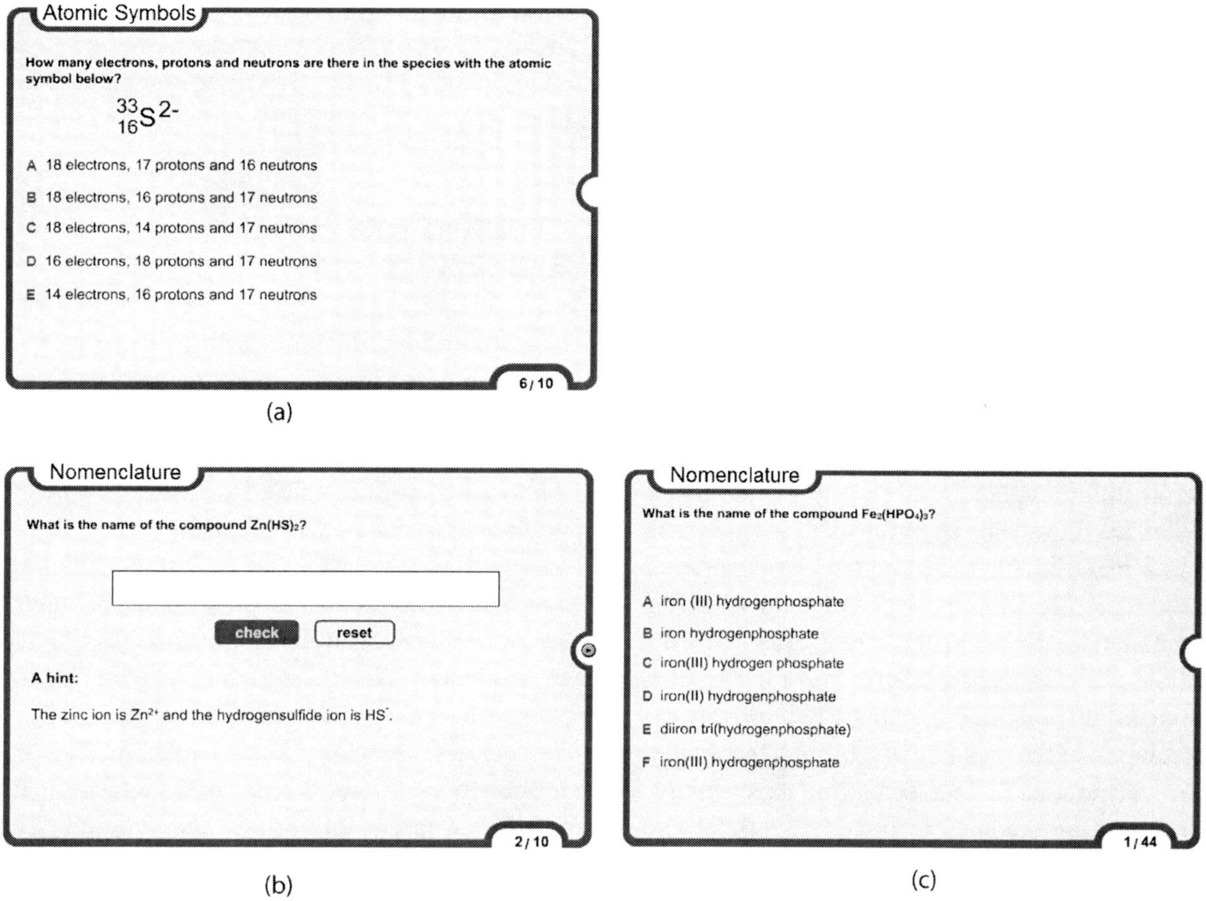

includes the common incorrect answers for use in the multiple choice quizzes and is thus able to provide at least some suggestions if one of these is entered by the students. It is also programmed to detect inappropriate use of spaces, capitalization and hyphens.

It is worth noting (D. M. Bunce, VandenPlas, & Havanki, 2006) that formative feedback quizzes and activities appear to give lead to superior student achievement in mid-semester tests than do those which utilize student response systems ('clickers') in lectures, possibly because the former allow for a certain amount of reflection and review. However, the long term benefit of online quizzes can be minimal as students do not spend time reviewing them before the end of semester examination.

To address this, the online post-lecture quizzes described and illustrated above were re-released in the study period between classes finishing and the examination with students able to recover lost marks through re-attempting the tutorials and quizzes.

Student feedback on the post-lecture quizzes was extremely encouraging. Although they required weekly work with each quiz contributing only 1% of the overall unit mark, students were very diligent in completing them. Around 80% of the cohort of 500 students completed all 10 quizzes and around 90% completed 9 quizzes. On average, students spent around 45 minutes on the quizzes each week. Such engagement contrasts very starkly with the effort most students put into their written homework. Feedback on the unit of study evaluation also indicated that students appreciated the motivation behind the quizzes: "I'd have been lost on this course without these quizzes" and "It was good to get loads of goes as I never get chemistry until I've done it a thousand times."

DATA ANALYSIS

In view of the difficulty students had defining 'significant' in the language survey described above and our separate desire to improve computer skills, we introduced an online data analysis task to our experimental program (A.J Bridgeman & Schmid, 2010). This involved students performing a detailed numerical analysis of the results of the whole class on a spreadsheet. From the results, students were guided towards a data-driven appreciation of the meaning of the terms 'precision' and 'accuracy' in real scientific experiments. By requiring students to consider outliers and experimental uncertainties and to justify their numerical results, they construct an appreciation of the meaning of significance in experiments.

IN CLASS ACTIVITIES

Even the most engaging educator will know that student attention will drift during a class and that this can be as dependent on the time of day and the weather as the lecture topic (McKeachie, 2002). A recent research study in chemistry classes by (D.M. Bunce, Flens, & Neiles, 2010) suggests that student attention spans can be as short as a few minutes. This section describes two student-centred pedagogies that we have found to work to improve student engagement and hence decrease attention lapses. The first describes our use of clickers to provide two-way feedback during class. As the use of clickers in chemistry classes is described in more detail elsewhere in this volume, it is deliberately short. The second describes the use of worksheets and group activities based on an inquiry based approach to student-centred learning.

STUDENT RESPONSE SYSTEMS (CLICKERS)

In a traditional, didactic lecture, there is little opportunity for students to ask questions, for the lecturer to gauge the level of understanding of the class or for students to organize their thoughts. Although a good didactic lecture can be inspiring, in general, students are not challenged; they do not have a chance to reflect on their understanding. and receive no feedback. Only the best students are actively engaged in the learning process and all students will have several attention lapses. Some may completely disengage.

A typical chemistry lecture often involves the introduction of one or more new concepts with associated terms, definitions and symbols. The consequences of these concepts are then explored through examples and problem solving. If the class has not understood the concept, they become 'lost' and any learning stops. Within a course, the concepts build upon those previously introduced: chemistry is a linear subject and a student who is lost early on can find it very difficult to recover.

Student response systems (SRS) can change the dynamics of the classroom by enabling real time feedback so that students can gauge their own understanding and instructors can monitor how well the students have mastered material. Australian (AUSSEAUSSE: Australian Survey of Student Engagement, 2008; K-L. Krause, Hartley, James, & McInnis, 2005) and international studies (Hounsell, McCune, Hounsell, & Litjens, 2008; Rust, 2002; Yorke & Longdon, 2008) point to low student satisfaction with the amount, quality, and promptness of feedback that they receive. These studies report that this area often scores lowest on student evaluations. In contrast to satisfaction in other areas, satisfaction with feedback provision is actually in decline. SRS offer in-class feedback during the learning process. Most importantly, the quality of the feedback is actually enhanced in large classes as each individual is provided with a statistical comparison of their performance with that of a large number of students.

This real-time statistical analysis of the level of understanding in the class is also very useful for the instructor. The ability to gauge the level of comprehension of a topic whilst the students are still present is a privilege. Just as with the online quizzes discussed above, multiple choice questions with well chosen distracters not only provide information on how well the class has understood but also precisely where the misconceptions and misunderstandings are. Of course, this information is useless unless the instructor is prepared to deviate from the lecture slides at this point. Often the most effective learning experiences occur when the presentation is put aside and the class becomes an ad-lib discussion.

The two questions below were used in a class on strong *vs* weak acids. This topic involves a number of language issues. Students will have varying levels of background knowledge on this topic based on lay definitions and on discussion in other courses, especially biochemistry. These ideas are very difficult to shift and must be confronted directly. The distractors were based on the answers received when the same questions were given as short answer problems and thus span the most common misconceptions.

1. Alongside H_2O, what are the major species present in a 1.0 M solution of HCl(aq)?
   A. HCl, H_3O^+ and Cl^-
   B. H_3O^+ and Cl^-
   C. HCl
   D. HCl, H_3O^+, OH^- and Cl^-
   E. H_3O^+, OH^- and Cl^-

2. Alongside $H_2O(l)$, what are the major species present in an aqueous solution of 1.0 M CH_3COOH?
   A. $CH_3COOH(aq)$, $H_3O^+(aq)$ and $CH_3COO^-(aq)$
   B. $H_3O^+(aq)$ and $CH_3COO^-(aq)$
   C. $CH_3COOH(aq)$
   D. $CH_3COOH(aq)$, $H_3O^+(aq)$, $OH^-(aq)$ and $CH_3COO^-(aq)$
   E. $H_3O^+(aq)$, $OH^-(aq)$ and $CH_3COO^-(aq)$

The correct answers are B and C respectively. The most common incorrect answers are C and B respectively. When a class gives these answers, it is clear that there is a misunderstanding of the terms strong and weak acid and the instructor must re-address this topic.

The introduction of short quizzes using SRS also helps to break up the lecture into small segments and it has been demonstrated that this can help prevent attention lapses (D.M. Bunce, et al., 2010). Although it is sometimes useful to introduce a rapid 'pop quiz' without allowing students to prepare, generally SRS segments should include a period to allow students to reflect on the material that has been covered before committing themselves to an answer. It is always worthwhile using a short period after a quiz and any instructor led discussion for students to reflect and even chat about the responses.

Indeed, SRS segments in class offer a very natural way to encourage peer discussion. By providing a goal (a successful response to the SRS question), the segment becomes an excellent way of encouraging even the most introverted to take part in group discussions. By combining individual and group questions, the segment can also be used to illustrate to even the high achieving students that group work has benefits (M.K. Smith, et al., 2009). Students have to discuss the concept and are forced to agree a group response. As well as building their comprehension, they must vocalize their conceptual understanding and develop key language and communication skills. Students will

Figure 3. The structure of the antihistamine, ZyrtecTM, is shown below in its 'stick' structure representation with the hydrogen atoms bonded to carbon not shown

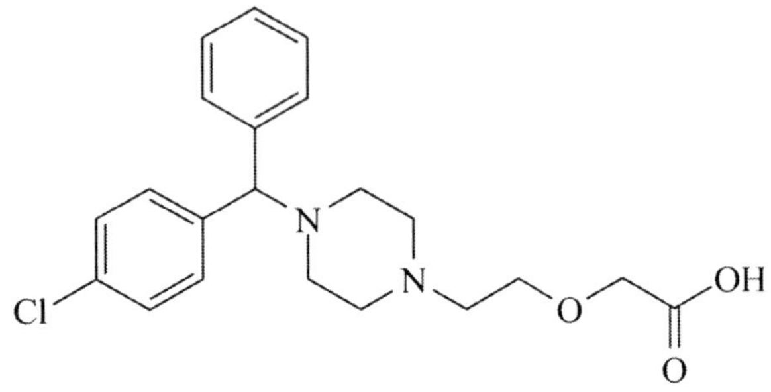

be much more willing to expose their misconceptions and challenge the ideas of other students in small groups than they will to the lecturer or in front of the whole class.

'Clickers', handheld SRS devices which use microwave technology to interact with software on the lecturer's computer, are commonly used in large classes. Alternatives include the 'VotApedia' system based on students own mobile phones which is described in detail elsewhere in this volume. However, it is possible to use distinctly low tech alternatives such as a show of hands and voting using red and green cards. We have found these to be *at least* as effective as electronic systems, at least when used in moderation. They also have the benefit of involving physical activity from the students and can also help to lighten the mood.

For example, at the start of a course on organic chemistry, students must learn to translate between the 'stick' structure representation of complex organic molecules and their formula (see Figure 3). This representation is used throughout chemistry and biochemistry, primarily to simplify the drawing and save time. This new symbolic representation can be difficult for students to pick up and yet, without this skill, they will be lost as the organic course moves rapidly onto organic isomerism and transformations. The students are asked to stand up and are shown the complex organic molecule below. They are then asked the series of questions below and remain standing or sit down depending on their response. Those answering correctly proceed to the next question until the whole formula has been derived and the winner(s) is announced. This exercise is repeated with 2 or 3 new structures.

1. If you think there are 2 nitrogen atoms in the formula, stay standing. Otherwise sit down. (Students answering correctly are still standing. The others are eliminated.)
2. If you think there are 2 oxygen atoms in the formula, stay standing. Otherwise sit down. (Students answering correctly are sat down. Those standing are eliminated.)
3. If you think there are 19 carbon atoms, stand up. (Those who stand up are eliminated.)
4. If you think there are 20 carbon atoms, stand up. (Those who stand up are eliminated.)
5. If you think there are 21 carbon atoms, stand up. (Those who stay sat down are eliminated.)

6 – 8 Repeat questions 3 – 5 with 27, 26 and 25 hydrogen atoms.

Students appreciate and enjoy SRS segments in lectures with feedback on evaluations generally very positive. They particularly like receiving feedback they receive from these segments as part of the learning process and the chance to interact with their peers during a lecture. Two comments from unit of study evaluations underline this: "liked working with my friends rather than listening/sleeping" and "getting feedback from clickers every lecture." The study by (D. M. Bunce, et al., 2006) suggests, however, that SRS questions do not significantly affect mid-semester quizzes or end of course examinations. These workers suggest that their benefits can be short term as they are not available out of class for students to use for revision or reflection.

Many publishers include sets of pre-formatted questions on textbook websites. (Asirvatham, 2009) has recently published an excellent set of 250 field-tested questions covering macroscopic and microscopic topics and symbolic representations and a pedagogical commentary on their use. Such resources will aid their adoption by more instructors (Towns, 2010). Most colleagues are aware of their potential but are put off by the perceived administrative difficulties involved in the distribution and collection of the devices. For their use to be sustainable and equitable, investment by universities to purchase the devices and / or the development of administration systems to loan them to students need to be investigated.

STUDENT-CENTRED, INQUIRY BASED APPROACH TO LECTURES AND TUTORIALS

In most lectures in science and chemistry, the vast majority of class time is taken up with the instructor. Depending on the nature of the class, there may be some student-instructor interaction although many large class make it very hard for this to happen and science students are often cautious and uncommunicative. Such an approach is an efficient way to cover a large amount of material but almost guarantees that students adopt a rote learning, "what is on the exam" approach and assessments involving memorizing algorithms (Mazur, 1997). This teaching style is not conducive and may even by incompatible with language development and improving graduate attributes such as communication skills and group work. To actively engage students, they must be given something to do!

To combat language issues, to facilitate conceptual development and to deal with misconceptions, it is important to know what is going on in the student's mind. As discussed above, a SRS segment may provide this although its focus is still the instructor. This section describes student-centred, inquiry-based approaches for large lectures and for tutorial classes. This is based on a version of the 'Process Orientated Guided Inquiry Learning' (POGIL) approach (Moog, 2008) but includes other student-centred activities that have been found useful. Its particular Australian implementation involves refinements for large classes with single instructors known as 'Active learning in University Science – Leading Change in Australian Science Teaching' (ALIUS) (Bedgood Jr, Gardiner, et al., 2010; Bedgood Jr, Yates, et al., 2010; Bedgood Jr., et al., 2008). This is part of the ALTC project 'Developing Leaders Of Change In The Teaching Of Large University Chemistry Classes' project.

In the ALIUS approach, a large proportion of the traditional lecture time is given over to student-centred activities. In some implementations, this can be a one or two whole lectures each week. In the author's class at The University of Sydney, each lecture is broken up into several segments typically involving 2 or 3 ALIUS activities interspersed with SRS sections, chemical demonstrations and traditional lecture segments. The classes generally begin with an instructor-led review of the previous class with perhaps some review questions for the students. In the ALIUS parts, students working in small groups use specially designed activities that guide them through new concepts. Students work in groups and they must take an active role: the individuals in the groups learn by reading, asking each other questions, listening to each other's explanations and discussing the concepts. The emphasis is on social interaction and cooperative learning. The worksheets used in these activities follow the learning cycle (Abraham, 2005) involving questions which guide them through:

- 'exploration' phase in which students read or look at data and develop a hypothesis,
- 'concept invention' phase in which new concepts and terms are introduced based on this understanding and
- 'application' phase in which these concepts are applied to new situations.

In a large class, these phases may be broken into 2 or 3 segments of 5 – 10 minutes in length. They are followed by review sessions to give feedback, sometimes involving SRS questions, and ensure that misconceptions are dealt with.

In the initial use of this format, the worksheets were tackled in longer timeslots. However, it was soon discovered that student attention lapses occur and off-task conversation can become disruptive if the timeslot is greater than 10 minutes. We have found that a combination of 2 – 3 minute and 8 – 10 minute segments work well. These problems are partly due to the large size typical first year lecture classes and to the physical limitations of tiered lecture theatres which make it difficult for the instructor to effectively facilitate and keep everyone on-task. As discussed below, longer exercises have been found to work well in our tutorial classes.

Appendix A contains two typical worksheets given out to the 'Fundamentals of Chemistry B' (CHEM1002) class in semester 2, 2010. Both sheets replaced 5 – 10 lecture slides and required twice the length of time that was taken to deliver the same material in a conventional style. Appendix B contains more similar worksheets on other topics. Adopting student centred approaches such as these unavoidably requires removal of some content from the lecture. In the absence of general agreement about a reduction of syllabus content, material which is purely information can be given as out-of-class reading.

The first, covering an 'Introduction to Acids and Bases', starts off with students reading a new definition of what an acid and bases are. They are then asked to critically analyze this definition and use this and their chemical knowledge to develop an understanding of the concept of proton transfer and the relationship between an acid and its conjugate base. This relationship is not formally given in the Model but rather students 'discover' it themselves in their groups. The first side ends with by asking the group to commit themselves to a definition of this relationship in a grammatically correct sentence. 10 minutes is given over to this activity. It is followed by a review session involving 2 or 3 groups communicating their answer to the full class with the best written up on the board as the agreed class definition. The second side reinforces this definition and the students have to then apply this knowledge to change the chemical formula, requiring them to work within the symbolism of chemistry. Finally, students have to apply their understanding to the novel liquid ammonia system, with which they have no experience. The students can then consolidate their learning by completing the set of exercises at the end. In a large class, the speed of different groups can vary considerably and these exercises provide some reward and incentive for keeping on task.

The second, covering 'Electronegativity and Bonding', the concept of electronegativity is introduced by students identifying and expressing in words the trends in data. The students then have to transform this knowledge into a graphical form and use it to make predictions about new systems. The sheet ends with each group reflecting on their understanding and agreeing with a description that they will communicate with the class. This kind of activity is very effective in developing students' ability to read and order tabulated and graphical data. These skills are developed in a subject-based context and are a natural part of communicating in a scientific way. Process and language skills such as information processing, critical thinking, communication, self-assessment and self-reliance, are built to assist in the mastery of the content and, at the same time, these graduate attributes are developed.

As this approach has been refined during 2009 and 2010, one of the biggest changes has been a tightening up of the group activity time. In a lecture theatre, 10 minutes appears to be maximum time for efficient work. Time limits are given and stuck to.

At the core of the success of this approach is that students are active in the lecture. They must communicate and must write down chemical explanations and symbols rather than simply copy down or passively read the words on the projector. In the author's implementation, the worksheets are only available in hard copy to those attending classes. With lectures available to download as documents and increasingly as podcasts and in video format, students are increasingly voting with their feet and not turning up to class. Perhaps because of the quality of the ALIUS classes and perhaps because they miss out on material otherwise, attendance at these classes is much higher than in classes taught in a traditional format.

Almost all students enjoy this style of lecture. On the unit evaluation, there was a marked improvement on the question "Overall I was satisfied with the quality of this unit of study") from a baseline of 3.94 in 2007 to 4.16 in 2008 and 4.23 /5 in 2009 following the phasing in and refinement of its implementation. As this is a terminating unit for 95% of the students, these scores and improvements are very pleasing. In comparison, this question averages about 3 – 3.5 /5 in other first year science units. From the lecturer's personal evaluation in 2009, on the question, 'what did you like most in this unit?', comments from students include:

"daily work sheets very effective in learning concepts"

[positive] "questions to reinforce understanding"

"we learnt concepts thrwough thinking. Better than being told"

"use of problem sheets engaged students. I'm used to falling asleep"

"made everyone involved to increase learning, especially with work sheets and clickers"

"interactive, made w do working/exercises during lectures!"

"I liked the sheets he gave. More like a tutorial than a lecture. Liked having prizes – especially chocolates"

"made learning interesting"

"the sheets and clickers were great for consolidation"

"interactive learning"

"critical thinking questions were awesome"

"make other subjects – and other chemistry lecturers – use critical thinking questions and clickers"

"liked working with my friends rather than listening/sleeping"

In 2010, this approach has been extended to tutorial work. Students are given homework to complete before tutorials and, traditionally, our tutorials have involved the tutor going over the answers to these problems at the board. Some of the tutors used group work and extension activities. Even in these small classes of 20 – 30 students, student engagement and activity could be minimal and was very dependent on the particular tutor. We have now introduced ALIUS style worksheets for these classes too. As flat rooms with moveable furniture are used for tutorials, more formal group roles can be used when appropriate. For example, one student is given the responsibility for leading the discussion whilst another is given the role of reporting back to the class. These roles are cycled over the semester so the communication skills of all students are developed.

The tutors now take on the role of facilitator and act to keep the students on task and to steer conversations where appropriate. The tutors are asked to limit the time they talk to the whole class and their use of the whiteboards. Instead, the onus is placed on the students to develop their own understanding and to use peer discussion to assemble and organize their thoughts. The tutors have to resist intervening, stand back and trust the interaction. This can be difficult for tutors of all levels and the implementation of this approach has meant (re-)training in student centred approaches and group work for the tutors.

It may be noted that despite reservations from a number of the lecturers and tutors involved, all reported that they had enjoyed the classes taught within this paradigm much more than those taught in a more instructor-led approach. Almost all also reported that they had overestimated students' prior knowledge and understanding and underestimated the ability of students to work problems out on their own.

A typical worksheet, taken from 'Chemistry 1A' (CHEM1101) is included in Appendix 1. It follows the same general pattern as those described above for use in lectures, requiring students to develop their understanding through group discussion. The topic is one which students are fairly confident on but, as their exam answers demonstrate, this is based on misconceptions from school. The group works together to develop the concepts, through guided exposure to graphical experimental data. At some point in the worksheet, the data is likely to counter their previous understanding and the students are forced to deal with their own misconceptions and those of the rest of the group. The worksheet ends with a deliberately open ended question which asks them to take and articulate a position based on their analysis.

This approach has many advantages over the traditional lecture and tutorial formats in delivering the outcome of improving language skills in first year chemistry. As discussed above, it is a student centred approach involving social interaction which encourages communication in both written and oral forms and seeks to replace surface and rote learning with deeper learning. Student become more active and engaged in classes and their ability to process chemical symbols and language is continually exposed and improved. At the end of the class, students have completed a worksheet which then becomes a useful learning, reflection and, of course, revision tool.

Table 6. Percentage of students answering multiple choice questions 19 – 30 correctly in the CHEM1002 examination in 2008 – 10 at The University of Sydney

Question	2008	2009	2010
19	35	41	71
20	56	60	68
21	79	82	92
22	68	66	79
23	91	92	91
24	54	83	83
25	80	84	83
26	75	78	78
27	92	91	91
28	78	86	84
29	32	47	46
30	92	93	91
Average	69	75	80

The ALIUS approach is now used in all of the author's first, second and third year lecture classes. As well as being used in all first year chemistry tutorials, for which the author is directly responsible, the approach is now being used in the majority of those second year tutorials taken by other lecturers. All of the worksheets are freely available on the project website ((“Active Learning in University Science,” 2010).

Extensive research in the US is consistent with improved student outcomes when this approach is adopted (M. D. Perry & Wight, 2008). In particular, there is encouraging evidence of deeper learning with improved knowledge retention in future units. At The University of Sydney, merit grades are norm referenced and the content of examinations is very different each year. This makes it difficult to directly compare examination results or merit grade distributions to assess the effect of an intervention such as the one described here. Only a number of the multiple choice questions are common in the final examinations before and after the change in teaching style was adopted in 2009 (see Table 6).

These questions covered various topics such as acids/bases chemistry, transition metal complexes, periodicity, phase changes. There is a sustained increase in performance after the intervention and this is particularly noticeable on those questions which score badly in 2008 and which presumably reflect the topics which student find more difficult.

CONCLUSION

The mix of unfamiliar scientific with everyday words used in a new way makes reading and comprehending chemistry difficult for many students. Success in chemistry also requires adoption of a new symbolic language to describe both abstract, molecular events and experimental observations. These aspects of language contribute greatly to the investment that students must make in order to genuinely understand chemical concepts and, hence, to the image of chemistry as a ‘hard subject’.

This chapter describes a number of online and in-class responses to the challenge of teaching chemistry at a first year level, to students with a wide range of abilities and levels of motivations. At their centre is a belief that students learn chemical concepts most effectively when they are actively engaged in doing and talking about it, rather than rote learning and listening to it.

The strategies described have been developed within the context of large classes and limited resources. They are fully adaptable to other topics beyond first chemistry and to other sciences. They are also necessarily designed to be scalable to large or small classes and to be sustainable. Online resources are useful for helping students become familiar with chemical language and symbolism and to provide them with a means of practicing their use. Online quizzes are an invaluable means of students self-assessing their progress and of providing meaningful assessment of their level of mastery. In-class activities involving student response systems and student-centred, inquiry based approaches are built around active learning and on-going formative assessment. To develop language skills, social interaction via peer instruction and group discussions can be utilized to ensure that classes are vibrant and engaging.

REFERENCES

Abraham, M. (2005). Inquiry and the learning cycle approach . In Pienta, N. J., Cooper, M. M., & Greenbowe, T. J. (Eds.), *Chemists' guide to effective teaching*. Upper Saddle River, NJ: Prentice Hall.

Active Learning in University Science. (2010). *Website*. Retrieved 11th November 2010, from http:// www.alius.edu.au/

Angell, C., Guttersrud, O., Henriksen, E. K., & Isnes, A. (2004). Physics: Frightful, but fun: Pupils' and teachers' views of physics and physics teaching. *Science Education*, *88*, 683–706. doi:10.1002/sce.10141

Arons, A. B. (1990). *A guide to introductory physics teaching*. New York, NY: Wiley.

Asirvatham, M. (2009). *Clickers in action: Increasing student participation in general chemistry*. New York, NY: W.W. Norton & Co.

AUSSE. Australian Survey of Student Engagement. (2008). *Engaging students for success*. Retrieved March 22, 2010, from http:// ausse. acer.edu.au/ images/ docs/ AUSSE_2008_Australasian_Student_Engagement_Report.pdf

Barrie, S. C., Buntine, M. A., Jamie, I. M., & Kable, S. H. (2001). *APCELL: Developing better ways of teaching in the laboratory*. Paper presented at the Research and Development into University Science Teaching and Learning Workshop.

Beall, H. (1994). Probing student misconceptions in thermodynamics with in-class writing. *Journal of Chemical Education*, *71*(12), 1056–1057. doi:10.1021/ed071p1056

Bedgood, D. R. Jr, Gardiner, M. B. Y., Bridgeman, A. J., Morris, G., Buntine, M., & Mocerino, M. (2010). The development of teaching skills to support active learning in university science (ALIUS). *Journal of Learning Design*, *1*(3), 10–19.

Bedgood, D. R. Jr, Yates, B., Buntine, M., Pyke, S., Lim, K., & Mocerino, M. (2008). Why are we still teaching the way we were taught in the 1980s? *Chemistry in Australia*, *11*, 22–23.

Bedgood, D. R. Jr, Yates, B., Buntine, M., Pyke, S., Lim, K., & Mocerino, M. (2010). Leading change in Australian science teaching. *Chemistry in Australia*, *7*(5), 18–19.

VerBeek, K. V., & Louters, L. (1991). Chemical language skills. *Journal of Chemical Education*, *68*(5), 389–392. doi:10.1021/ed068p389

Ben-Zvi, R., Eylon, B., & Silberstein, J. (1988). Theories, principles and laws. *Education in Chemistry*, *5*, 89–92.

Boud, D. (1993). Experience as the base for learning. *Higher Education Research & Development*, *12*(1), 33–44. doi:10.1080/0729436930120104

Bowen, C. W., & Bunce, D. M. (1997). Testing for conceptual understanding in general chemistry. *The Chemical Educator*, *2*(2), Bradley, J. D., & Brand, M. (1985). Stamping out misconceptions. *Journal of Chemical Education*, *62*(4), 318.

Bridgeman, A. J. (2005). Spectroscopy, mechanisms and calculations online. *Chemical Reviews*, *15*(3).

Bridgeman, A. J. (2010). *Chemical games*. Retrieved 11th November, 2010, from http:// firstyear. chem.usyd.edu.au/ games/ index.shtml

Bridgeman, A. J., & Schmid, S. (2010). *Collaborative laboratory for quantitative data analysis*. Paper presented at the Uniserve Science.

Brown, D. E. (1992). Using examples and analogies to remediate misconceptions in physics: Factors influencing conceptual change. *Journal of Research in Science Teaching*, *29*(1), 17–34. doi:10.1002/tea.3660290104

Bruner, J. S. (1986). *Actual minds, possible worlds*. Cambridge, MA: Harvard University Press.

Bunce, D. M., Flens, E. A., & Neiles, K. Y. (2010). How long can students pay attention in class? Astudy of student attention decline using clickers. *Journal of Chemical Education*, *87*(12), 1438–1443. doi:10.1021/ed100409p

Bunce, D. M., Vandenplas, J. R., & Havanki, L. L. (2006). Comparing the effectiveness on student achievement of a student response system versus online WebCT quizzes. *Journal of Chemical Education, 83*(3), 488–493. doi:10.1021/ed083p488

Burton, L. J., & Dowling, D. (2005). *In search of the key factors that influence student success at university*. Paper presented at the HERDSA 2005 International Conference: Higher Education in a Changing World. Retrieved from http:// eprints.usq.edu.au/

Cary, W. R. (1984). State of the art in the high school curriculum. *Journal of Chemical Education, 61*(10), 856–857. doi:10.1021/ed061p856

Chin, C., & Brown, D. E. (2002). Student-generated questions: A meaningful aspect of learning in science. *International Journal of Science Education, 24*(5), 521–549. doi:10.1080/09500690110095249

Cummings, K., Marx, J., Thornton, R., & Kuhl, D. (1999). Evaluating innovation in studio physics. *American Journal of Physics, 67*(S1), S38–S44. doi:10.1119/1.19078

Dupin, J. J., & Johsua, S. (1989). Analogies and modeling analogies in teaching: Some examples in basic electricity. *Science Education, 73*(2), 207–224. doi:10.1002/sce.3730730207

Eilks, I., & Byers, B. (2010). The need for innovative methods of teaching and learning chemistry in higher education – Reflections from a project of the European Chemistry Thematic Network. *Chemistry Education Research and Practice, 11*, 233–240. doi:10.1039/c0rp90004d

Ellem, G. K., & Mclaughlin, E. A. (2005). *Tales from the coalface: From tragedy to triumph in a blended learning approach to the teaching of 1st year biology*. Paper presented at the 2005 National UniServe Science Symposium: Blended Learning in Science Teaching and Learning. Retrieved from http:// sydney.edu.au/

Elrod, S. (2007). *Genetics concept inventory*. Retrieved from http:// bioliteracy.net/ Readings/ papersSubmittedPDF/ Elrod.pdf

Flavell, J. H. (1976). Metacognitive aspects of problem solving . In Resnick, L. B. (Ed.), *The nature of intelligence* (pp. 231–236). Hillsdale, NJ & New York, NY: Lawrence Erlbaum Associates.

Gardner, P. L., & Australian Science Education Project. (1972). *Words in science: An investigation of non-technical vocabulary difficulties amongst form I, II, III, & IV science students in Victoria, a report*. Melbourne, Australia: Australian Science Education Project.

Gardner, P. L. (1977). Logical connectives in science: A summary of the findings. *Research in Science Education, 7*(1), 9–24. doi:10.1007/BF02643108

Garvin-Doxas, K., & Klymkowsky, M. W. (2008). Understanding randomness and it impact on student learning: Lessons learning from building the biology concept inventory (BCI). *CBE Life Sciences Education, 7*(2), 227–233. doi:10.1187/cbe.07-08-0063

Gijbels, D., Dochy, F., Van Den Bossche, P., & Segers, M. (2005). Effects of PBL: A meta-analysis from the angle of assessment. *Review of Educational Research, 75*(1), 27–61. doi:10.3102/00346543075001027

Granville, M. F. (1985). Student misconceptions in thermodynamics. *Journal of Chemical Education, 62*(10), 847–848. doi:10.1021/ed062p847

Gunstone, R. F. (1994). The importance of specific science content in the enhancement of metacognition . In Fensham, P. J., Gunstone, R. F., & White, R. T. (Eds.), *The content of science: A constructivist approach to its teaching and learning*. London, UK: Routledge.

Halliday, D., Resnick, R., & Walker, J. (2008). *Fundamentals of physics* (8th ed.). Hoboken, NJ: John Wiley & Sons, Inc.

Halloun, I. A., & Hestenes, D. (1985). The initial knowledge state of college physics students. *American Journal of Physics*, *53*(11), 1043–1055. doi:10.1119/1.14030

Harris, K.-L., Krause, K., Gleeson, D., Peat, M., Taylor, C., & Garnett, R. (2007). Enhancing assessment in the biological sciences: Ideas and resources for university educators. Retrieved 1st April, 2007, from http:// www.bioassess.edu.au

Hestenes, D., & Wells, M. (1992). A mechanics baseline test. *The Physics Teacher*, *30*(3), 159–166. doi:10.1119/1.2343498

Hestenes, D., Wells, M., & Swackhamer, G. (1992). Force concept inventory. *The Physics Teacher*, *30*(3), 141–158. doi:10.1119/1.2343497

Hounsell, D., Mccune, V., Hounsell, J., & Litjens, J. (2008). The quality of guidance and feedback to students. *Higher Education Research & Development*, *27*, 55–67. doi:10.1080/07294360701658765

Itza-Ortiz, S. F., Rebello, N. S., Zollman, D. A., & Rodriguez-Achach, M. (2003). The vocabulary of introductory physics and its implications for learning physics. *The Physics Teacher*, *41*(6), 330–336. doi:10.1119/1.1607802

Jacobs, G. (1989). Word usage misconceptions among first-year university physics students. *International Journal of Science Education*, *11*(4), 395–399. doi:10.1080/0950069890110404

Kempa, R. F., & Ayob, A. (1995). Learning from group work in science. *International Journal of Science Education*, *17*(6), 743–754. doi:10.1080/0950069950170606

Khodor, J., Halme, D. G., & Walker, G. C. (2004). A hierarchical biology concept framework: A tool for course design. *Cell Biology Education*, *3*(2), 111–121. doi:10.1187/cbe.03-10-0014

Koch, A. (2001). Training in metacognition and comprehension of physics texts. *Science Education*, *85*(6), 758–768. doi:10.1002/sce.1037

Kolb, D. A. (1984). *Experiential learning: Experience as the source of learning and development*. Englewood Cliffs, NJ: Prentice-Hall.

Krause, S., Birk, J., Bauer, R., Jenkins, B., & Pavelich, M. J. (2004). *Development, testing, and application of a chemistry concept inventory*. Paper presented at the 34th ASEE/IEEE Frontiers in Education Conference. Retrieved from http:// fie-conference.org/ fie2004/

Krause, K.-L., Hartley, R., James, R., Mcinnis, C., & Centre for the Study of Higher Education, University of Melbourne. (2005). *The first year experience in Australian universities: Findings from a decade of national studies*. Retrieved from http:// www.dest.gov.au/

Lave, J., & Wenger, E. (1991). *Situated learning: Legitimate peripheral participation*. Cambridge, England & New York, NY: Cambridge University Press.

Lemke, J. L. (1998). *Teaching all the languages of science: Words, symbols, images, and actions.*

Lidbury, B., & Zhang, F. (2008). Comprehension of scientific language as a strategy to enhance learning and engagement for molecular biology students. *Australian Biochemist*, *39*(3), 10–13.

Maloney, D. P., O'kuma, T. L., Hieggelke, C. J., & Van Heuvelen, A. (2001). Surveying students' conceptual knowledge of electricity and magnetism. *American Journal of Physics*, *69*(S1), S12–S23. doi:10.1119/1.1371296

Marais, P., & Jordaan, F. (2000). Are we taking symbolic language for granted? *Journal of Chemical Education*, *77*(10), 1355–1357. doi:10.1021/ed077p1355

Markow, P. G. (1988). Teaching chemistry like the foreign language it is. *Journal of Chemical Education*, *65*(1), 57–58. doi:10.1021/ed065p57

Marton, F., Hounsell, D., & Entwistle, N. J. (Eds.). (1984). *The experience of learning*. Edinburgh, UK: Scottish Academic Press.

Martyn, M. (2007). Clickers in the classroom: An active learning approach. *EDUCAUSE Quarterly*, *30*(2). Retrieved from http:// www.educause.edu/.

Maskill, R., & Cachapuz, A. F. C. (1989). Learning about the chemistry topic of equilibrium: The use of word association tests to detect developing conceptualizations. *International Journal of Science Education*, *11*, 57–69. doi:10.1080/0950069890110106

Mason, D. S. (2006). Persistent misconceptions. *Journal of Chemical Education*, *83*(5), 681. doi:10.1021/ed083p681

Mayer, K. (2011). Addressing students misconceptions about gases, mass, and composition. *Journal of Chemical Education*, *88*(1), 111–115. doi:10.1021/ed1005085

Mazur, E. (1997). *Peer instruction: A user's manual*. Upper Saddle River, NJ: Prentice-Hall.

McDaniel, C. N., Lister, B. C., Hanna, M. H., & Roy, H. (2007). Increased learning observed in redesigned introductory biology course that employed Web-enhanced, interactive pedagogy. *CBE Life Sciences Education*, *6*(3), 243–249. doi:10.1187/cbe.07-01-0003

McDermott, L. C. (1991). Millikan lecture 1990: What we teach and what is learned-Closing the gap. *American Journal of Physics*, *59*, 301–315. doi:10.1119/1.16539

Mckeachie, W. J. (2002). *McKeachie's teaching tips* (11th ed.). New York, NY: Houghton Mifflin.

Mehmet, S. (2009). Exploring university students' expectations and beliefs about physics and physics learning in problem based learning context. *Eurasia Journal of Mathematics . Science & Technology Education*, *5*(4), 321–333.

Michael, J. (2007). Conceptual assessment in the biological sciences: a National Science Foundation-sponsored workshop. *Advan. Physiol. Edu.*, *31*(4), 389-391.

Millar, R. (1991). Why is science hard to learn? *Journal of Computer Assisted Learning*, *7*(2), 66–74. doi:10.1111/j.1365-2729.1991.tb00229.x

Miller, V., Gregg, G., & Kelly, B. (2000). *Peer Assisted Study Sessions (PASS) in a first year biology teaching program-strategies developed by undergraduate PASS leaders*. Paper presented at the Effective Teaching and Learning at University Conference. Retrieved from http:// www.tedi.uq.edu.au/ conferences/

Moog, R. S. (2008). *Process Oriented Guided Inquiry Learning (POGIL)* (*Vol. 994*). American Chemical Society.

Nakhleh, N. B. (1992). Why some students don't learn chemistry: Chemical misconceptions. *Journal of Chemical Education*, *69*(3), 191–196. doi:10.1021/ed069p191

Nussbaum, J., & Novick, S. (1983). Alternative frameworks, conceptual conflict and accommodation: Toward a principled teaching strategy. *Instructional Science*, *11*(2), 183–200.

O'Byrne, J. W., Britton, S., George, A., Franklin, S., & Frey, A. (2009). Using academic predictors to identify first year science students at risk of failing. *CAL-laborate International*, *17*(1), 15-25. Retrieved from http:// sydney.edu.au/

Odom, A. L., & Kelly, P. V. (2001). Integrating concept mapping and the learning cycle to teach diffusion and osmosis concepts to high school biology students. *Science Education*, *85*(6), 615–635. doi:10.1002/sce.1029

Oliver, J. S., & Simpson, R. D. (1988). Influences of attitude toward science, achievement motivation, and science self concept on achievement in science: A longitudinal study. *Science Education*, *72*(2), 143–155. doi:10.1002/sce.3730720204

Perry, M. D., & Wight, R. D. (2008). Using an ACS general chemistry exam to compare traditional and POGIL instruction. [American Chemical Society.]. *Process Oriented Guided Inquiry Learning*, *994*, 240–247. doi:10.1021/bk-2008-0994.ch020

Perry, W. G. (1988). Different worlds in the same classroom . In Ramsden, P. (Ed.), *Improving learning: New perspectives*. London, UK: Kogan Page.

Pickersgill, S., & Lock, R. (1991). Student understanding of selected non-technical words in science. *Research in Science & Technological Education*, *9*(1), 71–79. doi:10.1080/0263514910090107

Pipes, R. B., & Wilson, J. M. (1996). A multimedia model for undergraduate education. *Technology in Society*, *18*(3), 387–401. doi:10.1016/0160-791X(96)00020-6

Pollock, S. J. (2008). *Comparing student learning with multiple research-based conceptual surveys: CSEM and BEMA*. Paper presented at the Physics Education Research Conference 2008. Retrieved from http:// www.colorado.edu/

Redish, E. F. (1994). The implications of cognitive studies for teaching physics. *American Association of Physics Teachers*, *62*(6), 796–803.

Richardson, A., & Zhang, F. (2008). *Language support for statistics learners*. Paper presented at the OZCOTS 2008 - 6th Australian Conference on Teaching Statistics. Retrieved from http:// sky.scitech.qut.edu.au/ ~macgilli/ ozcots2008/

Richardson, A., Zhang, F., & Lidbury, B. A. (2008). *Activating multiple senses in learning statistics*. Paper presented at the Assessment in Science Teaching and Learning Symposium. Retrieved from http:// sydney.edu.au/

Ritchie, S. M. (2001). Actions and discourses for transformative understanding in a middle school science class. *International Journal of Science Education*, *23*(3), 283–299. doi:10.1080/09500690175006652

Roth, W.-M., & Roychoudhury, A. (1994). Science discourse through collaborative concept mapping: New perspectives for the teacher. *International Journal of Science Education*, *16*(4), 437–455. doi:10.1080/0950069940160405

Roy, H. (2003). Studio vs interactive lecture demonstration–Effects on student learning. *Bioscene: Journal of College Biology Teaching*, *29*(1), 3-6. Retrieved from http:// acube.org/ pdf/ bioscene/ v29-1.pdf

Rust, C. (2002). The impact of assessment on student learning: How can the research literature practically help to inform the development of departmental assessment strategies and learner-centred assessment practices? *Active Learning in Higher Education*, *3*, 145–158. doi:10.1177/1469787402003002004

Sanger, M. J., & Greenbowe, T. J. (1999). An analysis of college chemistry textbooks as sources of misconceptions and errors in electrochemistry. *Journal of Chemical Education*, *76*(6), 853–860. doi:10.1021/ed076p853

Schon, D. A. (1987). *Educating the reflective practitioner: Toward a new design for teaching and learning in the professions*. San Francisco, CA: Jossey-Bass.

Schulte, J. (2006). *Delivering first year physics assignments with limited resources - An Australian three-centre study*. Paper presented at the 2006 National UniServe Science Symposium: Assessment in Science Teaching and Learning. Retrieved from http:// sydney.edu.au/

Simpson, O. (2006). Predicting student success in open and distance learning. *Open Learning: The Journal of Open and Distance Learning*, *21*(2), 125–138.

Smith, M. K., Wood, B. W., & Knight, J. K. (2008). The genetics concept assessment: A new concept inventory for gauging student understanding of genetics. *CBE Life Sciences Education*, *7*, 422–430. doi:10.1187/cbe.08-08-0045

Smith, M. K., Wood, W. B., Adams, W. K., Wieman, C., Knight, J. K., Guild, N., & Su, T. T. (2009). Why peer discussion improves student performance on in-class concept questions. *Science*, *323*(5910), 122–124. doi:10.1126/science.1165919

Svinicki, M. D. (2004). *Learning and motivation in the postsecondary classroom*. Boston, MA: John Wiley and Sons.

Tinto, V., & Pusser, B. (2006). *Moving from theory to action: Building a model of institutional action for student success*. Paper presented at the National Symposium on Postsecondary Student Success. Retrieved from http:// nces.ed.gov/ npec/ papers.asp

Touger, J. S. (1991). When words fail us. *The Physics Teacher*, *29*(2), 90–95. doi:10.1119/1.2343227

Towns, M. (2010). Crossing the chasm with classroom response systems. *Journal of Chemical Education*, *87*(12), 1317–1319. doi:10.1021/ ed9000624

Wandersee, J. H. (1988). The terminology problem in biology education: A reconnaissance. *The American Biology Teacher*, *50*(2), 97–100. Retrieved from http:// www.jstor.org/.

Wellington, J. J., & Osborne, J. (2001). *Language and literacy in science education*. Buckingham, UK & Philadelphia, PA: Open University.

Wileyplus. (2010). *Website*. Retrieved from http:// www.wileyplus.com

Williams, H. T. (1999). Semantics in teaching introductory physics. *American Journal of Physics*, *67*(8), 670–680. doi:10.1119/1.19351

Wu, H.-K., Krajcik, J. S., & Soloway, E. (2001). Promoting understanding of chemical representations: Students' use of a visualization tool in the classroom. *Journal of Research in Science Teaching*, *38*(7), 821–842. doi:10.1002/tea.1033

Yeo, S., & Zadnik, M. (2001). Introductory thermal concept evaluation: Assessing students' understanding. *The Physics Teacher*, *39*(8), 496–504. doi:10.1119/1.1424603

Yore, L. D., Florence, M. K., Pearson, T. W., & Weaver, A. J. (2006). Written discourse in scientific communities: A conversation with two scientists about their views of science, use of language, role of writing in doing science, and compatibility between their epistemic views and language. *International Journal of Science Education*, *28*(2-3), 109–141. doi:10.1080/09500690500336601

Yore, L. D., Hand, B. M., & Prain, V. (2002). Scientists as writers. *Science Education*, *86*(5), 672–692. doi:10.1002/sce.10042

Yore, L. D., & Treagust, D. F. (2006). Current realities and future possibilities: Language and science literacy-empowering research and informing instruction. *International Journal of Science Education*, *28*(2-3), 291–314. doi:10.1080/09500690500336973

Yorke, M., & Longdon, B. (2008). *The first-year experience of higher education in the UK: Final report of a project funded by the Higher Education Academy*. York, UK: Higher Education Academy. Retrieved March 22 2010 from http:// www. heacademy.ac.uk/ news/ detail/ fye_final_report

Zhang, F., & Barber, B. (Eds.). (2008). *Handbook of research on computer-enhanced language acquisition and learning*. Hershey, PA: Information Science Reference. doi:10.4018/978-1-59904-895-6

Zhang, F., Lidbury, B., Schulte, J., Bridgeman, A., Yates, B., & Rodger, J. (2008). *Language difficulties in first year science - An interim report*. Paper presented at the 2008 National UniServe Science Symposium: Assessment in Science Teaching and Learning. Retrieved from http:// sydney.edu.au/

Zhang, F., Lidbury, B., Schulte, J., Bridgeman, A., Yates, B., & Rodger, J. (2009). *A cross-disciplinary approach to language support for first year students in the science disciplines*. Paper presented at the HERDSA 2009.

Zhang, F., & Lidbury, B. A. (2006). *It's all foreign to me: Learning through the language of genetics and molecular biology*. Paper presented at the 2006 National UniServe Science Symposium: Assessment in Science Teaching and Learning. from http:// sydney.edu.au/

Zhang, F., Lidbury, B. A., Schulte, J., & Bridgeman, A. J. B., Y., & Rodger, J. (2010). *Catering for the language needs of diverse first year science students*. Paper presented at the First Year in Higher Education.

Zhang, F., Lidbury, B. A., Schulte, J., Bridgeman, A. J., & Yates, B. (2008). *Language difficulties in first year science: An interim report*. Paper presented at the Uniserve Science.

Zhang, F. Z. (2006). *The teaching of Mandarin prosody: A somatically-enhanced approach for second language learners*. Unpublished PhD thesis, University of Canberra, Canberra.

Zoller, U. (1990). Students' misunderstandings and misconceptions in college freshman chemistry (general and organic). *Journal of Research in Science Teaching*, *27*(10), 1053–1065. doi:10.1002/tea.3660271011

ADDITIONAL READING

Ãœce, M. (2009). Teaching the mole concept using a conceptual change method at college level. *Education*, *129*(4), 683.

Bowen, C. W., & Bunce, D. M. (1997). *Testing for conceptual understanding in general chemistry. the Chemical Educator, 2(2)*. Retrieved from.

Chandrasegaran, A. L., Treagust, D. F., & Mauro, M. (2007). The development of a two-tier multiple-choice diagnostic instrument for evaluating secondary school students' ability to describe and explain chemical reactions using multiple levels of representation. *Chemistry Education Research and Practice*, *8*(3), 293–307.

Johnstone, A. H. (2000). Teaching of chemistry - logical or psychological? Chemistry Education Research and Practice in Europe, 1(1), 9-15. Retrieved from http:// www.uoi.gr/ cerp/

Kaya, O. N. (2008). A student-centred approach: Assessing the changes in prospective science teachers' conceptual understanding by concept mapping in a general chemistry laboratory. Research Science Education (. *Research in Science Education*, *38*, 91–110. doi:10.1007/s11165-007-9048-7

Krause, S., Birk, J., Bauer, R., Jenkins, B., & Pavelich, M. J. (2004). Development, testing, and application of a chemistry concept inventory. Paper presented at the 34th ASEE/IEEE Frontiers in Education Conference, Savannah, GA. Retrieved from http:// fie-conference.org/ fie2004/

Michael, J. (2007). Conceptual assessment in the biological sciences: A national science foundation-sponsored workshop. *Advances in Physiology Education*, *31*(4), 389–391. .doi:10.1152/advan.00047.2007

Nieswandt, M. (2001). Problems and possibilities for learning in an introductory chemistry course from a conceptual change perspective. Science Education, 85(2), 158-79. doi: 10.1002/1098-237X(200103)85:2<158::AID-SCE40>3.0.CO;2-3

KEY TERMS AND DEFINITIONS

Process Orientated Guided Inquiry Learning (POGIL): approach (Moog, 2008) is a classroom and laboratory technique that seeks to simultaneously teach content and key process skills such as the ability to think analytically and work effectively as part of a collaborative team. A POGIL classroom or lab consists of any number of students working in small groups on specially designed guided inquiry materials. These materials supply students with data or information followed by leading questions designed to guide them toward formulation of their own valid conclusions—essentially a recapitulation of the scientific method. The instructor serves as facilitator, observing and periodically addressing individual and classroom-wide needs. (Taken from http://www.pogil.org/about)

APPENDIX A

These three worksheets illustrate the exercises used in ALIUS-style lectures and tutorials.

Introduction to Acids and Bases

Model 1: A Definition of Acids and Bases

Brønsted–Lowry definition:

An **acid** is a substance that donates a proton, H^+, to another species.

A **base** is a substance that accepts a proton, H^+, from another species.

	Reaction
(i)	$HCl(g) + H_2O(l) \rightleftharpoons H_3O^+(aq) + Cl^-(aq)$
(ii)	$NH_3(aq) + H_2O(l) \rightleftharpoons NH_4^+(aq) + OH^-(aq)$
(iii)	$HCN(aq) + H_2O(l) \rightleftharpoons H_3O^+(aq) + CN^-(aq)$

Critical thinking questions

1. (a) What chemical species are the Brønsted-Lowry acids in the *forward* reactions in the table above?

   (i) (ii) (iii)

   (b) What chemical species are the Brønsted-Lowry bases in the *forward* reactions in the table above?

   (i) (ii) (iii)

2. Is it possible for a substance to act as both an acid and a base? Explain your reasoning.
3. (a) What species results from the loss of a proton from the Brønsted-Lowry acid in the forward reaction (i)

   (b) Does this species act as an acid or a base when the reverse of reaction (i) occurs?

   (c) What species results from the gain of a proton from the Brønsted-Lowry base in the forward reaction (i)

   (d) Does this species act as an acid or a base when the reverse of reaction (i) occurs?

   (e) Repeat questions (a) – (d) for reactions (ii) and (iii). Describe any general relationship you deduce using a grammatically correct sentence.

Model 2: Conjugate Pairs

As you discovered in Model 1, certain pairs of molecules are related through their acid and base properties. These pairs are described as a **conjugate acid–base pair**. They differ by *a single proton*. A base has *one less proton* than its conjugate acid. An acid has *one more proton* than its conjugate base.

Acid	Base
H_2CO_3	HCO_3^-
HCO_3^-	CO_3^{2-}
H_3O^+	H_2O
H_2S	HS^-

Critical thinking questions

4. (a) What is the conjugate base of H_2O?
   (b) What is the conjugate acid of NH_3?
5. There are two conjugate pairs in the following reaction. Identify *both* pairs.

$$H_2O(l) + H_2O(l) \rightleftharpoons H_3O^+(aq) + OH^-(aq)$$

6. Ammonia can also react with itself as an acid or a base. Complete the following acid–base reaction in which $NH_3(l)$ acts as both an acid and a base:

$$NH_3(l) + NH_3(l) \rightleftharpoons$$

Exercises

1. Write the formula of the conjugate bases of the following acids:
   (a) H_2SO_4 (b) $HClO_4$ (c) CH_3COOH
   (d) H_3PO_4 (e) $H_2PO_4^-$ (f) HPO_4^{2-}
2. Write the formula of the conjugate acids of the following bases:
   (a) CN^- (b) H_2S (c) HS^-
   (d) S^{2-} (e) CH_3COO^- (f) SO_4^{2-}
3. Amino acids have both an acidic (–COOH) and a basic (–NH_2) functional group. The simplest amino acid is glycine; $H_2N–CH_2–COOH$. Give the structure of the conjugate acid and the conjugate base of glycine.

Electronegativity and Bonding

Model 1: Electronegativitiy

When two bonded atoms are the same (e.g. C-C, O-O, H-H etc), the bonding electrons are shared equally. When the two atoms are different, one of the atoms attracts the bonding electrons more strongly. The ability of an atom to attract the electrons in a bond is called the *electronegativity*. It is possible to quantify this on a scale.

H						
2.30						
Li	Be	B	C	N	O	F
0.91	1.58	2.05	2.54	3.07	3.61	4.19
Na	Mg	Al	Si	P	S	Cl
0.87	1.29	1.61	1.92	2.25	2.59	2.87
K	Ca	Ga	Ge	As	Se	Br
0.73	1.03	1.76	1.99	2.21	2.42	2.69
Rb	Sr	In	Sn	Sb	Te	I
0.71	0.96	1.66	1.82	1.98	2.16	2.36

Critical thinking questions

1. Describe the trend in electronegativity as one moves from left to right across a period.
2. Describe the trend in electronegativity as one moves down a group.
3. Compare the trends in questions 2 and 3 to the corresponding trends in ionization energy and explain any relationship you observe.

Model 2: Relationship Between Electronegativity and Bonding Type

compound or metal	average electronegativity of atoms ($\overline{EN}$)	difference in electronegativity of atoms (ΔEN)	type of bonding
CsF(s)	2.42	3.53	ionic
NaCl(s)	1.87	2.00	ionic
NaI(s)	1.62	1.49	ionic
Cs(s)	0.66	0	metallic
Na(s)	0.87	0	metallic
CuZn (s; brass)	1.7	0.2	metallic
F_2(g)	4.19	0	covalent
CH_4(g)	2.42	0.24	covalent
C(s; diamond)	2.54	0	covalent

Critical thinking questions

4. What combination of $\overline{EN}$ and ΔEN values lead to metallic bonding (i.e. small, medium or large)?
5. What combination of $\overline{EN}$ and ΔEN values lead to ionic bonding (i.e. small, medium or large)?
6. What combination of $\overline{EN}$ and ΔEN values lead to covalent bonding (i.e. small, medium or large)?

Model 3: The Bond Type Triangle

The chart opposite uses the ideas explored in Model 2 and enables us to predict the properties of a compound. The bond type triangle is divided into regions which indicate the predominant type of bonding present. The compounds in the region marked "SM" are **semimetals** and have properties which are intermediate between the three bond types: metallic, covalent and ionic.

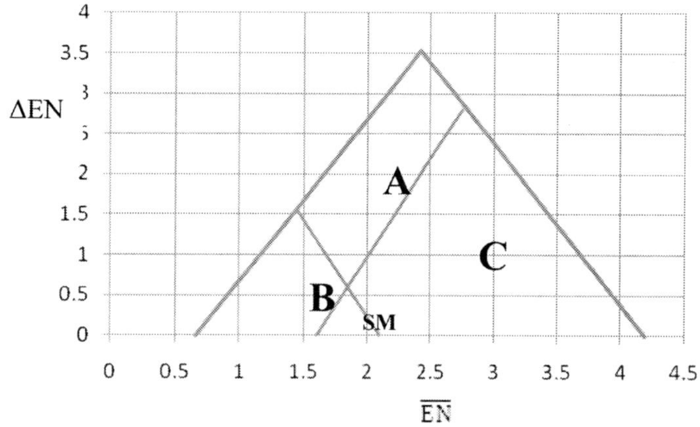

Critical thinking questions

7. Place a point on the chart for (i) NaCl (a compound with ionic bonding), (ii) Na (a metal), (iii) CH_4 (a compound with covalent bonding) and (iv) Si (a semiconductor).
8. Associate the regions (A, B and C) with bond types (metallic, covalent and ionic).
9. Quartz, SiO_2 is a very high melting and hard solid. Place a point on the chart for SiO_2 and predict the type of bonding you would expect to be predominant in quartz.
10. For each of the following compounds, classify their bonding type as metallic, covalent, ionic or semimetal.

(a)	Na_2O	(b)	MgO	(c)	Al_2O_3	(d)	SiO_2
(e)	P_2O_5	(f)	SO_3	(g)	Cl_2O_7		

11. In your group, describe how the changes in the bonding in oxides across the second row

INTERMOLECULAR FORCES

Information

Intermolecular forces are the interactions *between* rather than *inside* molecules. They are responsible for many of the physical properties of substances, including their melting and boiling points.

In pure substances, there are 3 important intermolecular forces which may be present:

- *Dipole – dipole forces*. The dipole moment in a molecule will tend to align with those in its neighbours. This type of interaction is only possible if the molecule possesses a dipole.
- *Hydrogen bonds*. This is a particularly strong dipole – dipole interaction involving the interaction between the δ+ H atoms in very polar bonds and lone pairs on very electronegative atoms. Hydrogen bonding therefore requires the presence of both δ+ H atoms *and* electronegative atoms.
- *Dispersion forces*. These forces are present in *all* molecules and atoms. At any moment in time, the electron density in a molecule or atom may not be symmetrical and this leads to a dipole moment. This momentary or *instantaneous* dipole moment *induces* a matching dipoles in neighbouring molecules or atoms by polarizing their electron density.

  Dispersion forces increase with the number of electrons in a molecule.

Model 1: Boiling Points Change Down a Group of the Periodic Table

Molecules are held in the liquid phase due to intermolecular forces so that boiling points are a good guide to their strength.

The figure opposite shows the boiling points of the Group 14 hydrides. All have the same shape but differ in the total number of electrons.

For example:

- C has 6 electrons and each H has 1 electron so CH_4 has $6 + 4 \times 1 = 10$.
- Sn has 50 electrons so SnH_4 has 54 electrons.

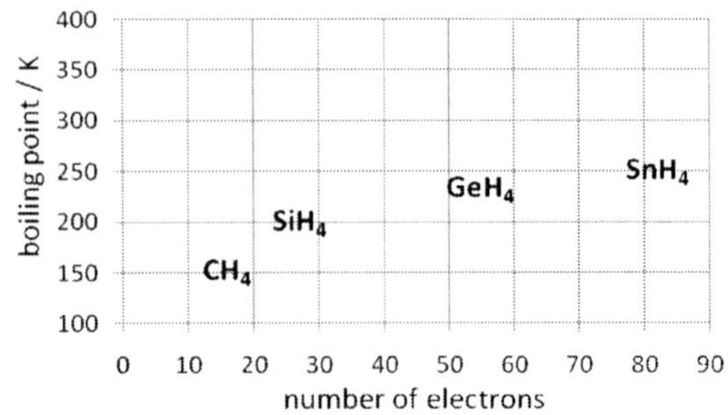

Critical thinking questions

1. What happens to the boiling point as the number of electrons increases?
2. What shape are the Group 14 hydrides?
3. Are dipole – dipole forces present in these molecules?
4. Is hydrogen bonding possible in these molecules?
5. What intermolecular force is present in these molecules?
6. Explain why the boiling points vary in the way you described in answer to Q1

Model 2: Boiling Points Change Across a Row of the Periodic Table

On the graph opposite, the boiling points for the other hydrides have been added:

- Group 14 SiH_4, GeH_4 and SnH_4
- Group 15 PH_3, AsH_3 and SbH_3
- Group 16 H_2S, H_2Se and H_2Te
- Group 17 HCl, HBr and HI

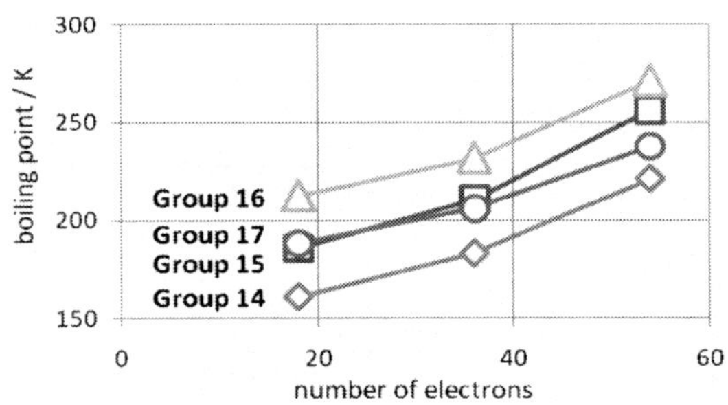

Critical thinking questions

7. Use a Periodic Table to confirm that SiH_4, PH_3, H_2S and HCl all have 18 electrons.
8. What happens to the boiling point as the number of electrons increases? Why?
9. What is the molecular shape of PH_3, H_2S and HCl?
10. Do PH_3, H_2S and HCl have dipole moments?
11. Why is the boiling point of SiH_4 *lower* than that of PH_3, H_2S and HCl?
12. Is the boiling point of SnH_4 (54 electrons) higher or lower than the boiling point of PH_3 (18 electrons)?
13. Explain your answer to Q12, making sure that it is consistent with your answers to Q8 and Q11.

Model 3: Anomalous Boiling Points of NH_3, H_2O and HF

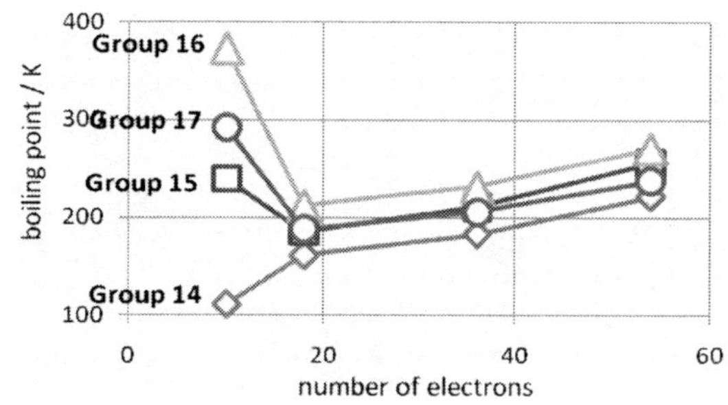

The graph opposite adds the boiling points of CH_4, NH_3, H_2O and HF to Model 3. N, O and F are very electronegative and N-H, O-H and H-F bonds are very polar.

Critical thinking questions

14. How do the boiling points of the Group 14 hydrides change down the group?

Re-read your answers to Q1 – 6.

15. How many δ+ H atoms are there on electronegative element in the molecules below?

    (a) NH_3 (b) H_2O (c) HF

16. How many lone pairs are there on electronegative element in these molecules?

    (a) NH_3 (b) H_2O (c) HF

17. Explain why the boiling points of NH_3, H_2O and HF (10 electrons) are *higher* than those of PH_3, H_2S and HCl (18 electrons) Refer to the **Information** if you are unsure.

18. Given your answer to Q17, suggest why the boiling point of NH_3 (10 electrons) is *lower* than that of SbH_3 (54 electrons).

19. Order the N-H, O-H and F-H bonds in terms of their polarity.

20. Predict the *relative* strength of the intermolecular forces between *two* NH_3 molecules, *two* H_2O and *two* HF molecules.

21. How many hydrogen bonds can each NH_3 molecule make on average in NH_3(l)? (*Hint*: re-read your answers to Q15 and Q16).

22. How many hydrogen bonds can each HF molecule make on average in HF(l)? (*Hint*: re-read your answers to Q15 and Q16).

21. How many hydrogen bonds can each H_2O molecule make on average in H_2O(l)? (*Hint*: re-read your answers to Q15 and Q16).

22. Use your answers to Q20 and Q21 – 23 to explain why the boiling points vary in the order NH_3 < HF < H_2O.

23. Many textbooks and websites state that dispersion forces are *weaker* than hydrogen bonding and dipole – dipole forces. Critically analyse this statement in the light of the evidence in Models 1 – 3.

APPENDIX B: ADMINISTERED AT THE UNIVERSITY OF SYDNEY AS AN INTERVENTION IN THE DISCIPLINE OF CHEMISTRY

Practical hint: The following clicker questions and group activities have been used extensively in tutorials in Chemistry.

A. Organic 'clicker' questions:

For questions 1 - 5, choose from **A - E** the term that best describes the isomeric relationship for each pair of compounds.
A. conformational isomers **B.** constitutional isomers **C.** enantiomers
D. diastereomers **E.** identical compounds

		A B C D E
		A B C D E
		A B C D E
	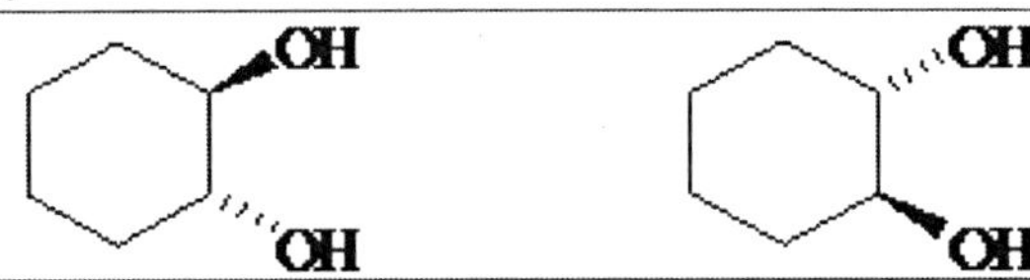	A B C D E
		A B C D E

In answering questions 6 - 12, choose from compounds **A - E**.

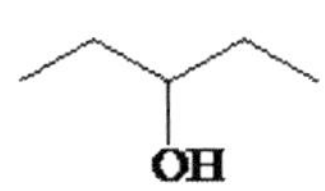

A

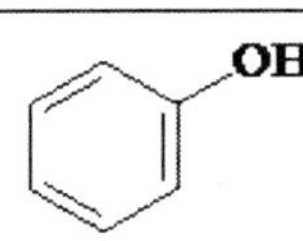

B

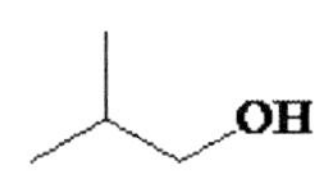

C

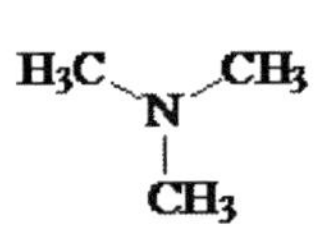

D

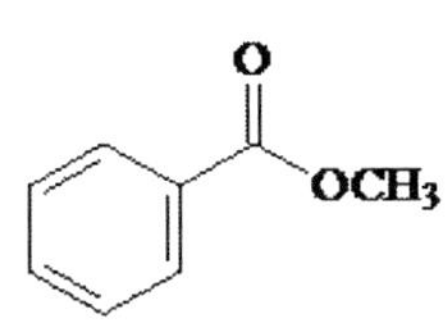

E

Which one of the compounds is a primary alcohol?	
Which one of the compounds will react with NaOH in an acid-base reaction?	
Which one of the compounds will react with HCl to form a	

B. Concept development group work:
HOW MANY ATOMS ARE THERE IN A UNIT CELL?
Model 1: The body centred cubic unit cell

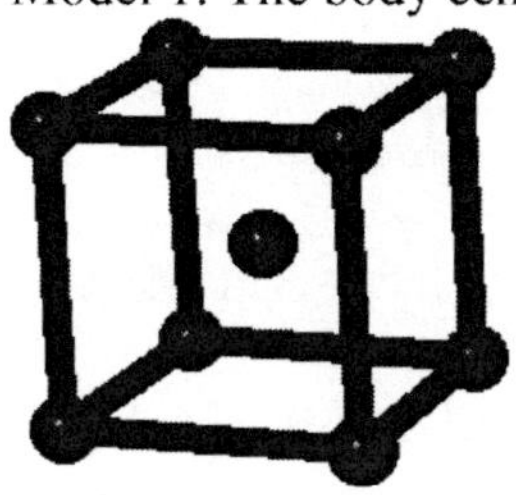

Critical thinking questions

1. Where are the atoms located in the unit cell (e.g. corners, edges, faces or centre) and how many of each type are there?

Model 2: The unit cell as a repeating unit

The unit cell is the simplest building block from which the whole solid can built.
To make the solid, the unit cell is repeated in all three dimensions an infinite (or, at least, a very, very large) number of times.
Because of this, atoms may be shared between more than one cell. The picture below shows 8 cells placed together to make a $2 \times 2 \times 2$ block.
One atom has been drawn in a different colour in this diagram to help answer the questions below.

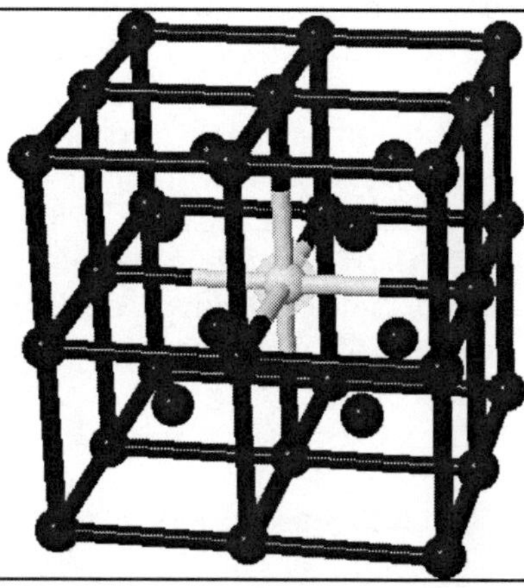

Critical thinking questions

2. Consider the atom at the centre of cell in Model 1. How many cells it is shared between?
3. Consider one of the atoms at the corners of cell in Model 1. One such atom has been coloured grey in Model 2. How many cells is it shared between?
4. What *fraction* of a corner atom is actually in one unit cell?
5. Complete the table below to work out the total number of atoms that are actually in one body centred cubic unit cell?

position	number in that position	fraction in one unit cell	overall contribution
at the centre			
on the faces			
on edges			
on corners			

6. What is the total number of atoms in a body centred cubic unit cell?

Model 3: The face centred cubic unit cell

APPENDIX C: LANGUAGE DIFFICULTY SURVEYS USED IN THE CHEMISTRY DISCIPLINE

Test 1: Questionnaire administered at the University of Sydney and the University of Tasmania in semester one, 2008 as a pre-intervention assessment in Chemistry.

1. "We carry out research to find out the answers to scientific questions."
What is meant by the word "research" in the sentence above? (There may be more than one correct answer)
After answering the question, you will be asked how confident you are that you understand this word.
a. Answering an exam question
b. Copying information out of a textbook
c. Looking for information in the library or on-line
d. Observing the results of a series of chemical reactions
e. Testing a hypothesis

1.1. confidence
Are you confident with your answer to this question on the meaning of the word "research"?
a. No, I do not understand the meaning of this word.
b. I have some idea of the meaning of this word.
c. Yes, I understand the meaning of this word.

1.2. "In most organic compounds, carbon combines with a power of four."
What is meant by the word "power" in the sentence above? (There may be more than one correct answer)
After answering the question, you will be asked how confident you are that you understand this word.
a. Always bonds to four other elements
b. Has a valency of four
c. Is four times more reactive than hydrogen
d. Forms four bonds
e. Makes bonds which are four times stronger than hydrogen

1.2. confidence
Are you confident with your answer to this question on the meaning of the word "research"?
a. No, I do not understand the meaning of this word.
b. I have some idea of the meaning of this word.
c. Yes, I understand the meaning of this word.
1.3. "Jane carefully mixed 5 grams of sodium chloride with 100 mL of water in order to get the exact concentration."
What is meant by the word "concentration" in the sentence above? (There may be more than one correct answer)
After answering the question, you will be asked how confident you are that you understand this word.
a. Paying attention in a lecture
b. The weight of sodium chloride that Jane added to the water
c. The number of grams per litre of sodium chloride
d. Evaporation of water
e. The amount of something per unit volume
1.3. confidence
Are you confident with your answer to this question on the meaning of the word "research"?
a. No, I do not understand the meaning of this word.
b. I have some idea of the meaning of this word.
c. Yes, I understand the meaning of this word.
1.4. "The reaction mixture was stirred for 20 minutes until it reached equilibrium."
What is meant by the word "equilibrium" in the sentence above? (There may be more than one correct answer)
After answering the question, you will be asked how confident you are that you understand this word.
a. When the rates of the forward and reverse reactions are equal
b. When the reactants and products reach a state of balance
c. When no more stirring occurs
d. When the reaction stops
e. When the mixture cools down
1.4. confidence
Are you confident with your answer to this question on the meaning of the word "research"?
a. No, I do not understand the meaning of this word.
b. I have some idea of the meaning of this word.
c. Yes, I understand the meaning of this word.

1.5 "The temperature was measured at 5 minute intervals and plotted on a graph."
What is meant by the word "graph" in the sentence above? (There may be more than one correct answer)
After answering the question, you will be asked how confident you are that you understand this word.
a. An XY plot of temperature versus time
b. A diagram expressing the connection between temperature and time
c. A table of numbers representing temperature and time
d. A picture of the experiment
e. A piece of paper with lines on it
1.5. confidence
Are you confident with your answer to this question on the meaning of the word "research"?
a. No, I do not understand the meaning of this word.
b. I have some idea of the meaning of this word.
c. Yes, I understand the meaning of this word.
1.6. "During the reaction, the test tube got colder and the energy of the system increased."
What is meant by the word "system" in the sentence above? (There may be more than one correct answer)
After answering the question, you will be asked how confident you are that you understand this word.
a. Everything in the universe
b. The test tube
c. The laboratory
d. The chemicals undergoing the reaction
e. The test tube and its contents
1.6. confidence
Are you confident with your answer to this question on the meaning of the word "research"?
a. No, I do not understand the meaning of this word.
b. I have some idea of the meaning of this word.
c. Yes, I understand the meaning of this word.

1.7. "The tutor helped the student balance the equation."
What is meant by the word "equation" in the sentence above? (There may be more than one correct answer)
After answering the question, you will be asked how confident you are that you understand this word.
a. A description of a reaction in terms of chemical symbols
b. A mathematical tool for calculating molecular mass
c. A representation of a chemical reaction
d. A statement that two things are the same
e. A problem to solve to find the answer
1.7. confidence
Are you confident with your answer to this question on the meaning of the word "research"?
a. No, I do not understand the meaning of this word.
b. I have some idea of the meaning of this word.
c. Yes, I understand the meaning of this word.
1.8. "In today's experiment, we will find out whether water and salt react by mixing them in a test tube."
What is meant by the word "experiment" in the sentence above? (There may be more than one correct answer)
After answering the question, you will be asked how confident you are that you understand this word.
a. A way of testing an idea
b. The test tube, water and salt
c. A way of proving what is in the textbook
d. The instructions in the laboratory manual
e. The idea that some chemicals react and some do not
1.8. confidence
Are you confident with your answer to this question on the meaning of the word "research"?
a. No, I do not understand the meaning of this word.
b. I have some idea of the meaning of this word.
c. Yes, I understand the meaning of this word.

1.9. "A spring is a model for a chemical bond." What is meant by the word "model" in the sentence above? (There may be more than one correct answer) After answering the question, you will be asked how confident you are that you understand this word. a. An analogy b. An exact visible representation of something too small to be seen c. A way of picturing something more complicated d. The theory that chemical bonds are springs e. A simple representation of a complex idea 1.9 confidence Are you confident with your answer to this question on the meaning of the word "research"? a. No, I do not understand the meaning of this word. b. I have some idea of the meaning of this word. c. Yes, I understand the meaning of this word. 1.10. "When using this instrument, only the first two numbers are significant." What is meant by the word "significant" in the sentence above? (There may be more than one correct answer) After answering the question, you will be asked how confident you are that you understand this word. a. Large b. Reliable c. Precise d. To be recorded e. Accurate 1.10. confidence Are you confident with your answer to this question on the meaning of the word "research"? a. No, I do not understand the meaning of this word. b. I have some idea of the meaning of this word. c. Yes, I understand the meaning of this word

Test 2: Cloze exercise
Language Test 2 administered at the University of Sydney and University of Tasmania in semester one 2008 as a pre-intervention assessment.
Instructions: Fill in the gaps in the story below using the following words (some words may be used more than once).
Words (in no particular order)
system, equation, experiment, model, significant, research, power, concentration, equilibrium, graph
Story
Last week in our chemistry laboratory a student carried out an ____________ (01) in which she diluted some acetic acid and then mixed it with sodium hydroxide. However she had not used the acid with the correct ____________(02) and there was a vigorous reaction which produced a large amount of heat. She waited until the ____________ (03) had cooled down and then tipped the mixture into the waste container and started again. This time her teacher suggested that she write down the ____________(04) for the reaction first and calculate the ____________(05) of acid required to react with the sodium hydroxide. To help her balance the ____________ (06), the teacher told her that hydroxide has a combining ____________(07) of one. After mixing the reagents the student used a digital thermometer to measure the temperature of the mixture to 3 ____________(08) figures. She continued to measure the temperature at 2 minute intervals and plotted the results on a ____________(09). The system reached ____________(10) after 10 minutes and there was no further change in the temperature. The student was asked to develop a ____________(11) to explain her results and then design an ____________(12) which could be carried out next time as part of her ____________(13) into the topic.
Key:
Last week in our chemistry laboratory a student carried out an experiment in which she diluted some acetic acid and then mixed it with ammonium hydroxide. However she had not used the acid with the correct concentration and there was a vigorous reaction which produced a large amount of heat. She waited until the system had cooled down and then tipped the mixture into the waste container and started again. This time her teacher suggested that she write down the equation for the reaction first and calculate the concentration of acid required to react with the ammonium hydroxide. ***To help her balance the equation, the teacher told her that hydroxide has a combining power of one***. After mixing the reagents the student used a digital thermometer to measure the temperature of the mixture to 3 significant figures. She continued to measure the temperature at 2 minute intervals and plotted the results on a graph. The system reached equilibrium after 10 minutes and there was no further change in the temperature. The student was asked to develop a model to explain her results and then design an experiment which could be carried out next time as part of her research into the topic.
List of words: experiment, concentration, system, equation, concentration, equation, power, significant, graph, equilibrium, model.

Chapter 3
First Year Chemistry Language Support at the University of Tasmania

Brian F. Yates
University of Tasmania, Australia

Michael G. Gardiner
University of Tasmania, Australia

ABSTRACT

A number of in-class and post-lecture learning strategies are described that have recently been incorporated into the first year chemistry program at the University of Tasmania, Hobart. These methodologies are centered on supporting the various language based challenges that students face in achieving a firm introduction to the discipline. A number of delivery strategies were trialed for in-class activities that overcame institutional specific limitations in access to technologies in an attempt to increase student involvement in lectures and tutorials. Activities included mobile phone-based technologies (and simple overhead transparency/show-of-hand alternatives) for in-class audience surveys, word matching exercises, and chemical songs. The post-lecture approaches also covered a range of delivery platforms, including (i) on-line based quizzes that focused on providing specific, tailored feedback to allow maximal self-reflection on the understanding of key terms and concepts prior to attempting weekly summative assignments where feedback is deliberately kept limited; (ii) developing a broader range of weekly assignment questions (beyond conventional multiple choice) to focus on language issues; (iii) chemical Sudoku puzzles, and (iv) crossword activities. Evaluations of these approaches are presented.

DOI: 10.4018/978-1-61350-062-0.ch003

INTRODUCTION

The introduction to the previous chapter has provided an excellent discussion on the general language based issues that confront students embarking on their pursuit to master chemistry. We do not wish to cover this ground again with any great overlap. Students of the discipline have faced many language based hurdles for centuries (Wellington & Osborne, 2001); however rapidly advancing technologies that have created new niche areas of specialization, changing requirements of graduates (not only in these new areas, but even in traditional vocations), the changing environments of tertiary education and even the level of part-time work students commit to during the academic year have brought new challenges.

The opening sentence of the previous Chapter, "Chemistry is the central science", is one that is becoming ever more relevant as the life sciences attain increasing levels of understanding at the molecular level and advances in physical sciences and engineering demand new materials (aka chemical substances) with improved properties and reliabilities to construct more sophisticated devices. With this, comes increasing challenges in the teaching of chemistry across a range of areas. Most obvious is in catering for many areas of service teaching that require tailored content (Barrie, Buntine, Jamie, & Kable, 2001). In parallel, delivering effective first year programs for students who intend on pursuing a major in chemistry is always undergoing modernization to adapt to changing needs, important breakthroughs and shifts in emphasis as society progresses. Enforced combined teaching of, previously, often separated streams in the face of budgetary/staffing pressures only adds to these issues. Such compromises can risk specialist programs undertaking the teaching of introductory to intermediate level chemistry within other schools, which has consequential organizational effects within chemistry departments and potentially can put at risk the students not mastering sufficient general grounding in the discipline to be effective in their profession. Ensuring that a student's first year of university study gets off on the right track is vital across many agendas (Burton & Dowling, 2005; Simpson, 2006; Tinto & Pusser, 2006).

In general, students of increasingly diverse prerequisite chemical knowledge and vocational interests are often being co-taught (Harris, et al., 2007). This broad range of students also tends to have a naturally wider instinctive range of numerical and language related abilities. Across this wide cohort of students with physical and health sciences interests, even personalities and favored learning approaches are, in general, quite disparate. Also adding to the challenge is that relatively few students pursue their first year chemistry study with the passion expected of someone intending on majoring in chemistry (Oliver & Simpson, 1988). This leaves many students seeing this aspect of their training as a necessary task to get through in order to achieve a qualification in their chosen vocation. As noted in the previous Chapter, such motivational issues are of major concern to mastering first year study in chemistry. An unrelated, additional, point relevant to this project is that tertiary institutions have increasingly larger proportions and wider range of diverse student nationalities in their programs, for whom English maybe a second language. This makes tertiary education institutions very interesting work places to work and study, but due to the many first languages and English language skills of the students this creates further challenges in relation to language support in the crucial first year of study for these students. For the institution, and discipline, at stake is retention and progression of students through their degree.

The upshot of all these issues is that (i) a range of significant, increasing language based challenges that all students face in achieving an adequate introduction to the discipline and (ii) teaching staff need to manage these challenges to meet the needs of their students and the many other stakeholders they interface with. It is a challenge to satisfy the needs heard internally within institutions, with secondary schools, government and relevant industries demanding well trained graduates that meet each of these needs.

We have always to find more effective ways to tackle the time old problems associated with the high usage of symbolism and word meaning in chemistry. Many of the issues are constantly evolving. New technologies offer innovative ways to approach them but there is a multitude of options available for students to choose independently and teaching staff to direct students to; some good and some bad. Textbook publishers can offer a wide variety of material in various formats, but some student direction is needed in trying to achieve a balanced set of activities to maximize effectiveness in the time that students have to devote to all their units of study. As with all aspects of life, this set of recommended activities may not be suitable to the learning styles of all students. Whatever intervention strategies are implemented, their design also needs to take into consideration such side influences as the student flexibility that they offer and the time commitment that is required given the level to which students fund their education by part-time employment during the academic year. This chapter outlines a number of initiatives that have been embedded in first year chemistry at the University of Tasmania that relate to the aims of the Australian Learning and Teaching Council project discussed in the Preface.

Context

The School of Chemistry at the University of Tasmania operates under the influence of most of the issues discussed above that continue to shape its unit offerings. The current structure of first year chemistry taught within the 3 year Bachelor degrees on the Hobart based campus has an intake that requires, beyond the standard University entry requirements, satisfactory completion of chemistry at the final year of pre-tertiary secondary school. If this prerequisite is not met, a single semester foundation unit is needed (and possibly a similar mathematics foundation unit) prior to enrolling in the main Chemistry 1A and 1B units (KRA113 and KRA114, respectively).

Unlike some offerings at other universities, those without secondary school leaving chemistry qualifications undertaking the foundation unit are not allowed progression into the main first year units mid-way through first year studies to allow progression to second year within their first two semesters of university study. With the exception of single semester first year units designed for Bachelor of Engineering and Bachelor of Applied Agriculture students, students of all other degrees take Chemistry 1A and 1B. Mid-year entry into these unit offerings has been possible for several of years and this now accounts for nearly 10% of the overall intake. This issue, as later discussed, has had impact on many aspects of syllabus design, including issues that relate to best practice for language support initiatives.

Whilst prerequisite of secondary leaving chemistry for entry into Chemistry 1A and 1B somewhat levels the playing field of background chemical knowledge, the minimum university entry level for the many degree programs of the students varies considerably and some students struggle to grasp the new topics that build on assumed knowledge. Enrolments in Chemistry 1A and 1B include students in degree programs administered within every Faculty of the University. Core degree requirements in chemistry range from only Chemistry 1A and 1B through to some second year units. Electives may include up to 3rd year level chemistry units in some streams. As an indication of the extent of service teaching within the School, enrolments in Chemistry 1A and 1B are typically between 200 and 250 whilst around only 5 to 10 students complete their degrees with a major in chemistry each year. Each module of Chemistry 1A and 1B contain three topics, each lasting four weeks, taught by different lecture staff at various career stages. These lecturers also all head highly active research groups.

Language Difficulty Surveys

A key language issue common to the study of all sciences is the situation where either subtle, but important, or major discrepancies between lay usage of a word and their meaning within the discipline can occur (Gardner, 1977; Gardner & Australian Science Education Project., 1972; Pickersgill & Lock, 1991). Two examples of this in first year chemistry not discussed previously are *spontaneous* and *rate constant*. These terms can cause significant problems in parts of physical chemistry for two quite different reasons (Markow, 1988). In lay usage, the adjective *spontaneous* refers to something that begins by itself without outside interference. In its chemical usage, a spontaneous process is one that is energetically favourable (in terms of the Gibbs free energy). A spontaneous process may not actually start to any significant extent by itself or, more confusing still, a process that is not spontaneous may actually occur quite rapidly (at least to some extent of forward progression). For this term, the issue is that most people attempting to adapt the word to a chemical context without specific knowledge of its correct usage would incorrectly build their logical guess around chemical kinetics (the rate of reactions) concepts rather than basing this correctly on thermodynamics.

The term *rate constant* can cause angst for quite different reasons. If taken from the view point of a student accustomed to modeling observations to mathematical expressions in the physical sciences, this refers to a physical *constant* (numerical value determined by experiment) that relates to the *rate* (or speed) of a reaction. However, if this line of analysis is not followed and the common, often very successful, approach of dissecting the term into the individual words *rate* and *constant* is attempted, one gets into a problem of being faced with two apparent contradictory words (i.e., rate is associated with speed of change and yet 'constant' suggests no change) and there is little chance of a correct understanding being expected to follow.

Student problems associated with the comprehension of scientific terms often extends beyond examples such as these quite discipline specific cases discussed above. Correct usage of very general scientific terms, such as those relating to scientific method, can present the worst case of misinterpretation as the more common terms are often assumed across all of a student's first year enrolment and may last through to higher research degree stage.

First year chemistry students at the University of Tasmania enrolled in KRA113 were invited to take part in the voluntary survey on comprehension of ten common terms that students will meet regularly in first year chemistry. The purpose and design of this survey was described in depth in the previous Chapter and a description will not be repeated here. The surveys were delivered through the on-line delivery system already in place for the unit approximately half way through the first semester of the first year program and were completed by 30 students, corresponding to 14.6% of the class. The results are summarized in Table 1 below.

The data in Table 1 map closely onto that collected at the University of Sydney in their various programs that varied regarding chemistry prerequisite requirements. Linking with The University of Sydney cohorts that have undertaken high school chemistry, we gratifyingly note the clustering of the mainly chemistry based terms "concentration", "equilibrium", and "equation" tend towards the easiest terms for the University of Tasmania students who have all satisfactorily completely high school chemistry. This is gratifying in relation to the entry level of the students beginning their first year studies. The generic term "experiment" is clearly the best understood term in all surveys that have been conducted. "Graph" is similarly quite well handled, which is in stark contrast to other generic terms associated with the scientific method and/or data interpretation, such as "significant", "model" and "system". The standout term

Table 1. Language survey results for KRA113 (all students with high school chemistry) at University of Tasmania

Terms	Confidence level (%)	Understanding (%)	Delusion index	Ranking (easiest=1, hardest=10)
Significant	48.40	3.20	15.13	10
Research	73.50	9.10	8.08	9
Model	48.40	12.90	3.75	8
System	58.10	19.40	2.99	7
Power	48.40	22.60	2.14	6
Concentration	87.10	25.80	3.38	5
Equilibrium	76.70	32.30	2.37	4
Equation	77.40	35.50	2.18	3
Graph	87.10	41.90	2.08	2
Experiment	83.90	71.00	1.18	1

causing most issues is clearly "significant". These trends are to be expected given the early stage of the students in their professional career where their exposure to experimental design and interpretation are fairly limited, though this was embedded in the year 12 chemistry syllabus in Tasmania at the time of the survey.

The most unexpected finding is the perceived meaning of "research", where many students recorded "testing a hypothesis" as their sole response to this question. This description of the term is reinforced at high school, which no doubt accounts for this trend. Perhaps strangely, the cohorts that have done high school chemistry tend more to not list "observing the results of a series of chemical reactions" as a response associated with research.

LANGUAGE METHODS EMPLOYED TO SUPPORT LEARNING IN FIRST YEAR CHEMISTRY

We wanted to develop useful strategies in both in-class activities and post-lecture resource material in this project. We gave some priority to in-class activities knowing that this could serve several benefits. Firstly, being mindful of increasing time pressures imposed on many students through part-time employment during the academic year we realized that many students could potentially opt out of post-lecture activities depending on the perceived usefulness of these activities or motivational prejudice if left uninspired by their face-to-face experiences. This added to our desire to make the non-compulsory and increasingly less attended lecture contact time an opportunity to deliver inspiring and unique material that maximizes learning outcomes.

Examples of such strategies include: (i) presenting interactive explanations and chemical demonstrations, (ii) presenting alternative explanations to the textbook and,(iii) providing an alternative way of accessing the content. With these goals, we hoped to instill improved levels of motivation that may also then lead to good participation levels in the non-assessable post-lecture activities. Both approaches included elements of games and/or some moments of unexpected fun and lighthearted sidetracks designed to bolster attention. A range of technology and nontechnology-based delivery platforms were utilized.

In-Class Activities

Concurrent with the implementation of language support interventions in first year chemistry which is the subject of this chapter, staff in the School of Chemistry have been active in the introduction of student centred active learning techniques associated with the *Active Learning In University Science – Leading Change in Australian Science Teaching* (ALIUS: www.alius.edu.au) project (Bedgood Jr, Gardiner, et al., 2010; Bedgood Jr, Yates, et al., 2010; Bedgood Jr., et al., 2008) arising from the Australian Learning and Teaching Council project *Developing Leaders Of Change In The Teaching Of Large University Chemistry Classes* within the Leadership for Excellence in Learning and Teaching Program. The benefits of active learning strategies are well documented, but can be a great challenge to implement in large classes with traditional tiered lecture theatres. This is particularly the case if an approach is attempted that occupies the entire student contact time, such as the POGIL (Process Orientated Guided Enquiry Learning: Moog, 2008) methodology. Detailed discussion of this type of strategy is given in the previous Chapter. With less ambitious initial implementation targets, we implemented a number of stand-alone activities designed to periodically break up the often quite didactic 50 minute sessions (particularly in lectures, which tend to follow this style) and in less formal tutorials. Both technology and nontechnology-based approached approaches were developed that coupled both our language support strategies and active learning goals. These are discussed below.

Use of VotApedia and Nontechnology-Based In-Class Student Surveys

Institutional circumstances prevented the trialing of clickers as a delivery method for in-class student surveys. We were conscious of the benefits of reliability and increased participation associated with student anonymity and automated response collection that technology-based approaches offer in conducting in-class surveys. Such data offers the students instant feedback on their understanding of the topic being discussed including their position relative to the rest of their participating class members. While for the teaching staff, the responses can be used to great effect to address often unforeseen issues that the class can be experiencing.

We trialed the mobile phone based VotApedia system (www.VotApedia.com.au, now www.urvoting.com) that has been discussed in previous Chapters. A basic description of this free to access on-line based system was included in the unit outline and its usage was demonstrated during the introductory lecture in the first week of semester. Limited mobile phone coverage in the lecture theatre was somewhat of a limitation, with only around 30% of the students subscribing to a suitable network provider. Nevertheless, typically high participation rates amongst those students gave reliable survey results. Those students who were prevented from participating could still benefit through knowledge of the whole class response and their own correctness of response. No negative comments on this inequity were given in student evaluation surveys of the unit.

Participation rates were of the order of that typically noted for show-of-hands class surveys, without the possibility of a student being identified when offering incorrect answer which often leads to embarrassment. It must be said that show-of-hands alternatives are quicker to administer and sufficient time must be allocated to conduct technology-based in-class student surveys. Depending on the question being asked, it can be beneficial (for learning or simply class enjoyment) to consider choosing the options of either having the responses display in bar graph format in real time while the survey is open or to only show this analysis after the survey is closed by the lecturer. For this latter option, the total number of responders is still shown on screen in real time and can aid teaching staff in deciding on the time needed for student answers. You can cut the feeling of suspense with a knife some days before the survey results are revealed!

An example VotApedia delivered class survey from Chemical Kinetics:

Question:
If you measure the rate of a reaction in terms of the disappearance of reactant A in the following example equation:

$$2A + B \rightarrow C + 2D$$

How will the rate of change in the amount of A relate to the reaction rate?

Responses:
1: The rate of change of A will be the same as the rate of the reaction
2: The rate of change of A will be half that of the reaction
3: The rate of change of A will be twice that of the reaction
4: The rate of change of A will have nothing to do with the rate of the reaction

The question is typical of a mid-lecture instant revision type question that is based on a concept that is key to the core understanding of the topic that may have been the subject of perhaps half a lecture discussion. The surveys offer a good chance for instant feedback on their understanding of an important concept. If the outcome of the survey shows a good level of student understanding, the mini-break in the lecture while conducting the survey provides a good spell before moving onto other material. The lecturer needs to be adaptable in devoting time to addressing sometimes unanticipated levels of misunderstandings that can be made apparent from the survey results.

In addition to the use of VotApedia during lecture time, often 1-3 times in a 50 minute lecture, they were found to be a highly effective revision tool (for more VotApedia questions, please refer to Appendix A). Surveys were developed for student responses before the lecture commences based on previously discussed material. This can be an effective use of an otherwise dead period of time while setting up for the formal lecture to commence. Other pre-lecture strategies are discussed in the following sections. An example of this type of question is given below from Structure and Bonding topics. A useful benefit of this type of revision survey question is the ease with which the class can be brought into focus following the closure of the survey and revealing of the students responses and the correct response.

Question: *Which of the following is not electromagnetic radiation?*

Responses:

1. *Microwaves*
2. *X-rays*
3. *Sound Waves*
4. *Radio Waves*

Similarly, a number of questions were developed as lecture warm ups that were used to serve as effective discussion starting points for the concepts to be discussed in the current lecture. This approach was particularly suited to language support issues where, typically, known misconceptions could be aired and corrected. An example of this is in the introduction to nomenclature of coordination complexes, where the question and multiple choice responses shown in Figure 1 were offered. Of note, the inclusion of an unexpected lighthearted response option naturally increased overall participation rates (and still allowed for useful class discussion of the other options!). Obviously, the attitude portrayed by the teaching staff must be carefully considered in order for the class not to believe there is any sense that he/she thinks there is truth to the final response.

Figure 1 is an example of a VotApedia delivered lecture warm up in-class student survey (taught in the Coordination Chemistry topic) which aimed at allowing students to use self enquiry to establish if all details of a coordination complex are portrayed in the complex name.

In this particular case of chemical nomenclature, the light-hearted option in the survey is very strategically included as the topic is inherently a quite dry subject to teach. The correct answer is "yes", though the particular example given is carefully chosen to elicit some students to choose "no" in relation to whether the name should specifically end with "dichloride" for correctness. This enables the comparison with the naming of simple transition metal salts, such

Figure 1. An example of a VotApedia exercise used in lectures

Question: Is the following name sufficient to fully describe the coordination complex shown?

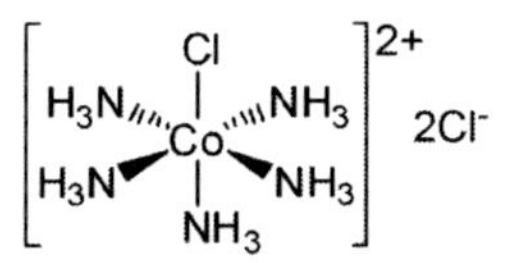

pentaamminechlorocobalt(III) chloride

Responses:

1: Yes

2: No

3: Not sure

4: Don't Care!

as $NiCl_2$ (nickel(II) chloride), to be discussed as an example of high school chemistry naming after the survey is closed. This serves as a good reminder that many of the nomenclature rules taught in the first year unit are already known to the students even if they may appear disguised in their new context. Thus, rather than being a new topic to students, this warm-up survey focuses attention on the new rules being introduced that relate only to the new features of the compounds being studied (the ligands) while most rules are already known to the students.

In order to correctly give the systematic name of even a simple coordination complex, around ten rules need to be introduced. This can be quite overwhelming and it can be perceived as though more than the minimum rules are being introduced and teaching methodologies are being used that do not soften the introduction in the same way that the students have known in their organic chemistry topics where naming issues can be gradually introduced.

A follow up from this VotApedia question is to create awareness of the distinction between a typical coordination complex and an organic compound studied in each topic. In this manner, it becomes very apparent that the complexity of the chemicals is very different, that the names thus also vary accordingly and that there must be a reasonable number of naming rules to cope with these differences. This strategy sets up the lecture that follows with typically good student attention to the material that can otherwise wane if this direction is not provided.

Word-Matching Lecture Warm Ups

Effective use of the five to ten minutes between arriving at the lecture theatre and commencing in the formal lecture can be achieved through lecture warm-up activities. Strategies must of course not be time consuming for the lecturer while they are typically setting up computer systems, demonstrations or talking to students. A number of overhead transparency word-matching puzzles were developed that allows students to opt into the challenge of matching terms/concepts/symbols with brief definitions. Figure 2 shows an example that was developed for the Thermodynamics topic relating to term definitions. This topic is often perceived as being difficult for a number of reasons. One of the reasons is that many term definitions that can appear to be only subtly distinguishable. Depending on lecture theatre location, room set-up, or even the amount of student movement into the lecture theatre, the success of this approach can be highly variable in regard to the number of students that pay attention to the activity.

Figure 2. Example of a completed word-matching exercise used as a lecture warm up that links terms with definitions (taught in the thermodynamics topic; created by Zhang, F, 2010 using HotPotatoes)

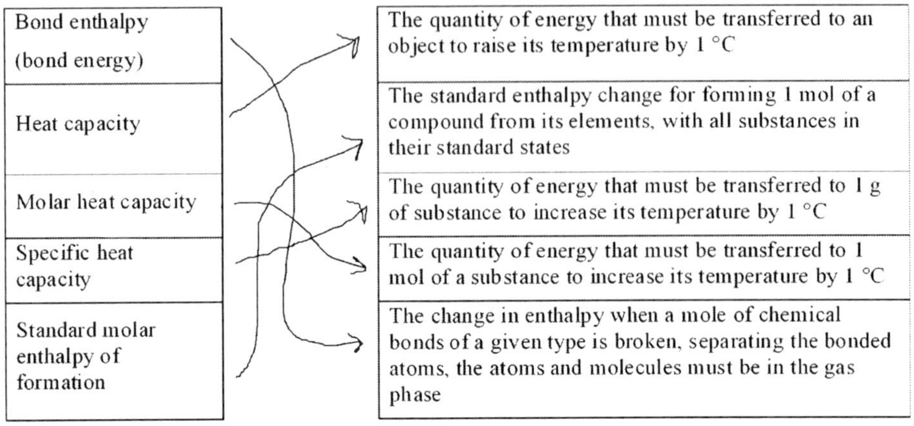

Figure 3. Example of a completed word-matching exercise used as a lecture warm up that links symbols with definitions (taught in the thermodynamics topic)

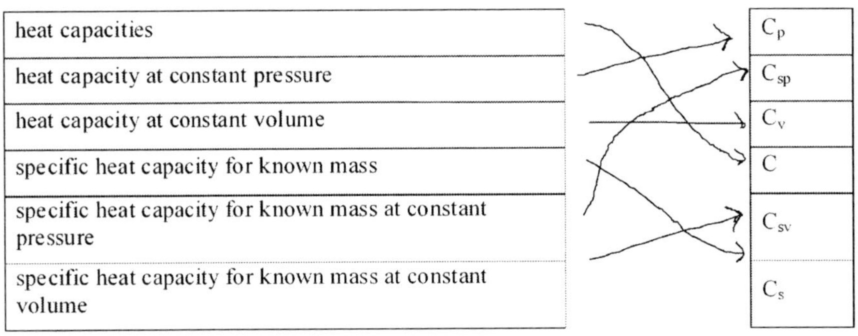

The further example in Figure 3, below, illustrates the adaptability of this approach to aiding the memory of the many symbols associated with specific terms in thermodynamics. Significantly, the symbols have been jumbled and appear on the right hand side of the puzzle, which aids students in noting the variations in the definitions on the left (that appear in a systematic order) and promotes a pattern recognition approach to completing the puzzle. This was also created with the computer freeware, Hotpotatoes.

Following the completion of the puzzle, it is of value to stress the commonality in the symbols regarding the systematic choice of the letters used, including the subscripts, and the link to the strict definitions of the terms and the conditions under which each are valid. For example, the matching puzzle is limited to heat capacities and specific heat capacities to allow for non-written attempts in the pre-lecture period, but after the puzzle is completed the symbols used for the various types of molar heat capacities (C_m, C_{mv}, C_{mp}) can also be revised and their systematic composition discussed. This exercise was designed to highlight the fact in a symbol like C_{sp}, 's' stands for specific heat capacity and in C_m, 'm' stands for molar.

There are clear limitations on the complexity of the puzzle that is set as students cannot benefit by drawing connections to complete that puzzle on paper. Around five matches were found to be ideal, otherwise the growing need to remember which matches remain can become a distraction and it is found that more students opt out in preference to talk with friends. In most cases, completion of the puzzle on the overhead with no initial student involvement or explanation is a good way to transition into the start of the lecture without taking much time. If a good proportion of the cohort has been focusing on the task, good extensions can include explaining the more difficult matches if close alternatives have been included. The students can be asked to comment on which matches they found hard. Alternatively, if the lecturer pays careful attention to student discussions while matches are being completed, the lecturer can tailor feedback to student reactions.

Such matching puzzles can be very effective in a number of ways. Firstly, they represent a good self-testing option substantially after the previous study of the students on this topic. In particular, the matching requires a significant dwell time of the puzzle (and the key terms) in the mind of the students while the full solution is completed and subtle distinctions in answers are thought through. Such thought time can be lacking in multiple choice questions that can be completed very quickly. More strategies that enhance dwell time advantages are discussed below in a number of post-lecture methods.

Chemical Vocabulary-Based Singing

There are a number of chemical songs available on the internet in which people have made up words to well-known tunes. These words are used to describe a chemical reaction or a chemical concept. By associating the words with a well-known tune, the students are better able to recall the chemistry information when it is needed. And it is entertaining!

We have used a number of available songs and in addition have written some of our own. For example, one of our own creations concerns the following reaction called 'alkylation' taught in Organic Chemistry topics. To add an Australian flavour, we set some words to the Australian national anthem, 'Advance Australia Fair', see below:

Advance alkyl amine fair
For amines all let us rejoice,
for they are basic and free.
They react with alkyl halides,
form salts as you can see.
Look out, here comes a base along
to take away a proton.
It makes some water and a sodium salt
in a process called alkylation.
And from this juice we at last produce
a new alkyl amine fair!
(Brian Yates, 28 September 2006)

The words to 'Advance Australia Fair' describe the mechanism of the alkylation reaction, Figure 4. A number of other songs are *performed* in lectures. The student feedback was very positive:

The songs were great. I'm glad you had the courage to sing them to us. They were a great way of learning some of the reactions.

I was quite positively surprised about the organic chemistry lectures. During the first audition I thought the singing was nice but not exactly useful but to my own surprise I have to admit that in fact I do now remember these reactions by singing the song myself...

I loved your songs and am wondering if you've got any more! I've printed them off and am taking them on my daily walks - singing as I go! (haven't been caught yet!) They're great for the reactions, which I find hardest to remember, so I'm hoping they'll stay in my head and that I don't burst out in song in the middle of exams!!!

Figure 4. Alkylation reaction between an amine and an alkyl halide (taught in Organic Chemistry topics)

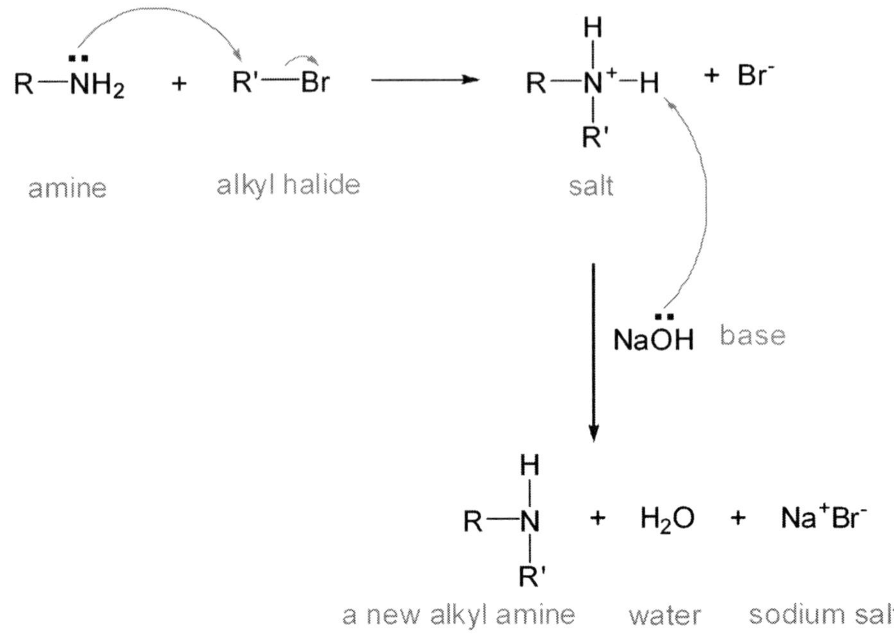

Additional chemistry songs are obtainable by web searching. The following provide a selection:

- http://www.departments.bucknell.edu/chemistry/courses/chem211/songs/,
- http://www3.delta.edu/slime/chemsongs.html, and
- http://www.wellesley.edu/Chemistry/Flick/mysongs.html.

Of wider familiarity, The Elements, written by Tom Lehrer, probably provides less benefits in terms of learning outcomes aimed at the aiding the memory of chemical terminology, but can be a useful extra to include in lectures as light entertainment. Accessible versions are available on-line from the original song writer, http://www.youtube.com/watch?v=DYW50F42ss8, or http://www.privatehand.com/flash/elements.html or a very recent partial version by Daniel Radcliffe (the actor that plays Harry Potter) is likely to have good appeal to students, http://www.youtube.com/watch?v=rSAaiYKF0cs.

Relating somewhat to the chemical songs developed with specific relevance to lecture material mnemonic sentences are used widely across the sciences as memory aids for key concepts and/or features. They represent a good blend of light entertainment and learning aids. In particular, when correct ordering of objects is important and is based solely on rote learning or when the terms are not based on word stems with lay meaning. A good selection of mnemonics relating to chemistry can be found at http://members.chello.nl/r.kuijt/index-chemistry.htm. An example relating to redox reactions is "Oxidation Is Losing electrons, Reduction Is Gaining electrons", whose underlined letters spell "OIL RIG". Other well known examples of mnemonic sentences aid in remembering the colours of the rainbow, biological classification and the planets of the solar system, and can span large variations in the age range of relevance to general and scientific learning.

Post-Lecture Activities

As previously discussed, the incorporation of any post-lecture activities for language support into the assessable or non-assessable materials for Chemistry 1A and 1B were considered very carefully in view of the workload aspect of the students. There is a large amount of material that can be recommended with either great thought given to its relevance and/or with relative ease and little thought from publisher sources. It is very easy to overwhelm the cohort with suggested exercises that may not be relevant to the syllabus and our may use different symbols for the many constants and variables in physical chemistry topics, for example, that can detract from comprehension and cause great confusion. This can be a major distraction to successful learning outcomes.

Concept Tests

The major post-lecture strategy adopted in this project was the introduction of what we termed 'Concept Tests'. These are weekly quizzes that do not contribute directly to assessment. They do not deliver examination type questions as such but deliberately focus on addressing common misconceptions or difficult concepts that students are known to have with each topic. These issues are often language based, but not exclusively. The design of the Concept Tests was based around the requirement of sustainability where the goal was their long term usage as a resource that is likely to largely survive any downstream staffing and syllabus changes by keeping content very much focused on core concepts. There were also particular local circumstances that influenced the inclusion of the Concept Test in both the Chemistry 1A and 1B units. Assessment in these units includes weekly on-line based assignments (10%), eight weeks of laboratory (20%), a mid-semester written test (10%), and final examination (60%).

The weekly on-line administered assignments were changed in response to the local circumstances from initially permitting two attempts to only allowing a single attempt. In addition, any existing feedback for these questions was closed off until the time period for the assignment was closed. These changes were justified in terms of attempting to increase the extent of individual performance in the assignments (as compared with "team efforts" aimed only at achieving 100% each week). The changes did lead to improvements in this regard. However the tasks had become counter to most principles of formative assessment which are behind on-going assessment items. These weekly on-line based assignments were designed to promote effective self-learning through the instant feedback options that are the hallmark of on-line assignment management in large class teaching.

In the face of the on-line assignment tasks that offered poor feedback, we implemented the non-assessable Concept Tests as a tool to provide considerable guidance to students in attempting the assessable assignments. Minimal time requirement, maximal feedback and the option to (almost entirely) opt out of this additional weekly task was the priority in design of these assignments. The focus was on understanding of the key terms and concepts in each unit. We were conscious of making students aware that the Concept Tests are primarily a learning tool and not for assessment.

We constructed the weekly on-line Concept Tests and Assignments to be linked, such that assessing the assessable weekly Assignment was conditional on students attempting the Concept Test initially and obtaining a score of at least 1 out of 10. This created an opt out option for students. This also set the bar quite low in terms of the minimal effort that could be devoted to the Concept Tests but, importantly, still required students to access them. Through monitoring of students grades for the Concept Tests, this minimum score could be raised substantially without excessive student negativity as the large majority of students achieve 50% scores for the Concept Tests throughout the year. We included an explanation of the conditional relationship and assessable status of both the Concept Test and Assignment at the header of each Concept Test in the way shown below:

Instructions

This Concept Test is provided to assist you in this week's assignment.

The Concept Test is not assessable. However, accessing the assessable Assignment is only possible by attempting this Concept Test with a minimum score of 1 out of 10. Obviously, we encourage an honest attempt in the Concept Test. It is designed to aid learning, direct your revision for the MyLO assignment and ultimately your assignment performance. Unlimited attempts are allowed. Immediate feedback is provided for the Concept Test in terms of the correct answers and extensive reasons for the incorrect responses are provided.

It may be useful to have a print out of the graded Concept Test available while doing the Assignment (which can be obtained even if all the Concept Test is not completed).

In addition, the Unit Outlines include similar information and a time suggestion and strategy guide to spend only 10 minutes on the Concept Tests and use this as a formative guide to preparedness for the assignment, see below. In this way, we have had less than 1% of students mistakenly only attempting the Concept Test thinking that they were the assessable assignment (which was addressed within the first weeks of semester through tracking of student grades). We are quite certain that prominent positioning of these messages was a key to not suffering from poor student remarks regarding workload issues for the unit. Indeed, the distribution of grades in the Concept tests throughout the whole year program showed that only 17% of students scored an average score in the Concept Tests less than 20% (or double the minimum requirement) and the vast majority of students attempt all 10 questions each week. The following information was provided to students in week 1 of the unit:

There is a non-assessable Concept Test must be attempted before accessing the weekly assignments in Weeks 2-13. A minimum score of only 1 out of 10 is required. The *Concept Tests* are designed to offer instant feedback on your understanding of the concepts underlying many of the assignment questions. Incorrect responses have extensive tips on addressing misconceptions. We suggest that you make a real effort on the Concept Tests. A suggestion is to *test yourself* by attempting them over just 10 minutes, respond with your first thought (long calculations won't be required) and then brush up on any problem areas. If you only opt for a token attempt at these, you can still benefit by referring back to the feedback given if you subsequently have a problem with an assignment question. Multiple attempts are also allowed for the Concept Tests.

An example Concept Test question illustrating the non-examination type question basis of the Concept Tests that focuses on understanding of key terms and concepts needed for assessment is shown below (from Coordination Chemistry and/or Acid/Base Chemistry). Where possible, each question included a note saying that this question was relevant to a particular Assignment question for that week. This further emphasized the benefit of attempting the Concept Test questions and/or reviewing the feedback that is available.

Question: What does a Lewis base (also called a ligand) donate?
(this question relates to Question 1 of the Week 19 Assignment)

(a)	an electron
(b)	an ^{-}OH group
(c)	a proton
(d)	a pair of electrons

The tailored feedback associated with this Concept Test question is given below. As the Concept Tests are non-assessable, were elected to give the correct answer in the feedback as well as tailored feedback relating specifically to why a student may have chosen each particular response:

1. **Incorrect.** The correct answer is D. Lewis bases do indeed donate 'electrons' but specifically they donate a pair of electrons. It is these electrons that form the bond between the ligand and the metal ion in a coordination complex.
2. **Incorrect.** The correct answer is D. This is the definition of an Arrhenius base. A Lewis base is a chemical species that can donate a pair of electrons. The hydroxide anion, ^{-}OH, is an example of a Lewis base however as the oxygen atom can be the donor atom using one of its lone pair of electrons to bind to metal centres.
3. **Incorrect.** The correct answer is D. Bases do not donate protons. Furthermore Lewis acids do not donate a proton; they are species that can accept a pair of electrons – such as the metal ion in a coordination compound.
4. **Correct.**

The removal of the grade quota system for the unit after the Concept Tests were implemented was a welcomed policy decision. We continue to use the Concept Tests in the format as described above, but we have been able to greatly improve the organizational issues associated with the online weekly Assignments. Unlimited attempts have been introduced and high level instant feedback has been written for all questions.

Given that we wished to maximize students undertaking repeat attempts at the Assignments with learning goals in mind (rather than adopting strategies to achieve full marks without any learning outcome) we have provided tailored feedback to the Assignments but not the correct answers at the time of grading. For example, if an incorrect response was given as the student selection in relation to a common error a specific tip is given to allow students to reattempt the problem using the correct methodology. For example, using temperature in degrees centigrade rather than Kelvin in a physical chemistry problem is a common error and an incorrect multiple choice response is often included for this purpose.

Concept Test and Assignment Questions Promoting Deeper Thought

In addition to the Concept Tests that have been created for the whole of the Chemistry 1A and 1B program, we have modified a number of the weekly on-line Assignment questions to further facilitate language support in some areas. We achieved this by introducing a broader range of weekly assignment questions beyond conventional multiple choice questions that have dominated the assignments prior to this project. Whilst this comprised the extent to which we included examination type questions (that are always received as popular by students if kept short in terms of the length of working required), the benefits to reinforcing the understanding of key terms was deemed beneficial. Jumbled sentence questions (created by using HotPotatoes software) were found to be particularly effective as they normally require an iterative process of elimination of the options provided as the question is completed that involves multiple rereading of the question.

This results in a significantly longer thought time for the terms and definitions, which is a clear memory aid. When carefully constructed to avoid vagueness, this question type can also highlight key aspects of definitions that can bypass notice if the only learning strategy adopted is to review a written definition from a text book or lecture slide. An example of a jumbled sentence question relating to the Thermodynamics topic is given below. Note how this question also reinforces the subject in the pre-lecture word marching exercise that was described earlier.

Question: *_________ relates the amount of heat energy given to something to the observed change in its _________. For pure substances it is possible to calculate the heat capacity per gram, which is* *then called the _________ or per mole, which is called _________ The units for specific heat capacity are _________ while the units of molar heat capacity are _________.*

Options: *heat capacity; temperature; molar hear capacity; specify heat capacity; J/°C/mol; J/°C/g*

The example below gives an example of a multiple choice question written in a style that achieves similar learning goals to that incorporated within the jumbled sentence question format (from Structure and Bonding). Depending on the question styles that are allowed within the on-line learning system being used, this can represent a close alternative.

Question: *Two atoms or ions are isoelectronic if they have the same number of _______. This also means that they have the same ___________. Differences in size across an isoelectronic size are due to the changing _________.*

1. *electrons, group number, nuclear charge*
2. *electrons, electronic configuration, nuclear charge*
3. *protons, charge, electronegativities*
4. *protons, electronic configuration, electronegativities*

Chemical Sudoku and Crossword Activities

We have also used a number of competitions, such as Sudoku puzzles and crosswords, in which we ask students to complete the puzzle and then go in the draw for a prize. An example of a Sudoku puzzle is shown in Figure 5. The letters in the puzzle are used to complete a statement of a chemical concept. The final aspect of the puzzle has a language, term, based purpose. Through attempting, and solving, the puzzle it is clear that the concept on which the puzzle is based must be

Figure 5. An example of a chemical Sudoku puzzle (taught in Organic Chemistry topics)

KRA114
Chemistry Sudoku #1

B				Be				N
	O	He	F					
	Be		C			H		
	H	O	B		1			
C				He				Li
					Li	C	F	
	2	H			F	4	C	
					B	Be	O	
He				H	3			B

Complete the puzzle using the first nine elements H-F of the periodic table.
(No element can be repeated in any one line either horizontally or vertically.
Each nine-square grid must also contain the nine elements without repetition.)

Please return to Brian Yates at the lecture on Friday 16 July 2010
if you wish to go in the draw for a text book prize.
Prize will be drawn at the lecture on Monday 19 July 2010

Aldehydes and ketones both contain a

1	A	R	2	3	4	Y	L

group.

(Fill in the missing letters using the solution to the Sudoku puzzle above.)

focused on for some time during the attempt and this has good memory benefits. However, it must be said that these post-lecture strategies, or 'extras' to the core unit material, are largely initiatives to awaken students interest (develop curiosity) and focus attention. The prizes are awarded during lectures in a strategic approach to help encourage lecture attendance.

Student feedback on the Sudoku puzzles was again very positive and a number of anecdotal comments on the outcome of the activities as a learning tool were provided:

the Sudoku was a great way to break up the lectures, just made things a bit more fun.

He seems to actually care if students understand.

He combines ...a genuine interest in the subject and an enjoyment in sharing knowledge.

The teacher ...sparked my interest.

Various analyses were performed to gauge the effectiveness of the Sudoku puzzles in relation to student performance. As noted above, the direct benefit of the puzzles is limited to aids in remembering a limited range of terms. However, the indirect motivational benefits are farther reaching. Figure 6(a) shows the correlation between performance in the final examination for the organic chemistry topic (in which the Sudoku puzzles were administered) with involvement in the Sudoku competitions. There are significant causal concerns with this analysis as the students who attempt the puzzles are likely to be the more diligent cohort. However Figure 6(b) provides a comparison of examination performance for the cohort that participated in the Sudoku exercises relative to the whole class average. As can be seen, the students performed markedly better in the topic where the puzzles were administered (organic chemistry).

The lecturer using this approach is shown as Section D.

Significantly, it can be seen from Figure 7 that the Sudoku puzzles engaged students with a wide range of ability. Although the data above provide no hard evidence for the effectiveness of the activities, further indications can be gleaned from anonymous feedback comments from the university instrument used to survey students at the end of each semester. The following provides a selection of these comments segregated into a two key indirect benefits:

Increased Concentration

whether or not the songs were relevant in helping me understand wasn't the point of them in my opinion. Having a laugh at the lyrics etc helped me get back to 100% concentration on what was happening in class, instead of losing concentration over the 50 minutes as happens in all lectures. Use them next year too!

I thought the songs were amusing as well as relevant. It added a bit of humour to keep me concentrating.

The lecture demonstrations "tend to wake people up and create some energy in the lecture rather than have one person up the front talking constantly and most of the students feeling like they would rather be asleep. Good on you for having a go with the songs."

Engagement and Motivation

The songs you sang in class were a new and interesting way to learn which proved helpful for my studies. The lecture demonstrations are good to have on a Friday afternoon as it motivates me to attend the lecture.

The Sudoku was extremely good. Glad you thought of a competition like that. Kept my mind on chem when I never thought I would be able to continue.

The use of crossword puzzles that are delivered as part of a competition share many of the same indirect benefits to the Sudoku puzzles described above. The additional benefit is a greater direct emphasis spent on recalling and/or revising to establish a number of terms that are central to the topic being taught. There is considerable dwell time associated with this process as the puzzle is completed, which is likely to have significant longer term memory benefits. An example crossword from Organic Chemistry is shown in Figure 8.

RESULTS OF IMPLEMENTATION AT THE UNIVERSITY OF TASMANIA

At the University of Tasmania, results on the final examination and the distribution of the final grades in semester one(Table 2) and two (Table 3) in 2008-9 are reported.

Figure 6. Analyses of Sudoku puzzle effectiveness, (a) Correlation of examination performance with Sudoku engagement for the lecturer who used this approach, (b) Comparison of relative performance in different sections of the examination (Sudoku cohort)

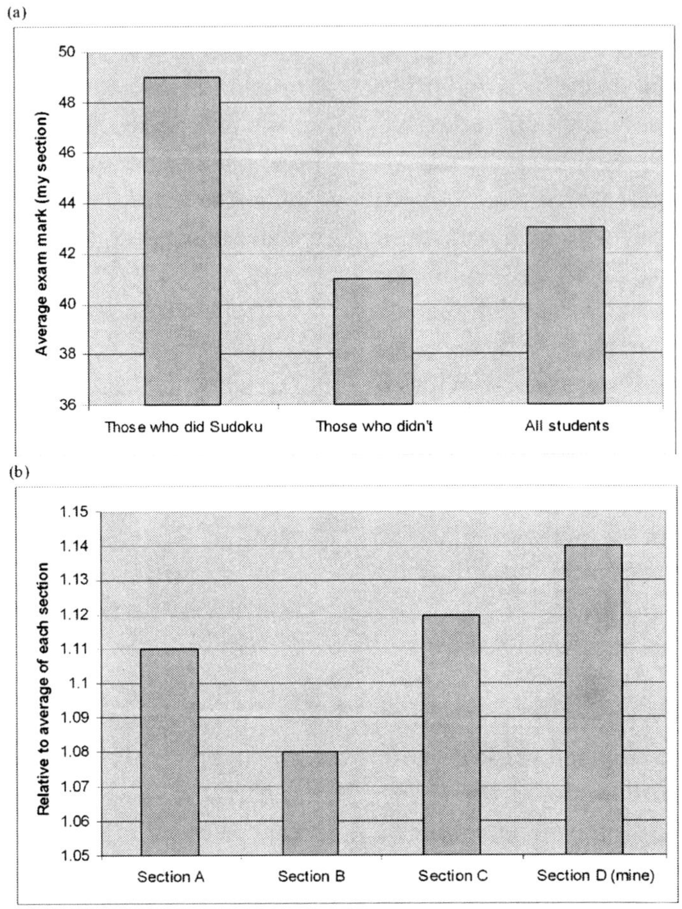

Figure 7. Analyses of Sudoku puzzle in relation to correlation of overall examination performance with Sudoku engagement

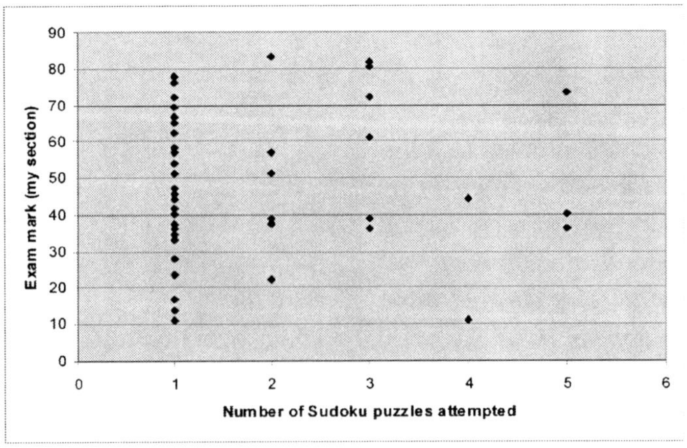

Figure 8. An example of a chemical crossword puzzle (taught in organic chemistry topics)

Name: ....

KRA114

Chemistry Puzzle #4 – Crossword

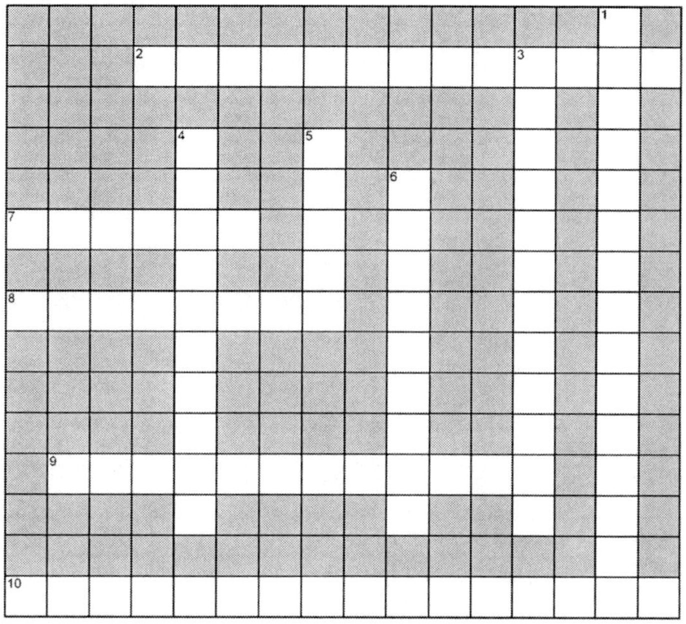

ACROSS	DOWN
2 Margarine is formed from the .... of vegetable oils.	**1** Containing more than one C=C bond.
7 Fats and oils are examples of complex	**3** Glycerol ester of three fatty acids.
8 Cholesterol and hormones are examples of	**4** The more general name for saponification.
9 This phospholipid covers nerve tissue.	**5** Esters formed from long-chain carboxylic acid and long-chain alcohol.
10 Cephalins and lecithins are examples of a	**6** The general name for the carboxylic acid that fats and oils are formed from (two words).

All the words appear on my summary pages for Lectures 7 and 8!

Please return to Brian Yates at the lecture on Friday 30 July 2010
if you wish to go in the draw for a text book prize.
Prize will be drawn at the lecture on Monday 2 August 2010.

Table 2 illustrates the distribution of the grades for semesters 1 in 2008 to 2009. It can be seen that the % of failures has increased slightly from 10.2% in 2008 to 12.2% in 2009, a rise of 2%; the % of Passes has dropped from 31.4% to 23%, a drop of 8.4%; the % of Credits has dropped slightly from 25.2% in 2008 to 23% in 2009, a drop of 2.2%; the % of Distinctions has increased from 10.2% in 2008 to 17.6% in 2009, an increase of 7.4%; and finally the % of High Distinctions has increased from 9.7% in 2008 to 9.9% in 2009, an increase of 0.2%. Distribution of grades for the continuation of the unit in semester two further reinforces this view as shown in Table 3.

According to Table 3, the % of failures decreased from 10.5% in 2008 to 8.62% in 2009, a fall of 1.9%; the % of Passes has dropped from 22.5% to 20.3%, a drop of 2.2%; the % of Credits dropped slightly from 29.2% in 2008 to 23% in 2009, a drop of 6.2%; the % of Distinctions increased from 13.9% in 2008 to 21.6% in 2009, an increase of 7.7%; and finally the % of High Distinctions increased from 10% in 2008 to 10.4% in 2009, an increase of 0.4%.

Table 2. Distribution of grades for the unit in semester one, 2008 and 2009 at the University of Tasmania

grades	% 2008	% 2009	Difference in %
HD	9.7	9.9	0.2
DN	10.2	17.6	7.4
CR	25.2	23.0	-2.2
PP	31.4	23.0	-8.4
TS	10.2	12.2	2

Table 3. Distribution of grades for the unit in semester two, 2008 & 2009 at the University of Tasmania

grades	% 2008	% 2009	Difference in %
HD	10	10.4	0.4
DN	13.9	21.6	7.7
CR	29.2	23.0	-6.2
PP	22.5	20.3	-2.2
TS	10.5	8.62	-1.9

Table 4. SETL survey results for Chemistry 1A and Chemistry 1B in 2009 at the University of Tasmania

Items	Subject	N	No answer %	Strongly agree %	Agree %	Neutral %	Disagree %	Strongly disagree %
1. I value the feedback on my understanding gained by the use of in lecture 'VotApedia' questions	Chemistry 1A	139	8	12	38	33	7	2
	Chemistry 1B	117	21	5	27	34	9	4
2. The non-assessable 'concept tests' conducted on MyLo helped me answer the weekly assignment questions	Chemistry 1A	139	0	23	42	20	13	2
	Chemistry 1B	117	0	16	53	18	11	2
3. The extensive feedback available when I make an incorrect response in the non-assessable 'concepts tests' are helpful	Chemistry 1A	139	0	37	46	11	5	1
	Chemistry 1B	117	3	31	48	16	1	1

The consistent increase in the number of students obtaining Distinctions in both semesters in chemistry at the University of Tasmania seemed to suggest that students who achieved above pass level tended to benefit from the language strategies implemented. This confirms the findings found in similar studies conducted at the University of Canberra in subjects such as genetics and molecular biology from 2005 to 2009.

In the Student Evaluation of Teaching and Learning (SETL) surveys conducted at the end of semester one and two, in 2009 (Table 4), the use of 'VotApedia' concept tests (Appendix A), the extensive feedback for the online non-assessable 'Concept Tests' (Appendix B) and the use of Hot potato exercises (Appendix C) were evaluated.

Most students agreed that VotApedia concept questions were useful during lectures. However, some 15% of the students would like more connection between the online concept questions and the Mylo assignments. This view was also confirmed by comments in the open-ended section of the SETL surveys.

CONCLUSION

The in-class and post-lecture strategies discussed in this chapter have been demonstrated to be amenable to large class first year teaching and are sustainable in their design. In a number of cases, improved student engagement and performance is demonstrated. Across the Chemistry 1A and 1B units, there are many other strategies for engagement and language support that lecturers adopt. Staff are encouraged to build on their own particular strengths in devising activities to motivate students. In most cases the methods described are clearly adaptable to a wide range of other disciplines and through the use of carefully chosen examples with descriptions is it hoped that this is clearly achievable by the interested reader.

ACKNOWLEDGMENT

The authors wish to thank the rest of the Chemistry 1A/1B lecturing staff for their support in the lead up to the development of the activities for this project and during their implementation. This enabled a whole of year coherence to the project that was very important to minimizing any backlash against the project leaders, who may otherwise have been perceived by a portion of the students as being the ones that "gave us more work to do that does not count towards assessment"! It has been rewarding that a number of the activities developed have continued to be used and developed further. Thanks are also due to the students that provided useful feedback to us and helped to improve the material that we developed. Particular thanks are due to Mr Bryce Lockhart-Gillett who did a fantastic job in writing many of the Concept Tests and VotApedia questions for the whole first year program. Bryce's familiarity with the unit material through private student tutoring and laboratory demonstrating was very helpful for his efficient completion of this large task that had seemingly endless rolling deadlines throughout 2008. But, more than anything, his deep understanding of the material and the goals of the Australian Learning and Teaching Council project, and good communication with the topic lecturers and project investigators led to an excellent set of materials having been developed. Team members of the ALIUS Australian Learning and Teaching Council project are also to be thanked in connect with some of the strategies from that project that meshed very well with the language support focus of this project.

Finally, both authors pay great appreciation to Felicia Zhang for the several visits that she made to Hobart that really helped to open our eyes to some of the language-based issues facing students, developing some starter materials that got us on track, proofing many of our lecture materials for language based vagueness that we had been completely blind too and keeping on top of the grant requirements.

REFERENCES

Barrie, S. C., Buntine, M. A., Jamie, I. M., & Kable, S. H. (2001). *APCELL: Developing better ways of teaching in the laboratory*. Paper presented at the Research and Development into University Science Teaching and Learning Workshop.

Bedgood, D. R. Jr, Gardiner, M. G., Yates, B. F., Bridgeman, A. J., Morris, G., & Buntine, M. (2010). The development of teaching skills to support active learning in university science (ALIUS). *Journal of Learning Design, 1*(3), 10–19.

Bedgood, D. R. Jr, Yates, B., Buntine, M., Pyke, S., Lim, K., & Mocerino, M. (2008). Why are we still teaching the way we were taught in the 1980s? *Chemistry in Australia, 11*, 22–23.

Bedgood, D. R. Jr, Yates, B., Buntine, M., Pyke, S., Lim, K., & Mocerino, M. (2010). Leading change in Australian science teaching. *Chemistry in Australia, 7*(5), 18–19.

Burton, L. J., & Dowling, D. (2005). *In search of the key factors that influence student success at university.* Paper presented at the HERDSA 2005 International Conference: Higher Education in a Changing World. Retrieved from http:// eprints.usq.edu.au/

Gardner, P. L., & Australian Science Education Project. (1972). *Words in science: An investigation of non-technical vocabulary difficulties amongst form I, II, III & IV science students in Victoria, a report*. Melbourne, Australia: Australian Science Education Project.

Gardner, P. L. (1977). Logical connectives in science: A summary of the findings. *Research in Science Education*, *7*(1), 9–24. .doi:10.1007/BF02643108

Harris, K.-L., Krause, K., Gleeson, D., Peat, M., Taylor, C., & Garnett, R. (2007). *Enhancing assessment in the biological sciences: Ideas and resources for university educators*. Retrieved 1st April, 2007, from http:// www.bioassess.edu.au

Markow, P. G. (1988). Teaching chemistry like the foreign language it is. *Journal of Chemical Education*, *65*(1), 57–58. doi:10.1021/ed065p57

Moog, R. S. (2008). *Process Oriented Guided Inquiry Learning (POGIL)* (*Vol. 994*). American Chemical Society.

Oliver, J. S., & Simpson, R. D. (1988). Influences of attitude toward science, achievement motivation, and science self concept on achievement in science: A longitudinal study. *Science Education*, *72*(2), 143–155. doi:10.1002/sce.3730720204

Pickersgill, S., & Lock, R. (1991). Student understanding of selected non-technical words in science. *Research in Science & Technological Education*, *9*(1), 71–79. doi:10.1080/0263514910090107

Simpson, O. (2006). Predicting student success in open and distance learning. *Open Learning: The Journal of Open and Distance Learning*, *21*(2), 125–138.

Tinto, V., & Pusser, B. (2006). *Moving from theory to action: Building a model of institutional action for student success*. Paper presented at the National Symposium on Postsecondary Student Success. Retrieved from http:// nces.ed.gov/npec/papers.asp

Wellington, J. J., & Osborne, J. (2001). *Language and literacy in science education*. Buckingham, UK & Philadelphia, PA: Open University.

ADDITIONAL READING

Ãœce, M. (2009). Teaching the mole concept using a conceptual change method at college level. *Education*, *129*(4), 683.

Bowen, C. W., & Bunce, D. M. (1997). Testing for conceptual understanding in general chemistry. the Chemical Educator, 2(2). Retrieved from Chandrasegaran, A. L., Treagust, D. F., & Mauro, M. (2007). The development of a two-tier multiple-choice diagnostic instrument for evaluating secondary school students' ability to describe and explain chemical reactions using multiple levels of representation. *Chemistry Education Research and Practice*, *8*(3), 293–307.

Johnstone, A. H. (2000). Teaching of chemistry - logical or psychological? Chemistry Education Research and Practice in Europe, 1(1), 9-15. Retrieved from http:// www.uoi.gr/cerp/

Kaya, O. N. (2008). A student-centred approach: Assessing the changes in prospective science teachers' conceptual understanding by concept mapping in a general chemistry laboratory. Research Science Education (. *Research in Science Education*, *38*, 91–110. doi:10.1007/s11165-007-9048-7

Krause, S., Birk, J., Bauer, R., Jenkins, B., & Pavelich, M. J. (2004). Development, testing, and application of a chemistry concept inventory. Paper presented at the 34th ASEE/IEEE Frontiers in Education Conference, Savannah, GA. Retrieved from http:// fie-conference.org/fie2004/

Michael, J. (2007). Conceptual assessment in the biological sciences: A national science foundation-sponsored workshop. *Advances in Physiology Education*, *31*(4), 389–391. .doi:10.1152/advan.00047.2007

Nieswandt, M. (2001). Problems and possibilities for learning in an introductory chemistry course from a conceptual change perspective. Science Education, 85(2), 158-79. doi: 10.1002/1098-237X(200103)85:2<158::AID-SCE40>3.0.CO;2-3

APPENDIX A: EXAMPLES OF VOTAPEDIA CONCEPT QUESTIONS ADMINISTERED AT THE UNIVERSITY OF TASMANIA IN CHEMISTRY

Contributors: Professor Brian Yates and Bryce Lockhart-Gillett

Practical hint: In each two-hour lecture segment, only three VotApedia concept questions were used to check understanding at appropriate junctures. Each question has been labeled according to the concept being studied.

Lecture 1

Concept: Carbonyl Group Polarization.

1.1) Which atom in the carbonyl group (C=O) has more electron density surrounding it?

1. C
2. O
3. They have the same electron density.
4. It depends on what the rest of the molecule is.

(Correct answer is 2)

Concept: Generalisation of Organic Naming

1.2) Which of the following hypothetical compound names is an aldehyde?

1. neoseshenol
2. para phenylic acid
3. dodecabeeral
4. benzoic ether

(Correct answer is 3)

Concept: Oxidation of Aldehydes

1.3) Which of the following pairs of compounds are related by oxidation?

1. aldehyde and carboxylic acid
2. aldehyde and ketone
3. ketone and carboxylic acid
4. 1 and 2

(Correct answer is 1)

Lecture 2

Concept: Nucleophilic Addition to Carbonyl Compounds

2.1) Where do the electrons in the C=O double bond go as the nucleophile attacks the carbonyl group?

1. They do not move.
2. They move onto the oxygen atom.
3. They move onto the nucleophile.
4. It varies depending on the nucleophile.

(Correct answer is 2)

Concept: Grignard Addition Reactions

2.2) The reaction of a Grignard reagent with a carbonyl compound followed by acidification yields an _______, and is an ________ reaction.

1. alcohol, irreversible
2. hemiacetal, reversible
3. imine, reversible
4. hydrazone, irreversible

(Correct answer is 1)

Concept: Keto-enol Tautomerism

2.3) Approximately what percent of acetone is present as the keto tautomer in solution?

1. .0001%
2. 1%
3. 50%
4. 100%

(Correct answer is 4)

Lecture 3

Concept: Carboxylic Acids and Derivatives All Contain a Carbonyl Group.

3.1) Which of the following functional groups does not contain a carbonyl group as part of the group?

1. carboxylic acid
2. acid halide
3. ester
4. ether
5. amide

(Correct answer is 4)

Concept: Acidity of Carboxylic Acids.

3.2) The O-H group of a carboxylic acid is more acidic than an alcohol because:

1. There are more oxygen atoms in the carboxylic acid group.
2. The carboxylate anion is more stable than the alkoxide anion.
3. They are called carboxylic acids.
4. They are more easily oxidised.

(Correct answer is 2)

Concept: Nucleophilic Acyl Substitution Reactions.

3.3) Where do the nucleophiles attack in a nucleophilic acyl substitution reaction?

1. At the C atom of the carbonyl group.
2. At the O atom of the carbonyl group.
3. Away from the carbonyl group.
4. It depends on the nucleophile.

(Correct answer is 1)

Lecture 4

Concept: Similarity of Reactions Between Acid Halides, Acid Anhydrides and Esters.

4.1) Acid halides, acid anhydrides and esters all undergo similar nucleophilic acyl substitution reactions, just at different rates. The fastest reactions occur with;

1. esters
2. acid anhydrides
3. acid halides

(Correct answer is 3)

Concept: Step-growth Polymerization Reactions.

4.2) What is the key bond-forming step in many step-growth polymerization reactions?

1. nucleophilic addition
2. nucleophilic acyl addition
3. nucleophilic acyl substitution
4. nucleophilic esterification

(Correct answer is 3)

Lecture 5

Concept: Amines Contain a Lone Pair of Electrons.

5.1) How many electron domains are there around the nitrogen atom in a non-heterocyclic amine, and how many of these are bonding domains?

1. 3, 3
2. 4, 4
3. 4, 3
4. It varies

(Correct answer is 3)

Concept: Basicity of Amines.

5.2) If amine A is a stronger base than amine B, what do we know about their pK_b values?

1. $pK_b(A)$ will be larger than $pK_b(B)$
2. $pK_b(B)$ will be larger than $pK_b(A)$
3. both pK_b values will be smaller than for water
4. 2 and 3

(Correct answer is 2)

Concept: Utilising Amine Salts in Purifying Amines.

5.3) Amines can be purified by extracting them into water after adding what to them?

1. ether
2. base
3. acid
4. it varies depending on the amine

(Correct answer is 3)

Lecture 6

Concept: Amines Reacting as Nucleophiles (Synthesis of Amines)

6.1) Amines readily react with alkyl halides via an S_N2 reaction because amines are excellent;

1. bases
2. leaving groups
3. electrophiles
4. nucleophiles

(Correct answer is 4)

Concept: Similarity Between Amines and Alcohols

6.2) Just as alcohols can be formed from ketones or carboxylic acids, amines can be formed from amides and nitriles by;

1. acyl substitution reactions
2. nucleophilic addition reactions
3. reduction
4. oxidation

(Correct answer is 3)

Concept: Aromatic Nitrogen Heterocycles: Hybridisation of the N Atom

6.3) What is the hybridisation of the nitrogen atom in an aromatic heterocycle?

1. sp^2
2. sp^3
3. sp^4
4. It depends on the molecule.

(Correct answer is 1)

Lecture 7

Concept: Complex Lipids Structure: Tri-ester of Glycerol.

7.1) When a complex lipid is hydrolysed the products are glycerol and three;

1. fatty acids
2. fatty alcohols
3. steroids
4. saturated fat molecules

(Correct answer is 1)

Concept. Saturated and Unsaturated Fatty Acids – Melting Points.

7.2) If you take an unsaturated fatty acid and remove all the double bonds from it, will the melting point become higher or lower?

1. higher
2. lower
3. it will not change
4. it is impossible to say for sure

(Correct answer is 1)

Concept: Saponification.

7.3) Crude natural soap can be formed by reacting what type of biological molecule with alkali?

1. waxes
2. sterols
3. triglycerides
4. 1 and 3

(Correct answer is 3)

Lecture 8

Concept: Phospholipids

8.1) Phospholipids are most closely related to which type of lipid?

1. waxes
2. steroids
3. triglycerides
4. oils

(Correct answer is 3)

Concept: Steroids.

8.2) All steroids contain the same backbone of 4 fused rings. How many of which size rings are there in the steroid backbone?

1. 2 six-membered, 2 five membered
2. 1 six-membered, 3 five membered
3. 3 six-membered, 1 five membered
4. It varies.

(Correct answer is 3)

Concept. Steroids: Biological Function

8.3) What function do steroids most commonly perform in humans?

1. energy storage
2. cell wall tissue
3. nerve tissue
4. hormones

(Correct answer is 4)

Lecture 9

Concept: Carbohydrates : Poly-hydroxylated Ketones and Aldehydes

9.1) All carbohydrates are poly-hydroxylated;

1. aldehydes and carboxylic acids
2. aldehydes and ketones
3. ketones and carboxylic acids

(Correct answer is 2)

Concept: D and L Sugars

9.2) Which chiral carbon is used to determine if a sugar is the D or the L form in the Fischer projection?

1. The one that is closest to the carbonyl group.
2. The one that is farthest from the carbonyl group.
3. The one that contains a hydroxyl group.
4. The one that does not contain a hydroxyl group.

(Correct answer is 2)

Concept: Cyclic Forms of Monosaccharides

9.3) What ring sizes are common in the hemiacetal form of cyclic monosaccharides?

1. 3&4 membered
2. 4 &5 membered
3. 5&6 membered
4. 6&7 membered

(Correct answer is 3)

Lecture 10

Concept: Mutarotation

10.1) Why does the optical rotation of a pure sample of alpha-D-glucopyranose decrease from its starting point of 112.2° in solution?

1. The compound is unstable in solution.
2. It is converting into a different sugar.
3. It is converting into a different enantiomer.
4. It is forming a polysaccharide.

(Correct answer is 3)

Concept: Reduction of Monosaccharides

10.2) When a reducing reagent is added to a solution of a monosaccharide, which functional group(s) are reduced?

1. the alcohols
2. the carbonyl
3. the hemiacetal
4. 2 and 3

(Correct answer is 2)

Concept: Starch vs. Cellulose

10.3) The difference in structure between the two common polysaccharides starch and cellulose is the configuration at which carbon (determining if the glycoside bond is alpha or beta)?

1. C1
2. C2
3. C3
4. C4

(Correct answer is 1)

Lecture 11
Concept: Zwitterion
11.1) What is the overall charge of a zwitterion?
1. +1
2. 0
3. -1
4. It varies depending on the pH.

(Correct answer is 2)
Concept: pH Changes Affecting the Charge of Amino Acids
11.2) If the pH is less than the isoelectric point of an amino acid, what will the overall charge c the amino acid be?
1. positive
2. negative
3. neutral
4. It varies.

(correct answer is 1)
Concept: Peptides
11.3) Peptides and proteins are amino acid polymers where the amino acid monomers or 'residues' are linked by;
1. ether bonds
2. imide bonds
3. disulfide bonds
4. amide bonds

(Correct answer is 4)

Lecture 12
Concept: Primary Structure of Peptides and Proteins
12.1) The primary structure of a peptide is;
1. the shape of it
2. the order of the sequence of the amino acids
3. amount of each amino acid residue in it

(Correct answer is 2)
Concept: Sequencing of a Peptide: Edman Degradation
12.2) What level of protein structure does the Edman Degradation process reveal?
1. primary
2. secondary
3. tertiary
4. quaternary

(Correct answer is 1)
Concept: Secondary Protein Structure
12.3) Alpha-helix and beta-pleated-sheet are both types of geometry observed in which level of protein structure?
1. primary
2. secondary
3. tertiary
4. quaternary

(Correct answer is 2)

APPENDIX B: EXAMPLES OF PRE-LECTURE OR TUTORIAL ONLINE QUIZZES USED AT THE UNIVERSITY OF TASMANIA IN CHEMISTRY

Contributed by Professor Brian Yatesand Bryce Lockhart-Gillett

Practical hints: these online quizzes have been created according to the concepts under study. Full feedback has been provided with each question.

Topic 1: Organic Chemistry

1. **Concept.C=O group:** Polarisation due to electronegativities. Since oxygen is more [*electronegative*] than carbon, the double bond in the carbonyl group will be [*polarised*]. There will be more electron density around the [*oxygen atom*] and less around the [*carbon atom*].
2. **Concept.C=O group:** Shape and hybridization. The carbon atom of the carbonyl group forms three sigma bonds. It is therefore ___ hybridised and the shape predicted for the C=O group is ______.
   A. **sp^3, tetrahedral.** *Incorrect. (The correct answer is D. The fourth valence electron on carbon remains in a p orbital and forms a pi bond to oxygen. Therefore the carbon atom is sp^2 hybridised, and it is not possible for an sp^2 carbon to form a tetrahedral geometry.)*
   B. **sp, linear.** *Incorrect. (The correct answer is D. The three sigma bonds arise from mixing one s and two p orbitals on carbon and thus the carbon atom is sp^2 hybridised. An sp carbon atom would have only two sigma bonds.)*
   C. **sp^2, pyramidal.** *Incorrect. (The correct answer is D. The carbon atom is indeed sp^2 hybridised. However a pyramidal geometry occurs when an atom is sp^3 hybridised and the atom has a non-bonding pair of electrons (e.g. nitrogen).)*
   D. **sp^2, planar.** *Correct.*
3. **Concept. C=O group:** Electrophilic carbon. Because the carbon in the C=O group has less electron density around it, it is more ________ and hence more reactive towards ________.
   A. **electrophilic, electrophiles.** *Incorrect. (The correct answer is B. The carbon is more electrophilic, which means that it will be reactive towards nucleophilic attack.)*
   B. **electrophilic, nucleophiles.** *Correct.*
   C. **nucleophilic, nucleophiles.** *Incorrect. (The correct answer is B. The carbon is more electrophilic, which means that it will be reactive towards nucleophilic attack.)*
   D. **nucleophilic, electrophiles.** *Incorrect. (The correct answer is B. The carbon is more electrophilic, which means that it will be reactive towards nucleophilic attack.)*
4. **Concept. Naming aldehydes:** If an organic compound has a name that ends in –al we know that it is what type of compound?
   A. **Ketone.***Incorrect. (The correct answer is B. Ketones have names that end in -one.)*
   B. **Aldehyde.***Correct.*
   C. **Alcohol.***Incorrect. (The correct answer is B. Alcohols have names that end in -ol.)*
   D. **Carboxylic acid.***Incorrect. (The correct answer is B. Carboxylic acids have names that end in -oic acid.)*

5. **Concept. Naming Ketones:** If an organic compound has a name that ends in –one we know that it is what type of compound?
   A. **Ketone.***Correct.*
   B. **Aldehyde.***Incorrect. (The correct answer is A. Aldehydes have names that and in –al.)*
   C. **Alcohol.***Incorrect. (The correct answer is A. Alcohols have names that end in -ol.)*
   D. **Carboxylic acid.***Incorrect. (The correct answer is A. Carboxylic acids have names that end in -oic acid.)*
6. **Concept. Synthesis of aldehydes:** Aldehydes are synthesised by oxidising ______ alcohols, however it is also possible to form the ________ if the oxidation goes too far, which is why PCC is usually used in aldehyde synthesis.
   A. **primary, ketone.** Incorrect. The correct answer is C. Aldehydes are formed from primary alcohols, however it is not possible to oxidise a primary alcohol to a ketone. The product from further oxidation would be a carboxylic acid.
   B. **secondary, carboxylic acid.** Incorrect. The correct answer is C. Aldehydes are formed from primary alcohols.
   C. **primary, carboxylic acid.** Correct.
   D. **secondary, ketone.** Incorrect. The correct answer is C. Aldehydes are formed from primary alcohols. It is not possible to form an aldehyde from a secondary alcohol; the product would be a ketone after the first stage of oxidation.
7. **Concept. Synthesis of ketones:** Which of the following is not a method of preparing ketones?
   A. **Oxidation of a secondary alcohol.***Incorrect. (The correct answer is D. This is a method of preparing ketones (only secondary alcohols can be oxidised to give ketones))*
   B. **Hydrolysis of terminal alkynes.***Incorrect.(The correct answer is D. This is a method of preparing ketones.)*
   C. **Friedel-Crafts acylation.***Incorrect. (The correct answer is D. This is a method of preparing ketones.)*
   D. **Oxidation of an aldehyde.***Correct. (This will yield a carboxylic acid.)*
8. **Concept. Oxidation of tertiary alcohols:** What is the product of oxidation of a tertiary alcohol?
   A. **An aldehyde.***Incorrect. (The correct answer is D. Aldehydes can only be formed by oxidation of primary alcohols, and it is not possible to oxidise a tertiary alcohol.)*
   B. **A ketone.***Incorrect. (The correct answer is D. Ketones are formed by oxidation of secondary alcohols. It is not possible to oxidise a tertiary alcohol.)*
   C. **A carboxylic acid.***Incorrect. (The correct answer is D. Carboxylic acids are formed by oxidation of primary alcohols or aldehydes. It is not possible to oxidise a tertiary alcohol.)*
   D. **A tertiary alcohol.***Correct. (It is not possible to oxidise a tertiary alcohol.)*

9. **Concept. Oxidation (of alcohols, ketones and aldehydes):** Which of the following compounds is unreactive towards oxidation?
   A. **Aldehydes.***Incorrect. (The correct answer is E. Aldehydes oxidise to give carboxylic acids. This is usually done using Tollens' reagent.)*
   B. **Primary and secondary alcohols.***Incorrect. (The correct answer is E. Typically alcohols are reactive towards oxidation (tertiary alcohols are the exception).)*
   C. **Ketones.***Incorrect. (The correct answer is E. Ketones are not reactive towards oxidation, but neither are carboxylic acids.)*
   D. **Carboxylic acids.***Incorrect. (The correct answer is E. Carboxylic acids are not reactive towards oxidation, but neither are ketones.)*
   E. **C & D.***Correct.*
10. **Concept. Tollens' reagent:** Tollens' reagent is used as a test for aldehydes. If the test is positive a shiny metal deposit is observed as the silver ions in solution are reduced back to metallic silver. What compounds are typically oxidised using Tollens' reagent?
   A. **Ketones.***Incorrect. (The correct answer is B. Ketones are unreactive towards oxidation.)*
   B. **Aldehydes.***Correct.*
   C. **Alcohols.***Incorrect. (The correct answer is B. Tollens' reagent is a mild reducing agent and is used to oxidise aldehydes to carboxylic acids.)*
   D. **All of the above.***Incorrect. (The correct answer is B. Tollens' reagent is used to oxidise aldehydes to carboxylic acids.)*

APPENDIX C: HOT POTATO EXERCISES FOR CHEMISTRY

Hint: answers are provided after the key in each exercise

Representations of Molecules
1. With the exception of (1)__________, the element further to the (2)__________ in the periodic table appears (3)__________.
2. If hydrogen is present, it appears (4)__________ except when the other element is from group (5)__________ or (6)__________ of the periodic table.
3. If both elements are from the (7)__________ of the periodic table, the (8)__________ one appears first.
4. If the compound is (9)__________, we write the (10)__________ first, followed by the (11)__________.

List: 16 17 anion cation first hydrogen ionic last left lower same group

----------Key----------

Representations of molecules
1. With the exception of hydrogen, the element further to the left in the periodic table appears first.
2. If hydrogen is present, it appears last except when the other element is from group 16 or 17 of the periodic table.
3. If both elements are from the same group of the periodic table, the lower one appears first.
4. If the compound is ionic, we write the cation first, followed by the anion.

Nomenclature 2

These rules are used to name compounds containing only two nonmetallic elements:
1. The element closer to the (1)______________ of the periodic table appears first. If both elements are from the (2)______________ group of the periodic table, the (3)______________ one appears first.
2. The element that appears (4)______________ retains its elemental name.
3. The second element begins with a root derived from its (5)______________ and ends with the (6)______________ -ide.
4. When there is more than one atom of a given element in the formula, the name of the element usually contains a (7)______________ that specifies the (8)______________ of atoms present.

List: elemental name first left lower number prefix same suffix

----------Key----------

Nomenclature 2

These rules are used to name compounds containing only two nonmetallic elements:
1. The element closer to the left of the periodic table appears first. If both elements are from the same group of the periodic table, the lower one appears first.
2. The element that appears first retains its elemental name.
3. The second element begins with a root derived from its elemental name and ends with the suffix -ide.
4. When there is more than one atom of a given element in the formula, the name of the element usually contains a prefix that specifies the number of atoms present.

Nomenclature 3

These rules are used to name anions containing oxygen and at least one other element.

1. the name has a root taken from the name of the (1)__________ atom.
2. When an element forms two different (2)__________, the one with (3)__________ oxygen atoms ends in -ite and the other ends in -ate.
3. Chlorine, bromine and iodine each form four different (4)__________ that are distinguished by prefixes and suffixes.
4. A (5)__________ anion with a charge more negative than 1- may add H^+ to give another anion. These anions are names from the parent anion by adding the word hydrogen.

List: central fewer oxoanions oxoanions polyatomic

----------Key----------

Nomenclature 3

These rules are used to name anions containing oxygen and at least one other element.

1. the name has a root taken from the name of the central atom.
2. When an element forms two different oxoanions, the one with fewer oxygen atoms ends in -ite and the other ends in -ate.
3. Chlorine, bromine and iodine each form four different oxoanions that are distinguished by prefixes and suffixes.
4. A polyatomic anion with a charge more negative than 1- may add H^+ to give another anion. These anions are names from the parent anion by adding the word hydrogen.

Ionisation Energy

Ionisation energy of an (1)____________________ or (2)____________________ is the (3)____________________ energy required to remove an electron from the (4)____________________ of the (5)____________________ or (6)____________________.

Ionisation energies (7)____________________ for an element as successive electrons are (8)____________________.

A sharp (9)____________________ in ionisation energy occurs when an (10)____________________ is removed.

List: atom ground state increase increase inner-shell electron ion ion isolated gaseous atom minimum removed

----------Key----------

Ionisation energy

Ionisation energy of an atom or ion is the minimum energy required to remove an electron from the ground state of the isolated gaseous atom or ion.

Ionisation energies increase for an element as successive electrons are removed.

A sharp increase in ionisation energy occurs when an inner-shell electron is removed.

The Lowering of Freezing Point of a Solution

When a solution freezes, a few (1)______________ molecules cluster together to form a pure (2)______________ thus forming a dynamic (3)______________ between (4)______________ and (5)______________ solvent. (6)______________ (7)______________ with this equilibrium (8)______________ the liquid. To (9)______________ this effect, (10)______________ of solvent and solute molecule should be (11)______________. This results in the (12)______________ being lowered.

List: counteract equilibrium favouring freezing point interfere liquid lowered mobility solid solid Solutes solvent

----------Key----------

The lowering of freezing point of a solution

When a solution freezes, a few solvent molecules cluster together to form a pure solid thus forming a dynamic equilibrium between liquid and solid solvent. Solutes interfere with this equilibrium favouring the liquid. To counteract this effect, mobility of solvent and solute molecule should be lowered. This results in the freezing point being lowered.

Definition of the Mole

The mole is a standard (1)________________. It is the amount of any substance with the (2)________________ number of atoms or molecules as there are atoms in 12 g (exactly) of ^{12}C. (3)________________, NA, is 6.022x 10 23 $mole^{-1}$ and gives the (4)________________ of specified entities in 1 mole of a substance. The (5)________________, M, is the mass of 1 mole of a substance.

List: Avogadro's constant molar mass number same unit

----------Key----------

Definition of the mole

The mole is a standard unit. It is the amount of any substance with the same number of atoms or molecules as there are atoms in 12 g (exactly) of ^{12}C. Avogadro's constant, NA, is 6.022x 10 23 $mole^{-1}$ and gives the number of specified entities in 1 mole of a substance. The molar mass, M, is the mass of 1 mole of a substance.

Electrons in Atoms

(1)____________________ occupy regions of space called (2)____________________ in (3)____________________. Each (4)____________________ has a characteristic (5)____________________ and (6)____________________. For example, the (7)____________________ situation for a hydrogen atom, the (8)____________________, occurs when the single (9)____________________ occupies an (10)____________________ in which its most probable distance from the (11)____________________ is 5.29×10^{-11} m. This (12)____________________ has a (13)____________________ electron distribution. If the ground state hydrogen atom absorbs a specific amount of (14)____________________, the electron can be (15)____________________ to a higher energy orbital to form an (16)____________________ in which the (17)____________________ lies, on average, (18)____________________ from the nucleus. Such a process is called an (19)____________________ and the electron distribution in the (20)____________________ energy orbital is (21)____________________ shaped.

List: atoms dumbbell electron electron electron distribution electronic transition Electrons energy energy excited state further ground state higher lowest energy nucleus orbital orbital orbital orbitals promoted spherical

----------Key----------

Electrons in atoms

Electrons occupy regions of space called orbitals in atoms. Each orbital has a characteristic electron distribution and energy. For example, the lowest energy situation for a hydrogen atom, the ground state, occurs when the single electron occupies an orbital in which its most probable distance from the nucleus is 5.29×10^{-11} m. This orbital has a spherical electron distribution. If the ground state hydrogen atom absorbs a specific amount of energy, the electron can be promoted to a higher energy orbital to form an excited state in which the electron lies, on average, further from the nucleus. Such a process is called an electronic transition and the electron distribution in the higher energy orbital is dumbbell shaped.

Periodicity of Atomic Properties

1. Which of the following statements is not true of the periodic table?

a) Atomic size decreases from left to right and increases from top to bottom of the periodic table.

b) Electron affinity tends to become more negative from left to right across a period of the periodic table.

c) As the atomic number (Z) increases, it becomes progressively more difficult to remove an electron.

d) Atomic anions are restricted to elements on the right side of the table: the halogens, oxygen and sulfur.

e) All the elements in the s, d and f blocks form compounds containing atomic cations.

f) The elements that can form anions relatively easily are metals.

----------Key----------

(f)

Thermodynamics 2

Table 5.

----------Key----------

Bond enthalpy (bond energy)	The change in enthalpy when a mole of chemical bonds of a given type is broken, separating the bonded atoms, the atoms and molecules must be in the gas phase.
Caloric value	The energy of complete combustion of a stated size sample of a food, usually reported in Calories (kilocalories).
Energy density	The quantity of energy released per unit volume of a fuel.
Fuel value	The quantity of energy released when 1 g of a fuel is burned to form carbon dioxide and water.
Heat capacity	The quantity of energy that must be transferred to an object to raise its temperature by 1 °C.
Molar heat capacity	The quantity of energy that must be transferred to 1 mol of a substance to increase its temperature by 1 °C.
Specific heat capacity	The quantity of energy that must be transferred to 1 g of substance to increase its temperature by 1 °C.
Standard molar enthalpy of formation	The standard enthalpy change for forming 1 mol of a compound from its elements, with all substances in their standard states.
Standard state	The most stable form of a substance in the physical state in which it exists at 1 bar and a specified temperature.

Table 6.

Properties of electrons

Louis de Broglie (1892-1987)	I summarized that the more accurately we know the position of a particle-wave, the more uncertain we are about its momentum and vice versa. This is due to the particle-wave duality of electrons.
Heisenberg (1901-1976)	This indexes the angular momentum of an atomic orbital. Electron density in an orbital can be concentrated along preferred axes. I can be zero, or any positive integer smaller than n. I indicates the preferred axes.
quantum number	This is a number that indexes restrictions on the possible orientations of electron distribution.
principal quantum number (n)	These are numbers that specify the values of the electron's quantised properties.
azimuthal quantum number (l)	This is a number that indicates how electrons behave when placed in a magnetic field. The two possible values are +1/2 or -1/2.
magnetic quantum number	This is the quantum number that indexes energy for any atom containing only a single electron. I can only have positive integer value and the energy I indicate is correlated with orbital size. As n increases, the energy of the electron increases, its orbital gets bigger and the electron is less tightly bound to the atom.
spin quantum number	I was the first one to suggest that electrons display wave-particle duality like that exhibited by photons.
The Pauli exclusion principle	A complete description of an atomic electron requires a set of four quantum numbers which must meet all the restrictions.

----------Key----------

Louis de Broglie (1892-1987)	I was the first one to suggest that electrons display wave-particle duality like that exhibited by photons.
Heisenberg (1901-1976)	I summarized that the more accurately we know the position of a particle-wave, the more uncertain we are about its momentum and vice versa. This is due to the particle-wave duality of electrons.
quantum number	These are numbers that specify the values of the electron's quantised properties.
principal quantum number (n)	This is the quantum number that indexes energy for any atom containing only a single electron. I can only have positive integer value and the energy I indicate is correlated with orbital size. As n increases, the energy of the electron increases, its orbital gets bigger and the electron is less tightly bound to the atom.
azimuthal quantum number (l)	This indexes the angular momentum of an atomic orbital. Electron density in an orbital can be concentrated along preferred axes. I can be zero, or any positive integer smaller than n. I indicates the preferred axes.
magnetic quantum number	This is a number that indexes restrictions on the possible orientations of electron distribution.
spin quantum number	This is a number that indicates how electrons behave when placed in a magnetic field. The two possible values are +1/2 or -1/2.
The Pauli exclusion principle	A complete description of an atomic electron requires a set of four quantum numbers which must meet all the restrictions.

Table 7.

Molecular shapes

Tetrahedral	SF_5
Octahedral	NO_3^-
Trigonal Planar	Water
Bent	SF_6
Square Pyramidal	CH_4

----------Key----------

Tetrahedral	CH_4
Octahedral	SF_6
Trigonal Planar	NO_3^-
Bent	Water
Square Pyramidal	SF_5

Table 8.

Electronic configuration

Neon	$1s^2 2s^2 2p^6 3s^2 3p^6 4s^2 3d^1$
Sodium	$1s^2 2s^2 2p^6 3s^2 3p^6 4s^2 3d^{10} 4p^1$
Gallium	$1s^2 2s^2 2p^6$
Aluminium	$1s^2 2s^2 2p^6 3s^1$
Scandium	$1s^2 2s^2 2p^6 3s^2 3p^1$

----------Key----------

Neon	$1s^2 2s^2 2p^6$
Sodium	$1s^2 2s^2 2p^6 3s^1$
Gallium	$1s^2 2s^2 2p^6 3s^2 3p^6 4s^2 3d^{10} 4p^1$
Aluminium	$1s^2 2s^2 2p^6 3s^2 3p^1$
Scandium	$1s^2 2s^2 2p^6 3s^2 3p^6 4s^2 3d^1$

Table 9.

Chemistry Nomenclature_1

carbon monoxide	1
silicon dioxide	4
nitrogen triiodide	2
tin tetrachloride	7
phosphorus pentachloride	5
sulfur hexafluoride	3
iodine heptafluoride	6

----------Key----------

carbon monoxide	1
silicon dioxide	2
nitrogen triiodide	3
tin tetrachloride	4
phosphorus pentachloride	5
sulfur hexafluoride	6
iodine heptafluoride	7

Table 10.

Chemical reactions and stoichiometry

reactants	This is defined as the amount of solute dissolved in a particular volume of solution.
products	This is used to calculate the mass of 1 mole of any substance or to convert between amount and mass.
reversible reactions	These are the substances that are formed in the reaction.
stoichiometry	These are reactions in which both the forward and reverse reactions occur.
stoichiometric coefficients	These are numbers in front of the formulae and they indicate the number of molecules of each kind among the reactants and products.
solute	This is any substance dissolved in the solvent.
solution	This is concerned with the relative amounts of reactants and products in a chemical.
solvent	These are the substances present before the reaction begins.
molarity	This is a homogeneous mixture in which the molecules or ions of the components freely intermingle.
molar mass	This is usually taken to be the component present in largest amount and is the medium into which the solutes are mixed or dissolved.
concentration	This is when the amount is given in moles and the volume in litres. It has the unit mol L^{-1} (often abbreviated M).

----------Key----------

reactants	These are the substances present before the reaction begins.
products	These are the substances that are formed in the reaction.
reversible reactions	These are reactions in which both the forward and reverse reactions occur.
stoichiometry	This is concerned with the relative amounts of reactants and products in a chemical.
stoichiometric coefficients	These are numbers in front of the formulae and they indicate the number of molecules of each kind among the reactants and products.
solute	This is any substance dissolved in the solvent.
solution	This is a homogeneous mixture in which the molecules or ions of the components freely intermingle.
solvent	This is usually taken to be the component present in largest amount and is the medium into which the solutes are mixed or dissolved.
molarity	This is when the amount is given in moles and the volume in litres. It has the unit mol L^{-1} (often abbreviated M).
molar mass	This is used to calculate the mass of 1 mole of any substance or to convert between amount and mass.
concentration	This is defined as the amount of solute dissolved in a particular volume of solution.

Table 11.

The Structure of the Atom

alpha particles	We are a happy family of subatomic particles.
protons	I have a radioactive nucleus.
nucleus	This is the number of protons in the nucleus.
neutron	This is the number of protons in the nucleus plus the number of neutrons (N) in the nucleus.
nucleons	We live in the nucleus and we have a positive charge.
isotopes	We are atoms of an element with the same number of protons but different numbers of neutrons.
radioactive	These are a stream of positively charged particles.
Nuclide	I have a mass close to or equal to that of proton but I have no electric charge.
Radionuclide	In me, my atomic number also equals to the number of electrons.
atomic number (Z)	This word means that the nucleus is unstable.
mass number (A)	I am the central core of an atom and i am positively charge and I possess most of the mass of the atom.
neutral atom	I am a normal atomic nucleus.

----------Key----------

alpha particles	These are a stream of positively charged particles.
protons	We live in the nucleus and we have a positive charge.
nucleus	I am the central core of an atom and i am positively charge and I possess most of the mass of the atom.
neutron	I have a mass close to or equal to that of proton but I have no electric charge.
nucleons	We are a happy family of subatomic particles.
isotopes	We are atoms of an element with the same number of protons but different numbers of neutrons.
radioactive	This word means that the nucleus is unstable.
Nuclide	I am a normal atomic nucleus
Radionuclide	I have a radioactive nucleus.
atomic number (Z)	This is the number of protons in the nucleus.
mass number (A)	This is the number of protons in the nucleus plus the number of neutrons (N) in the nucleus.
neutral atom	In me, my atomic number also equals to the number of electrons.

Table 12.

Atoms, Molecules, Ions, Elements and Compounds

atoms	These are substances containing two or more elements in a definite and unchanging proportion.
molecules	These are collections of one type of atom only.
covalent bonds	These are different ways of arranging (bonding) an element's atoms to make particles. (diamond vs graphite)
ions	These are positively charged ions.
cations	These are collections of atoms with a definite structure held together by chemical bonds.
anions	No detectable gain or loss of mass occurs in chemical reactions. Mass is conserved.
elements	These are chemical species that have either a positive or negative electric charge.
compounds	This is a notational system used to describe a chemical reaction.
law of conservation of mass	These are negatively charged ions.
Chemical equation	These are discrete chemical species comprising a centrally positively charged nucleus surrounded by one or more negatively charged electrons.
allotropes	Bonds formed by sharing of electrons between neighbouring atoms

----------Key----------

atoms	These are discrete chemical species comprising a centrally positively charged nucleus surrounded by one or more negatively charged electrons.
molecules	These are collections of atoms with a definite structure held together by chemical bonds.
covalent bonds	Bonds formed by sharing of electrons between neighbouring atoms
ions	These are chemical species that have either a positive or negative electric charge.
cations	These are positively charged ions.
anions	These are negatively charged ions.
elements	These are collections of one type of atom only.
compounds	These are substances containing two or more elements in a definite and unchanging proportion.
law of conservation of mass	No detectable gain or loss of mass occurs in chemical reactions. Mass is conserved.
Chemical equation	This is a notational system used to describe a chemical reaction.
allotropes	These are different ways of arranging (bonding) an element's atoms to make particles. (diamond vs graphite)

Table 13.

Atomic Energy Levels

frequency	This refers to the starting position of a wave with respect to one wavelength.
wavelength	This shows how the energy of light depends on its frequency and intensity.
amplitude	This is determined by the amplitude of a light wave.
intensity	I am the most stable state with the lowest energy state for an atom.
phase	This is the number of wave crests passing a point in space in 1 second.
photoelectric effect	This is the distance between successive wave crests and is measured in units of distance such as metres or nanometres.
photons	I am an unstable state of higher energy state for an atom.
ground state	I measure the frequencies of the photons that an atom absorbs.
excited state	These are 'packets' or 'bundles' from which light comes from as postulated by Einstein.
absorption spectrum	I am produced by excited atoms losing all or part of, their excess energy by emitting photons.
emission spectrum	This is the maximum displacement of a wave from its centre.

----------Key----------

frequency	This is the number of wave crests passing a point in space in 1 second.
wavelength	This is the distance between successive wave crests and is measured in units of distance such as metres or nanometres.
amplitude	This is the maximum displacement of a wave from its centre.
intensity	This is determined by the amplitude of a light wave.
Phase	This refers to the starting position of a wave with respect to one wavelength.
photoelectric effect	This shows how the energy of light depends on its frequency and intensity.
photons	These are 'packets' or 'bundles' from which light comes from as postulated by Einstein.
ground state	I am the most stable state with the lowest energy state for an atom.
excited state	I am an unstable state of higher energy state for an atom.
absorption spectrum	I measure the frequencies of the photons that an atom absorbs.
emission spectrum	I am produced by excited atoms losing all or part of, their excess energy by emitting photons.

Chapter 4
Language Focus for Genetics and Molecular Biology Students

Brett A. Lidbury
Australian National University, Australia

ABSTRACT

This chapter examines the role of scientific language comprehension and confidence for senior undergraduate students in Genetics and Molecular Biology, and the impact of language-centred learning strategies to assessment outcomes. A number of online and tutorial language exercises and strategies are described that were designed to promote scientific language competence and subsequent genetics learning. The effect of these interventions was analysed through grade and assessment performance comparisons with earlier traditionally taught Genetics cohorts. While no significant grade improvements were found for cohorts taught via language, deeper statistical analysis revealed that motivation to adopt new learning strategies was crucial for best student performance. Language was found to be most influential for middle range performing students. Despite at least a year of tertiary education, students still had difficulty interpreting some everyday words in a science context. The study also encourages a greater evaluation of student motivation in adopting new learning techniques.

INTRODUCTION

The majority of the study described and discussed within this volume concerns the experience of first year science undergraduates in the fields of biology, chemistry, physics and statistics, and the consideration of a role for field-specific language competence in performance for the first contact of students with tertiary science education. The studies conducted identified language difficulties and then set about addressing such difficulties via particular teaching and learning interventions focused on language. The importance of language to science education has been discussed and studied

DOI: 10.4018/978-1-61350-062-0.ch004

previously (Wellington & Osborne, 2001). One earlier study suggested that first year university biology students encounter more new words than their counterparts studying foreign languages in their first year (Yager, 1983; Wandersee, 1988). Given this evidence, the centrality of language to successful science learning has not been generally embraced, and is at best an auxiliary issue.

This chapter examines the role of language in learning in the advanced undergraduate subject of *Genetics* (with some reference also to the associated but separate subject *Molecular Biology*). The students involved in these studies were experienced university students with at least a year of tertiary level study when they entered *Genetics*. Also different to the other studies contained here, language support had been introduced three years earlier than for first year biology, chemistry and physics, and as such was a standard aspect of teaching once the formal study began (Zhang & Lidbury, 2006). This chapter describes the language interventions used in teaching genetics, the results of language comprehension (and confidence) testing using everyday words that also have usage in science (the same analysis as for first year biology, chemistry and physics, described in other chapters) and a statistical analysis of student data to reveal which learning factors, including language, are associated with student performance in undergraduate Genetics.

LANGUAGE METHODS EMPLOYED TO SUPPORT LEARNING IN GENETICS

Paper-Based Methods and Tutorial Activities

Language focused methods adopted for teaching genetics did not rely solely on technology, but actively borrowed approaches from foreign language education. The primary idea was to enhance genetic language comprehension, but also diversify and energise tutorial sessions and lectures provided during the semester of teaching. "Traditional" and transmissive styles of teaching were used in teaching genetics and molecular biology until 2005, after which language-centred teaching was developed and embedded routinely into subject instruction. Traditional/transmissive tutorial styles involved, generally, a reading or a set of statistics problems and so on, with the expectation that students would do the pre-reading and/or problems and the bring them to the tutorial for discussion. The subsequent tutorial often had the tutor or lecturer speaking most of the time with little engagement from the students, who often did not prepare adequately, often due to not understanding the content.

Group Language Mix and Match

To energise the group and shift the focus from the lecturer/tutor/professor to student-directed learning, "warm-up" exercises were employed based on language comprehension associated with a topic covered in lectures. The warm-up used a "mix and match" approach (inspired by a *Hot Potatoes*™exercises - see next section), with each student given a word or term that then required a "match" to another student holding the correct word or term. This exercise was performed at the very beginning of the tutorial and was given 5 - 10 minutes. It forced students from their chairs and into the wider group to speak with other students, and furthermore through language, primed the group for the tutorial topic. An example is based on the concepts of plasmid cloning of recombinant DNA and the *polymerase chain reaction* (PCR), a laboratory technique used to specifically amplify a DNA sequence. The words/terms employed were:

- **Plasmid (DNA cloning) exercise:** *Ori*, MCS, Amp^R, transcription control region, *Eco*RI, Restriction Enzyme, buffer.
- **PCR Exercise:** Primer 1, Primer 2, *Taq*, target sequence, dNTPs, $MgCl_2$, thermo-

stable DNA polymerase (interchangeable with *Taq*).

With one word/term each, the students were then invited to make a restriction digest for a plasmid or a PCR reaction. A recent lecture introduced both plasmids/restriction enzymes and PCR. This approach stimulated discussion and debate with little effort by the tutor, while reinforcing specialist language in genetic engineering. This method also encouraged students speak up on any misunderstandings or difficulties associated with the topic of discussion, helped by group support formed via the exercise.

Drawing Genetic Concepts

The early phase of the language project for genetics education identified that students no longer drew diagrams or free-hand pictures when grappling with complex scientific language and concepts. The approach commonly detected by the language education expert involved in the project was for students to simply underline words that they didn't understand, and come back to these words later. This resulted in reading text with up to half of the meaning missing, exacerbated by persisting with the reading when not comprehending the introductory and background concepts. An exercise was introduced for the first tutorial that focused on complex genetic language from a primary research article on *Drosophila* (fruit fly) genetics and a mutation associated with altered fly behaviour. Central to translating the research paper were a series of questions focused on key words and supported by collection of drawings that represented the concepts of the "gene", "protein" and "organism" (Figure 1). These drawn concepts were connected to the theme of the research paper; namely, a mutated gene results in a dysfunctional protein, which leads to an abnormal or disease phenotype in the fly (the "organism"). Anecdotally, students appear not to cross categorise concepts, for example, we have a "gene" that encodes normal genetic information, but a mutated gene that carries faulty information is a separate concept. This approach was referred to as a template for other tutorials that followed for the semester. The exercise also included the breaking down of complex words into separate root meanings, an approach that was often used in lectures to reinforce language comprehension. For example, for the above-mentioned fruit fly study, an *aminopeptidase* was identified as the faulty protein resulting from a genetic mutation. The word was broken into three parts; *amino-*, pertaining to the NH_3 (amino) entity of an amino acid chain, *peptid(e)*, several amino acids, a peptide chain, and *-ase*, pertaining to an enzyme. In translation, therefore, the abnormal gene found in the fruit flies was an enzyme that degraded protein, or peptide, chains from the amino terminus. This enzyme function was then linked to the broader organism function and behaviour phenotype.

Scaffolded Readings in Molecular Biology to Assist Comprehension

To assist with the complex language of the subject *Molecular Biology*, a graded series of readings on a particular topic, for example gene expression and RT-PCR (reverse transcriptase polymerase chain reaction), were set starting with the least difficult paper (e.g. a report from *New Scientist*) and finishing with a specialised research article from a peer-reviewed, high impact discipline journal. The reading/language scaffold comprised four layers of increasing conceptual difficulty and language challenge ("*VP35 blocks virus-induced IRF-3 phosphorylation and subsequent IRF-3 dimerization and nuclear translocation*") (Basler, Mikulasova, Martinez-Sobrido, Paragas, Mühlberger, Bray, Klenk, Palese P, & García-Sastre, 2003). These readings were discussed in lectures (a "lectorial" format) and were often supported by a group *Cloze* (gap-fill) exercise (see next section for an example).

Figure 1. Encouraging science language competency through association with symbols and pictorial representations. The concept of the "chromosome", "gene", "protein" and "organism". Exercise context is Drosophila genetics.

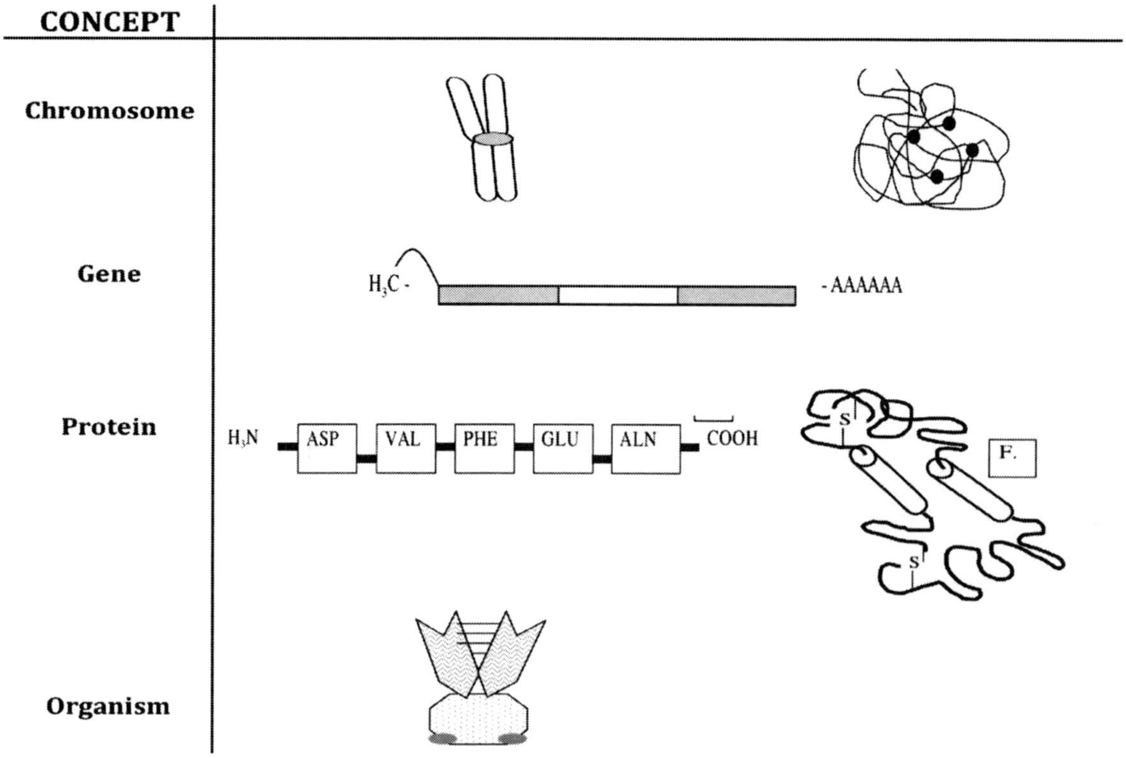

Of the approaches discussed, scaffolded readings were the least successful in terms of student engagement with the material. Clearly, articles pitched at an educated lay audience still contain unfamiliar concepts that have no connection with student experience, and therefore remain inherently difficult. The timing of such paper exercises to coincide with laboratory practical work may assist this learning approach. This experience, however, also stimulated thinking on student motivation in relation to new teaching styles and support; some students obviously enjoyed and benefited from this style of instruction, but they were the minority. The topic of motivation will be discussed in-depth later in this chapter.

Technology Centred Language Support; Adaptation of the Hot Potatoes™ Format to Genetics and Molecular Biology Education

Mix and Match

In the same spirit as described-above for mix-and-match exercises in tutorials, early lecture warm-up exercises were developed using the *Mix and Match* feature of *Hot Potatoes*™ (http://hotpot.uvic.ca/). Figure 2 shows examples of these exercises used in the early weeks of an undergraduate *Genetics* unit. The lecturer guides the students on this exercise at the very beginning of the lecture, and the concepts covered pertain to the previous

Figure 2.

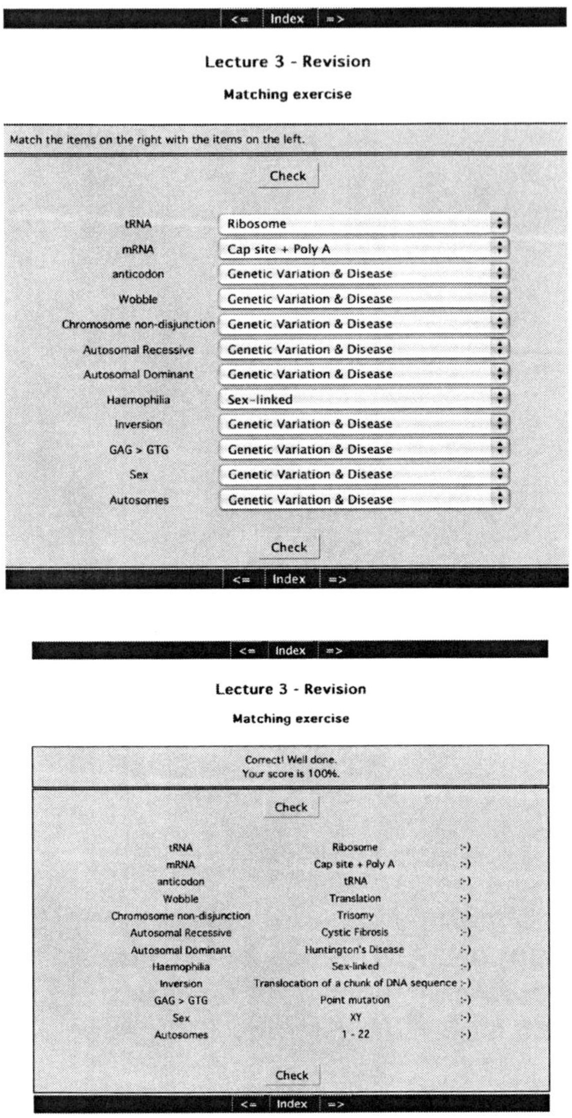

weeks' lecture. This mix and match exercise is projected onto the lecture theatre screen and the group asked to contribute ideas and answers related to the word match options.

The exercises are relatively simple, with a column of fixed words or terms in the left column, and the matching word options in the right column. Figure 2(a) shows an incomplete exercise; a box in the right column, when clicked on, will drop down all of the matching word options. Figure 2 (b) shows a completed exercise. While not all students participate in the exercise, a significant minority normally contributes answers, ideas or questions that helps generate a short discussion of the previous lecture, lasting 5 to 10 minutes. It is also usual not to achieve correct matches for all words at the first attempt, leading to further discussion of the exercise. The students always request access to these exercises afterward via the subject LMS (*WebCT* or *Moodle* in this case).

Cloze (Gap-Fill) Exercises

Using the *Hot Potatoes*™ gap-fill tool, language exercises were designed to support laboratory practical work in DNA cloning using plasmid vectors. The specialist language associated with recombinant DNA technology and genetic engineering is particularly onerous for students learning the discipline for the first time. Of most difficulty for the uninitiated is the *Materials and Methods* section of articles describing the strategies, reagents and techniques used to clone DNA fragments into plasmid vectors, precisely the area laboratory focus central to understanding these concepts in molecular biology and genetics.

Figure 3 shows and example of a gap-fill exercise in plasmid-based cloning. The passage on the left is a reading chunk copied directly from the published article (Gao, Peng, Xie, Xie, Gao & Yao, 2004), containing the specialist language of DNA cloning (For example, *"Pre-S2/S fragment of HBV was obtained using pTZ19U-HBV as a template, and HBV-P1 and HBV-P2 as primers. The purified product was then cloned into pUC-18 at the site between the restriction endonucleases Hind III and Eco RI to generate recombinant plasmid pUHB"*). This exercise was chosen as it had relevance to another subject taken by many students in the same semester (Immunology) and used restriction enzymes utilised in the student laboratory sessions (*Eco*RI and *Bam*HI). On the right of the exercise (Figure 3) is a brief introduction on the article and its aims described

Figure 3.

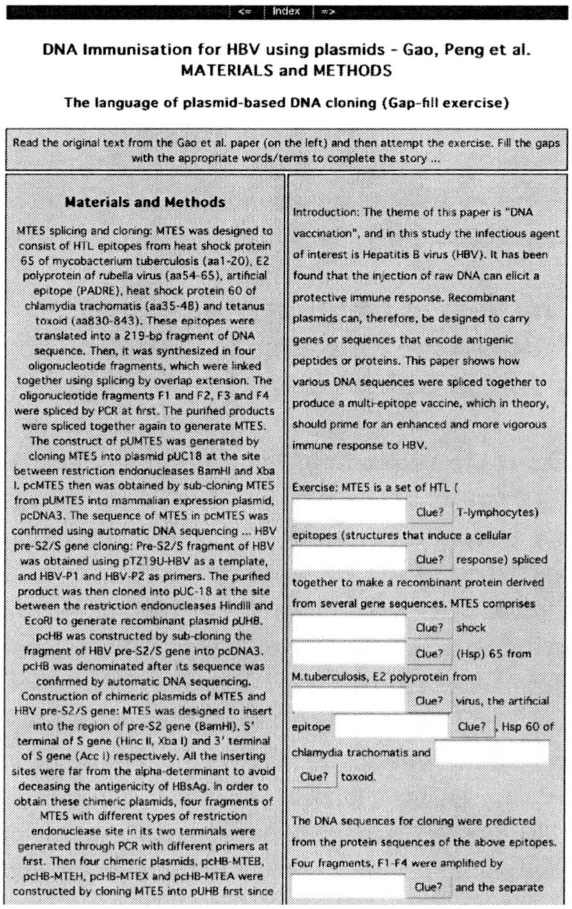

Figure 4.

in simpler language, then a passage below this comprising the same *Materials and Methods* chunk taken directly from the article, but written with simpler language and including embedded explanations. Amongst this passage are word/term gaps that require correct input to make sense of the recombinant DNA techniques employed by the scientists. The correct word or term required to complete this task can be found by reading the text taken directly from the adjacent *Materials and Methods* text.

Crossword

A number of on-line crossword exercises were designed to support tutorials, or as optional extension tasks for motivated students on key concepts (Figure 4). These were designed for individual use and were not used in-group settings due to their greater complexity. As for the other on-line exercises adapted from *Hot Potatoes*™, the intention of crosswords were to enhance scientific language comprehension and thereafter build confidence in understanding concepts in *Genetics*. The impact of crosswords on genetics learning were not monitored, but were placed as a permanent feature on student LMS sites.

The crosswords embedded in the student websites were a leading aspect to a number of non-assessable and non-compulsory on-line exercises that were evaluated at the end of the teaching semester as a guide to student motivation, that is, students who would do extra work and study on genetic concepts without being forced by assessment or participation demands required to pass the subject. Cloze exercises, like that described above, were also embedded and assessed for this purpose. Through tracking the student LMS activity at the end of the semester, a motivation score was assigned (based on total semester hits on exercises) and was included as a predictor variable in the analysis of final examination results and overall performance across all assessment. Other features found in LMS formats like *Moodle* and *WebCT*, for example number of discussion posts and total students hits on the site, were also considered in these analyses. An analysis of these results in relation to student motivation is discussed later in this chapter.

IMPACT OF SCIENTIFIC LANGUAGE INTERVENTION ON STUDENT LEARNING IN GENETICS

From 2001 to 2004 teaching undergraduate students *Genetics* involved what from now on will be referred to as "traditional" (transmissive style). Namely, teaching that was lecturer/professor focused, with an assumption that through previously studied university subjects adequate background knowledge and discipline language had been acquired. This style resulted in a content driven teaching philosophy that continued in spite of reduced student engagement through not understanding the early concepts. In 2005, language-centred teaching was introduced to *Genetics* through collaboration with a language education expert (Dr. Zhang) that involved her attendance at lectures and tutorials. Through these observations, the interventions discussed above were decided and designed for future teaching semesters.

The impact of the language focus for genetics was evaluated as a longitudinal comparison between traditional cohorts (2001 - 2004) and language cohorts (2006 - 2009). The 2005 cohort was not included in the longitudinal analysis because this was the year where language pedagogy was first tried for the undergraduate genetics subject (Table 1). The observations from 2005 directed the embedded language methods utilised routinely from 2006 - 2009.

Analysis of Language Impact on Student Performance in Genetics at a Whole Cohort Level

The student data from the transmissive teaching years of 2001 - 2004 was pooled (n = 94) as was student data from 2006 - 2009 representing pooled data from the language-centred genetics student cohorts (n = 346). A summary of overall student performance as reflected by the number and percentage of *High Distinction* (HD), *Distinction* (DI), *Credit* (CR) and *Pass* (P) grades, as well as number who failed or did not complete the subject, are summarised in Table 1. Pearson's Chi-square (χ^2) analysis showed that in terms of final grade, the language focus did not significantly influence (p = 0.52) performance for the 2006 - 2009 cohort when compared to the transmissive cohorts (2001 - 2004). Additionally, comparison of the mean final theory examination performance and mean overall performance (all assessment items) results by t-test showed no significant differences between language and transmissive cohorts (Table 1).

Overall performance is calculated from adding the weighted scores for each assessment item together for a final mark from 100%. For *Genetics* the assessment items included (weighting in parentheses); Final theory examination (50%), Bioethics Essay (20%), on-line Mid-Semester exam (multiple choice, short answer, mix and match questions) (15%) and Tutorial participation (15%).

For the traditional/transmissive cohorts, the mean final examination performance for pooled

Table 1. (A) Final student grades achieved for cohorts exposed to embedded language-centred teaching compared to traditional/transmissive teaching styles in undergraduate Genetics at the University of Canberra (2001 - 2009)

(A) Teaching Style *	Final Grade Category					
	HD	DI	CR	P	FAIL	Did not complete
Traditional (n = 93) Number in each grade	8	19	26	27	8	5
% of Total	8.5	20.2	27.7	28.7	8.5	5.3
Language focus introduced (n = 53) Number in each grade	3	13	17	12	5	3
% of Total	5.7	24.5	32.1	22.6	9.4	5.7
Language methods embedded (n = 346) Number in each grade	21	85	96	78	32	34
% of Total	6.1	24.6	27.8	22.5	9.3	9.8

* Traditional and transmissive styles were employed for undergraduate *Genetics* teaching during the years 2001 to 2004. Language-centred methods were introduced in 2005. Embedded language support was employed routinely in undergraduate *Genetics* teaching from 2006 to 2009.

Final grade categories: HD = *High Distinction* (85 - 100%); DI = *Distinction* (75 - 84%); CR = *Credit* (65 - 74%); P = *Pass* (50 - 64%); Fail (< 49%). To pass *Genetics* overall, a minimum of 50% was required in the final theory examination at the end of semester.

Table 1. (B) Mean overall performance (all assessment items) and mean specific theory examination performance for transmissively taught Genetics cohorts in comparison to cohorts receiving embedded language support

(B) Student performance	Traditional/Transmissive				Language Methods Embedded			
	2001	2002	2003	2004	2006	2007	2008	2009
Mean Final Exam (%)	60.1	64.4	64.0	65.6	65.3	62.0	66.1	68.1
Mean Overall (%)	67.4	67.1	67.4	68.8	68.1	67.2	68.4	71.7

results from 2001 - 2004 was 63.5%, with the mean overall student performance pooled for the same cohort equal to 67.7%. The pooled final examination mean for the language cohort (2006 - 2009) was 65.4%, with mean pooled overall performance for the same cohort of 68.8%. For the language introduction year 2005, the mean final examination mark was 64.5% and the mean overall mark 69.2%.

Student's t-test (unpaired, equal variance) showed no significant difference between the pooled traditional/transmissive final examination results compared to the pooled results for the language cohorts ($p = 0.32$). The same t-test analysis also showed no significant difference between the pooled overall performances for both cohorts ($p = 0.30$).

At the cohort level, the introduction of language pedagogy did not lead to a statistically significant increase in higher grades, nor a change to the proportion of students failing or not completing the subject. This finding confirms the analysis of the initial results from the 2005 introduction of language methods, with direct input from the language expert, when compared to previous years' student results in genetics from 2001 to 2004 (Zhang & Lidbury, 2006). Again using χ^2 analysis, no statistical differences in rates of movement from lower grades to higher grades were found, and no statistical differences found for final examination results or overall performance by t-test. However, the Zhang and Lidbury (2006) study did reveal subtle differences at the individual student level, particularly when considering the individual student grade point average (GPA) that reflects the total student performance across all subjects studied at university up to the semester in which they enrolled in *Genetics*. The GPA analysis (Pearson's correlation) showed that *Distinction* level students in the 2005 language cohort maintained their general higher level of performance in *Genetics*, as also achieved for earlier subjects reflected by their GPA (a correlation coefficient of $r = 0.64$ was calculated that was significantly different [$p < 0.05$] to independence, or $r = 0$). This association between GPA and genetics performance was not seen for *Distinction* level students in the transmissive 2001 - 2004 cohorts, nor was it found for other grade groups in the 2005 language cohort (Zhang & Lidbury, 2006). The suggestion here is that the introduction of new learning approaches suited the *Distinction* students most, and therefore were adopted in their studies to their advantage in terms of their performance in *Genetics*. This effect was also observed for the more advanced subject, studied by the same student cohorts, *Molecular Biology*. The molecular biology results showed that the effect stradled the Credit - Distinction band of student results (for both *Genetics* and *Molecular Biology*, because of lower numbers, the *High Distinction* students were included in the *Distinction* category).

Anecdotally, students who most needed help (those on the Pass/Fail boundary) and for who the introduction of language methods were intended, did not embrace the new methods, and in casual feedback, often said that they preferred transmissive teaching. By extension, this type of student probably preferred surface learning and did not welcome having innovative teaching/learning methods imposed upon them as this was seen as an extra burden. Such students could be thought of not being interested in "learning", *per se*, and sometimes resented the introduction of new techniques. While not causation, the correlation analyses of results across the two subjects examined were consistent for the Pass/Fail cohort across the years of the study, and on reflecting on these findings, which stood in strong contrast with the *Distinction* cohort, pointed to the role of student perceptions of learning and intrinsic motivation for study at university. From inspection of Table 1, while no statistically significant effect was observed, the trends support the above assertions on student motivation. For the traditional and language cohorts, the *Credit* grade was steady with 27.7% and 27.8% of students respectively achieving this final grade for *Genetics*. For the 2006 - 2009 language cohort, there was possibly a movement of some *Pass* students into the *Credit* grade and a further flow into the *Distinction* grade, as reflected by comparing percentages with the traditional cohort. However, this effect on the *Pass* cohorts also involved, for both the 2005 language introduction year and the embedded language cohorts of 2006 - 2009, a slight increase in failure rate, but more dramatically for the embedded language cohort, a clear increase in the "Did not complete" students. Therefore, a *Pass* student could go in either direction and this would result in poor correlations with GPA. In fact, examination of the scatter plots for the Pass/Fail groups did often show some individual students exceeding their GPA standard in *Genetics* or *Molecular Biology*. Clearly for others though, introducing non-traditional styles of university instruction may have exacerbated their study difficulties, not assisted them, and as alluded to above the attitude of such students may be the central issue, not the teaching itself.

In summary, the introduction of innovative teaching and learning methods based on language pedagogy to students studying subjects rife with specialised and unique language, genetics and molecular biology, did not result in statistically significant improvement of final grade attainment or final theory examination performance when analysed at a whole cohort level. However, clues emerged from these analyses that point to individual student attitudes and motivation to university study, which are subtle and not readily revealed by evaluation of gross student performance data such as presented in Table 2. The consideration of these subtle effects in association with embedded language learning methods, are discussed in the following sections.

ASSESSING LANGUAGE CAPACITY IN EXPERIENCED SCIENCE STUDENTS: A FOCUS ON THE 2008-2009 *GENETICS* COHORT

The obvious factor that underpins the adoption of language-based teaching methods, and therefore engagement with subject content, is the student's scientific language capacity. The studies described herein were with students with at least one year of university education on entering the subject *Genetics*. Opportunities to assess the development of scientific language skills from first year at university (where the students were exposed to concepts in biology, genetics, inorganic and organic chemistry, and basic statistical/quantitative methods) were therefore available. The results of the study with *Genetics* students can be contrasted with the results reported by identical studies with first year science students in biology, chemistry and physics (see other chapters). A primary aim of the language capacity assessment of university science students was to analyse words that in addition to their scientific meaning, also have assumed or known everyday meanings in English (e.g. model, significant).

In light of these language requirements, an online exercise was designed that used words with everyday meanings, but also currency in biology, chemistry and physics. The format and design of this survey is described in chapter 1. As well as assessing scientific language comprehension, the exercise also measured confidence associated with the comprehension of each word. Ten words were included in the study, and they were: *Equation, Experiment, System, Research, Model, Significant, Equilibrium, Power, Concentration and Graph.*

The language assessment also used a *Cloze* (gap-fill) exercise that comprised a scientific paragraph (mouse model of diabetes) with missing words, with each of the above-listed words a candidate to fill at least once gap in the paragraph. Students were assessed at the beginning of semester and in the final week of the semester (12 - 13 weeks after the initial test).

To enhance statistical power, the language data for 2008 and 2009 students were pooled into one dataset (n= 90). Prior to pooling the data, homogeneity of results between the two separate *Genetics* cohorts were analysed by F-test and Student's t-test. Final theory examination results and final overall mark (all weighted assessment items) were not significantly different for the 2008 compared to the 2009 cohort ($p = 0.126$ and $p = 0.09$ respectively). Language comprehension and language confidence were significantly higher for the 2009 cohort on the language pre-test (week 2) exercise compared to the 2008 cohort, but were not significantly different for the week 15 language post-test. For both 2008 and 2009 cohorts, mean language comprehension and language confidence increased from the week 2 pre-test to the week 15 post-test, indicating enhanced language competence and confidence over the course of the semester. It must be noted that the students only had one attempt at the beginning of the semester (10 minutes) and no other access to the language exercise until 12 weeks later. There were no practice exercises and none of the coursework or content of *Genetics* was connected to this exercise, or made reference to the individual words used.

In further analysis of the 2009 student data, correlation (Spearman method) analysis also suggested that student judgment on language improved significantly by the completion of a semester of second year genetics. For the week 2 language

Table 2. Leading factors and predictors associated with superior, middle range and poor performance across assessment tasks in Genetics

Level of Performance	Superior	Middle Range	Poor
Key factors and predictors of performance in Genetics	- GPA > 5.5 - Essay > 12.25/20 - Motivation - Tutorial participation	- Language competence and confidence (If GPA < 5.25)	- GPA < 5.25 - Essay < 12.25/20 - Semester LMS hits < 325 or > 500

"Superior" is defined as students achieving > 80% on either the final theory examination or for a combined overall assessment mark (final theory exam, mid-semester on-line exam, bioethics essay, tutorial participation). "Middle Range" is taken as students achieving overall or specific exam marks of high Pass range (60%) to low Distinction range (77%). "Poor" performance is defined as either a Fail mark (45 - 49%) for the final exam or for overall assessment performance. LMS is "learning management system" (i.e. Moodle or WebCT). GPA scores are a cumulative calculation of individual weighted subject grades up to that point of the degree programme. The maximum GPA at the University of Canberra is 7.0 (High Distinction = 7.0, Distinction = 6.0, Credit = 5.0, Pass = 4.0). Motivation was measured for each student by the number of LMS hits on non-assessed and non-compulsory on-line learning activities and additional exercises or readings in Genetics.

pre-test, only one word (equilibrium) showed a significant (positive) correlation between word comprehension and confidence, while at the end of the semester, five words (experiment, model, concentration, graph, equation) had a significant ($p < 0.05$) positive correlation between comprehension and confidence in their judgment on whether they knew the correct definition, or not. This judgment in word comprehension does not just relate to getting the word definition or meaning correct, and knowing its' correct, but also relates to not knowing a word and recognising that their understanding of word in the scientific context is inaccurate. This conclusion is supported by the inclusion of the word "model" in the list of five words. In both the early semester pre-test and the end of semester post-test language exercise (2009), "model" was the worst understood word of the ten listed. This was also found for the 2008 genetics cohort. "Equation" and "Experiment" also regularly featured in the top three least comprehended words. Therefore, language judgment, and confidence in that judgment, improved over the semester of genetics in the early second year period of their studies. This may be associated with the general focus on language during the subject.

In spite of this increase in language confidence, particular words were particularly problematic for second year genetics students. As discussed above, "model" was consistently the worst understood word when applied to a scientific context. For the 2009 language pre-test, 63.8% of students scored zero for the question (partially correct answers were allowed). While "model" comprehension improved for the language post-test, 42.2% of students still scored zero on this question (but confidence in understanding the word, or not, improved). Based on the proportion of students who scored zero for the question, and a mean or median score less than or equal to 5/10, other problematic words were; "Equation", "Power" and "Equilibrium", while the consistently best understood by the same criteria were; "Graph" and "Concentration". "System" and "Experiment" also were also well understood in one pre- or post-test language exercise, but not consistently. Interestingly, for both "Graph" and "Concentration", the percentage of 2009 students scoring zero increased for the post-test (end of semester) compared to the early semester pre-test.

Regardless of the trends revealed by the study of language comprehension and confidence in second year science students, in spite of a year of experience at university, language comprehension problems remained, although the problem words did vary when compared to first year students studying biology, chemistry or physics (see other appropriate chapters). Given the background of students enrolled in *Genetics*, generally having successfully passed first year biology (including genetic concepts), chemistry and so on, it could be reasonably expected that all ten words examined by the language comprehension and confidence exercises would be confidently comprehended; this was not found. This suggests ongoing issues with language in science programmes, and as revealed here, problems with successfully understanding some "everyday" words in the science context. This clearly has implications for successful science learning at university. How language capacity impacts on student performance is considered in the following section.

ANALYSIS OF LANGUAGE AND NON-LANGUAGE FACTORS THAT CONTRIBUTE TO STUDENT PERFORMANCE IN GENETICS

To understand the impact of varied assessment items, use of on-line learning tools and engagement with the technology, and language competence and confidence in student performance for *Genetics*, a data-mining strategy was employed for student data that captured the effect of these factors on overall performance in *Genetics* (final mark, all assessment items included) and final theory examination performance. Assessment items (% weighting overall) were;

- An end-of-semester theory examination (50% - 2 hour invigilated paper),
- A bioethics essay (20%),
- A mid-semester (on-line) exam (15% - short-answer questions, multiple choice with one or more correct answer, mix and match exercises. Questions shuffled for each student, internal MCQ choices shuffled),
- Tutorial participation (15% - completion of readings, statistical problems and exercises; engagement with tutorial discussions).

The influence of on-line tools were measured simply by the number of hits per student over the semester on the unit/subject specific LMS (2008, WebCT and 2009, Moodle), the number of posts per student to the LMS over the semester, and the engagement with additional non-assessed on-line exercises and content (also used as a measure of individual student motivation). Overall performance at university up to the semester where students enrolled in *Genetics* was included in the study as a grade point average (GPA) score. The measured influence of individual student scientific language competence and confidence, as described in the previous section, was also included in the analyses to assess the association between language and performance. Because of the change in LMS platform during the study, the 2008 and 2009 cohorts were analysed separately, and then pooled for further analysis; because of the change in LMS during the study, not all parameters were included in the pooled data analysis. Multiple linear regression and recursive partitioning (Decision Tree) methodologies that used the public domain statistical and data-mining software *R* (version 2.9.1) was employed for these studies.

For the analysis of the 2008 cohort, the assessment items listed above were included in addition to; GPA, "motivation" (total hits on non-assessed on-line exercises), total LMS hits over the semester, on line pages read and discussion posts to the LMS. The pre- and post language test scores and confidence measures were included (see previous section), as well as the student results on an early and late genetic concepts inventory quiz (Elrod, 2007). Like the language assessment, the same concept inventory quiz was run in the first week of the semester, and then 14 weeks later in the final week to gauge entry-level conceptual understanding and whether this improved by the completion of the semester in genetics. The 2009 cohort had similar parameters, but did not have concept inventory data, nor was the measure of motivation the same as for 2008 due to the change in LMS platform.

Figure 5.

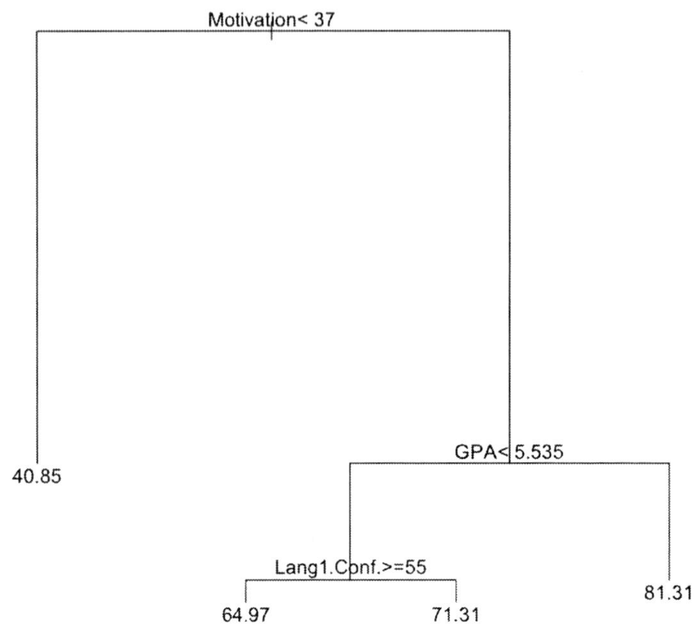

Overall Performance Across all Assessment in Genetics

When analyzing overall performance, not surprisingly, individual assessment items are highly significant predictors of overall performance. Apart from the role of specific assessment item performance on overall performance, for the 2008 cohort GPA (> 5.54/7.0) and "motivation" (as measured by engagement with non-assessed on-line activities) were the leading factors contributing to highest overall marks (> 80%), as analysed by decision trees (Figure 5). If student GPA was less than 5.54, confidence on the first language exercise (described in the previous section) was a factor to obtain an overall credit level mark of 71%. In general, language confidence and competence was not a leading factor in overall performance across all assessment items for Genetics, and this was confirmed by linear regression analysis that pointed again to GPA and motivation as the leading predictors of performance for 2008 students.

For the 2009 cohort with all assessment items included, a strong regression model ($R^2 = 0.84$) predictably showed each individual assessment item as a strong predictor to overall performance, but in amongst these obvious predictors GPA and confidence on the first language exercise were also significant ($p < 0.01$). When removing all assessment items from the model ($R^2 = 0.60$), GPA dominated as the leading factor of overall performance, with the student score on the end-of-semester language exercise also significant ($p < 0.05$). Language capacity was more influential for the 2009 cohort, and this was confirmed by decision tree analysis. For the highest overall mark of 84%, an essay score of > 12.25/20 and a GPA of > 5.46 were the leading factors. For middle to high-ranking marks (68.3% / 77.7%), student performance on the first (early semester) language exercise (> 54.5/70 for an overall mark of 77.7%) was also influential for overall performance in Genetics.

The pooled 2008 and 2009 data were analysed exactly as for the individual year cohorts. With all assessment items and other variables included, a strong regression model was achieved (R^2 = 0.81), with all assessment items and GPA highly significant predictors ($p < 0.0001$). Early semester language confidence was also a significant predictor ($p = 0.008$) of overall performance in Genetics from this model. With the removal of GPA from the regression model, early semester language confidence remained significant ($p = 0.024$). Other language factors were not significant either with or without GPA included in the model ($p > 0.35$). Decision tree analysis confirmed the regression findings with GPA and tutorial participation as particularly strong factors in superior overall performance. Interestingly, student early language score (i.e. understanding language in a scientific context when entering the subject) was a factor associated with middle-ranking students (credit range of 65.8% - 72.5%). This middle-range effect was also noted above for the 2009 cohort. The evidence suggests that for students with a weaker GPA (< 5.4), scientific language capacity may help compensate and assist the student in maintaining overall grades in the credit - lower distinction range. A summary of significant predictors for superior, middle range and poor academic performance in Genetics is presented in Table 2.

Performance in the Final Theory Examination for Genetics

For the 2008 cohort, GPA and "motivation" were the leading factors for the best theory examination performance. Motivation was measured by student engagement with non-assessable, non-compulsory additional on-line exercises embedded into the unit LMS (greater than 37 hits on these sites over the semester was found by decision tree analysis). For a middle-range final exam mark (36.6/50), a score of > 40.8/50 was needed for the final (end of semester) *Genetics Concept Inventory* score (Elrod, 2007) in addition to a tutorial participation mark of > 10.25/15, as determined by decision trees. The linear regression analysis of 2008 final theory examination performance again pointed to motivation as a significant predictor ($p < 0.003$) of final exam performance, and overall LMS hits over the semester ($p < 0.02$). Further investigation found that the role of LMS hits was nuanced in that too many hits were associated with poor performance; a positive influence was found for < 472 hits over the semester, suggesting a lack of study and/or engagement direction from the poorer students.

For the 2009 cohort, decision tree analysis of final theory examination performance echoed overall performance with a GPA > 5.46 and a bioethics essay mark of 12.25/20 the leading contributors to the best final exam mark. For middle-ranking exam marks, the student performance in the early language comprehension quiz was a factor. With GPA removed from the decision tree model, essay score (> 12.25/20) in combination with both early and late semester language competence were factors for high to middle range (31-41/50) final exam performance. An essay mark of < 12.25/20 was a factor in achieving a fail mark of 23.4/50. The linear regression analysis of the same data on 2009 final exam performance confirmed GPA ($p < 0.001$) as a leading exam predictor with early language confidence ($p < 0.01$) also a significant predictor (model $R^2 = 0.70$). With the removal of GPA from the model, there were no significant predictors found, including the previously identified early language confidence. Overall, while GPA is the most powerful predicting factor for examination performance, essay and language ability also featured as significantly influential on performance in Genetics; the language factor again appeared to be most relevant to middle-ranking students. Success in essays requires a solid language competence and judgment, thus suggesting for the 2009 Genetics cohort a greater role for language in exam achievement. Motivation was not measured for the 2009 cohort.

For the combined 2008 and 2009 data, the best final examination performance (42/50) had a GPA > 5.5 and early language confidence (> 6.5/10)

Figure 6.

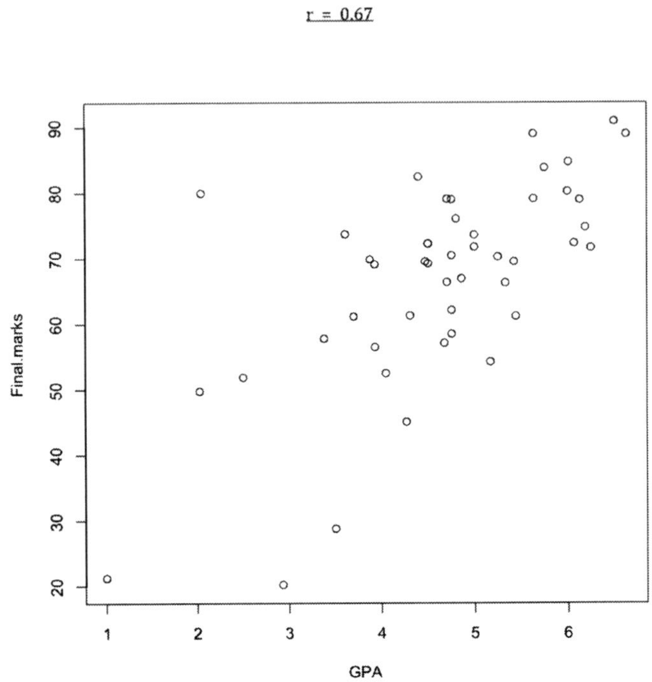

as leading factors as determined by decision tree analysis. Students with a GPA < 5.5, an essay mark of> 12.25/20 in combination with language competence and confidence were important for Pass to Credit range marks (29.6/50 - 36.4/50). Conversely, for an exam fail mark (22.8/50), a GPA < 5.5, < 321 total LMS hits and an essay mark < 12.25/20 were the leading factors. Linear regression analysis ($R^2 = 0.60$) confirmed GPA and the essay as significant predictors ($p < 0.001$), along with early language confidence ($p < 0.05$).

For final theory examination performance in general, as for overall performance, the student GPA was consistently the most power predictor of student performance. GPA measures success across the students total university education, and is not surprising as a prominent factor in student achievement in individual subjects (Figure 6). Underneath this GPA impact, language competence and confidence, as determined via the measures described in that section, were also a pronounced factor in success overall (all assessment) and specifically in the final theory examination. This language influence was particularly noted for middle-range performance (Pass to Credit), and may have compensated for the lack of other effective study skills or engagement. This often also linked with essay performance, which also involves enhanced language capacity for superior performance.

ROLE OF MOTIVATION

Implicit in the prominence of GPA as a primary factor in student performance in Genetics, is the role of motivation. In fact for the 2008 cohort where a measure of motivation was employed, this rated highly as a predictor to assessment performance. The teaching interventions and learning strategies developed and applied around science language comprehension were initially conceived with the intention of assisting less able students and helping them achieve higher grades. While this was seen for some individuals, there was no overall trend uncovered statistically, or anecdotally, that identified consistent enhanced performance for students in the Fail and low Pass range.

A science motivation questionnaire has been developed (Glynn, Taasoobshirazi & Brickman, 2009), which identified five dimensions to motivation to learn science; (i) intrinsic motivation and personal relevance, (ii) self-efficacy and assessment anxiety, (iii) self-determination, (iv) career motivation, and (v) grade motivation. Another study (Knight & Smith, 2010) has shown higher motivation, as measured by study time, interest and so on, for biology majors versus non-majors for an introductory genetics unit, linked to life relevance through career aspiration. "Deep" approaches to learning at university are consistently and positively associated with assessment outcome, while "achieving" approach changes with time over the course of study (Zeegers, 2001). This observation is supported by the consistency of GPA with individual subject performance for Distinction level students.

Reference is made to the "mid-range" performing students (Credit - 65 - 74%), both in terms of language intervention shifting students either into the higher Distinction or lower Pass range, and in relation to the importance of language competence and confidence as a factor in grade outcome for these students. A study of "meaningful" learning approaches to science proposed that while "instructional treatment" had little relationship to meaningful understanding, for students between "meaningful" and "rote" learning approaches, instructional treatment was important (Cavallo & Schafer, 1994). The introduction of language focussed teaching and learning support, as a style of instructional treatment, appeared to have an impact on Genetics students situated between the motivated and deep learning higher GPA Distinction students, and the less motivated surface learners who are more likely to be represented in the Pass/Fail range (or not complete the subject).

Based on such evidence, the introduction of new instructional styles cannot be assumed to be a wholesale positive development for all students, and may actually impede lower performance students in subjects where innovative styles are employed. The irony is that such innovation educational initiatives are primarily designed to help the lower level student. However, without the motivation to adopt new learning techniques and methods, such new approaches to teaching may provide an extra disincentive, while further stimulating superior students to maintain their high academic performance level.

CONCLUSION

It is difficult to merge the worlds of language ("arts") and science largely due to the persistent insouciant stance of scientists and science educators on the primacy of language comprehension for scientific learning (Yore & Treagust, 2006). Through genuine cross-collaboration with a colleague from the humanities, science academics have bridged this traditional gap and have effectively embedded language into tertiary science education. While not a panacea to the challenges of teaching science, for motivated students the benefits are clear as they look to enhance their deep and meaningful understanding of complex scientific concepts and principles. To ensure an impact across the broad range of student attitudes and ability, instructional methods like those described herein may now also need to examine how to augment motivation to study, with enhanced comprehension and practice via the multiple languages of science (Lemke; Pickersgill & Lock, 1991).

REFERENCES

Basler, C. F., Mikulasova, A., Martinez-Sobrido, L., Paragas, J., Mühlberger, E., & Bray, M. (2003). The Ebola virus VP35 protein inhibits activation of interferon regulatory factor 3. *Journal of Virology*, *77*, 7945–7956. doi:10.1128/JVI.77.14.7945-7956.2003

Cavallo, A. M. L., & Schafer, L. E. (1994). Relationships between students' meaningful learning orientation and their understanding of genetics topics. *Journal of Research in Science Teaching*, *31*, 393–418. doi:10.1002/tea.3660310408

Elrod, S. (2007). *Genetics concepts inventory*. Retrieved from http:// www.bioliteracy.net/ Readings/ papersSubmittedPDF/ Elrod.pdf

Gao, W. J., Peng, X. M., Xie, D. Y., Xie, Q. F., Gao, Z. L., & Yao, J. L. (2004). Construction of exogenous multiple epitopes of helper T lymphocytes and DNA immunization of its chimeric plasmid with HBV pre-S2/S gene. *World Journal of Gastroenterology*, *10*, 2979–2983.

Glynn, S. M., Taasoobshirazi, G., & Brickman, P. (2009). Science motivation questionnaire: Construct validation with nonscience majors. *Journal of Research in Science Teaching*, *46*, 127–146. doi:10.1002/tea.20267

Knight, J. K., & Smith, M. K. (2010). Different but equal? How nonmajors and majors approach and learn genetics. *CBE Life Sciences Education*, *9*, 34–44. doi:10.1187/cbe.09-07-0047

Lemke, J. L. (2010). *Teaching all the languages of science: Words, symbols, images, and actions*. Retrieved on the 30th August, 2010, from http:// www-personal.umich.edu / ~jaylemke/ papers/ barcelon.htm

Pickersgill, S., & Lock, R. (1991). Student understanding of selected non-technical words in science. *Research in Science & Technological Education*, *9*, 71–79. doi:10.1080/0263514910090107

Potatoes, H. ™. (2010). *Half-Baked Software Inc: University of Victoria, Canada*. Retrieved on the 30th November, 2010, from http:// hotpot.uvic.ca/

Wandersee, J. H. (1988). The terminology problem in biology education: A reconnaissance. *The American Biology Teacher*, *50*, 97–100.

Wellington, J., & Osborne, J. (2001). *Language and literacy in science education*. Buckingham, UK: Open University Press.

Yager, R. E. (1983). The importance of terminology in teaching K-12 science. *Journal of Research in Science Teaching*, *20*, 577–588. doi:10.1002/tea.3660200610

Yore, L. D., & Treagust, D. F. (2006). Current realities and future possibilities: Language and science literacy empowering research and informing instruction. *International Journal of Science Education*, *28*, 291. doi:10.1080/09500690500336973

Zeegers, P. (2001). Approaches to learning in science: A longitudinal study. *The British Journal of Educational Psychology*, *71*, 115–132. doi:10.1348/000709901158424

Zhang, F. Z., & Lidbury, B. A. (2006). It's all foreign to me: Learning through the language of genetics and molecular biology . In Johnston, I., & Peat, M. (Eds.), *Proceedings Assessment in Science Teaching and Learning, UniServe Science* (pp. 153–159). Australia: The University of Sydney.

ADDITIONAL READING

Bowling, B. V., Wang, L., Myers, M. F., Dean, G. E., & Markle, G. C. (2008). Development and evaluation of a genetics literacy assessment instrument for undergraduates. *Genetics*, *178*(1), 15–22. doi:10.1534/genetics.107.079533

Chen, Y., & Hoshower, L. B. (2003). Student evaluation of teaching effectiveness: An assessment of student perception and motivation. *Assessment & Evaluation in Higher Education*, *28*(1), 71–88. doi:10.1080/02602930301683

Fagen, A. P. (2003). *Assessing and enhancing the introductory science course in physics and biology: Peer instruction, classroom demonstrations, and genetic vocabulary*. Cambridge, Massachusetts: PhD, Harvard University.

Fazey, D. M. A., & Fazey, J. A. (2001). The potential for autonomy in learning: Perceptions of competence, motivation and locus of control in first-year undergraduate students. *Studies in Higher Education*, *26*(3), 345. doi:10.1080/03075070120076309

Lidbury, B., & Zhang, F. (2008). Comprehension of scientific language as a strategy to enhance learning and engagement for molecular biology students. *Australian Biochemist*, *39*(3), 10–13.

Ryan, R. M., & Deci, E. L. (2000). Intrinsic and extrinsic motivations: Classic definitions and new directions. *Contemporary Educational Psychology*, *25*(1), 54–67. doi:10.1006/ceps.1999.1020

Smith, M. K., Wood, B. W., & Knight, J. K. (2008). The genetics concept assessment: A new concept inventory for gauging student understanding of genetics. *CBE Life Sciences Education*, *7*, 422–430. doi:10.1187/cbe.08-08-0045

Zhang, F., Lidbury, B. A., Schulte, J., Bridgeman, A., Yates, B., & Rodger, J. (2008). Language difficulties in first year science - an interim report. Paper presented at the 2008 National UniServe Science Symposium: Assessment in Science Teaching and Learning, University of Sydney. Retrieved from http:// sydney.edu.au/

Zhang, F., Lidbury, B. A., Schulte, J., Bridgeman, A., Yates, B., & Rodger, J. (2009). In-discipline language support for first year university students in the sciences: Does it work? Paper presented at the 2009 International Conference of East-Asian Science Education, Taipei, Taiwan.

Zhang, F., Lidbury, B. A., Schulte, J., Yates, B., Bridgeman, A., & Rodger, J. (2010). Integrating Language Learning Practices in First Year Science Disciplines. *The International Journal of Learning*, *17*(4), 481–502.

KEY TERMS AND DEFINITIONS

Engagement: Purposeful and active interaction with learning tasks and willing adoption of new learning ideas and styles.

Genetics: A branch of the life sciences concerned with the inheritance of biological traits.

"Genetic" Language: The specialised language of the Genetics discipline that is unintelligible to the uninitiated. Comprises unique technical words and terms, as well as everyday English words with a different meaning.

Language Strategy/Approach/Intervention: Approaches to teaching and learning emphasising the comprehension of specialist scientific language as a first step to enhanced learning and greater grasp of complex concepts.

Molecular Biology: The scientific discipline associated with the manipulation and engineering of nucleic acids, as well as the study of proteins and gene expression.

Motivation: The intrinsic drive to achieve an outcome, in this case deep learning of genetics and/or molecular biology.

Traditional (Transmissive) Teaching: Academic (professor) centred educational style involving transmission of facts, ideas and concepts without involvement of students present in the lecture, tutorial or practical.

Chapter 5
Evaluating a Genetics Concept Inventory

Felicia Zhang
University of Canberra, Australia

Brett A. Lidbury
Australian National University, Australia

ABSTRACT

This chapter examines the reliability and validity of a subset of the Genetics concept inventory (S. Elrod) to discriminate good students from poor performing students in the undergraduate units Genetics and Molecular Biology taught at the University of Canberra, Australia. These two units went through a series of reforms since 2005. These reforms included the implementation of a number of online and tutorial language exercises and strategies designed to promote scientific language competence and subsequent genetics learning. The effect of these interventions was analyzed through grade and assessment performance comparisons with earlier traditionally taught Genetics cohorts as well using the genetic concept inventory. The genetic concept inventory questions used at the University of Canberra have been found to be reliable and valid according to a number of statistical tests.

INTRODUCTION

Molecular Biology and *Genetics* are taught to undergraduate students at the second and third year level of their University of Canberra degrees. The assumption is, particularly for the third year students who have passed previous units/subjects in fundamental chemistry, biology and biochemistry that the concepts and many of the details confronted in the more specialised discipline areas like genetics/molecular biology will be familiar. Anecdotally, this has not been true and the specialised language, for a majority of students, leads to a loss of engagement with the content of the unit and hence the learning outcomes. While there are issues around retention of knowledge from previously studied foundation units, it seems a bigger problem is that students

DOI: 10.4018/978-1-61350-062-0.ch005

find the lecture material, readings and other study material impenetrable.

Researchers have acknowledged the importance of language in science education for over 40 years. The history of thinking in this field has been reviewed by Wellington and Osborne, and these authors go so far to suggest that language is of primary significance to learning science, a situation that may have been deflected traditionally by the focus on practical/laboratory work in science curricula. Furthermore, previous research in looking at the language of science (Gardner; Gardner and Australian Science Education Project.; Pickersgill and Lock) suggested students have problems with both technical and non-technical vocabulary especially with the logical connectives. While these studies did not set out to examine the retention of scientific vocabulary, there was recognition that the comprehension of difficult scientific articles necessitates the retention and understanding of both types of vocabulary. These studies promoted the idea that language understanding, as well as reflective and active reading, are skills that must be taught.

The aim of this study was to investigate whether learning of the genetics/molecular biology unit content, and associated concepts, was enhanced through the application of educational techniques normally found in language learning. To determine the efficacy of this mode of teaching, quantitative and qualitative analyses were performed to compare the 2005 - 2008? (experimental) cohort with the 2001- 2004 student cohorts who studied *Genetics* under a traditional transmissive mode of delivery.

SURVEY PROCEDURES AND STUDENT PROFILE

In order to establish what students do know when reading scientific articles, a pre-project questionnaire was administered to 32 students consisting of 24 females and 8 male students in the 2005 cohort. The average age of the student body was 23 years of age. Questionnaire data reviewed that 21/32 students highlight the difficult words in the article and try to read and read again if they encounter something they do not understand. Most of the difficulties encountered were: (1) not understanding the terms, and (2) having difficulty in maintaining focus.

Students were also asked to make suggestions to improve their reading. Many of them requested shorter articles and a provision of terms, definitions and abbreviations. The pre-experiment questionnaire results suggested that this group of students might be using too few strategies in reading; and secondly when reading, too few modalities were used. In other words, students in this unit were rather passive in the way they learned. Not only did they appear to learn passively, it was also difficult to ascertain whether there was any active construction of knowledge involved at all. Learning seemed to involve a process of reading with a pen.

For the student body involved in this experiment, there were a large number of students who were school leavers. Guided by the research of Gardner (1972, 1977) and Pickersgill and Lock, (1991), a range of techniques were used to teach or reinforce active reading and subsequent reflection skills to students who were set a number of readings from the primary literature in genetics and molecular biology. Underpinning much of this process was a focus on translating scientific words/terms as if they were from a foreign language. Specific techniques employed in this study are outlined in the next section.

EXPERIMENTAL TECHNIQUES APPLIED TO ENCOURAGE LANGUAGE LEARNING

To encourage active learning, a number of techniques used in language teaching and education

Table 1. The language learning techniques used in the experiment

1. Small group discussion in tutorials
2. Breaking down long words to aid memory
3. Warming up activities such as match definitions to concepts
4. Online exercises
5. Providing lecture notes and exercises on *WebCT or Moodle (LMS)*
6. Online assessment
7. Connecting different concepts
8. Consistently revising previously learned concepts in tutorials through group work
9. Using the structure of scientific articles to guide the understanding of what is important and what is not in readings and tutorials through group work
10. Giving students opportunities to put forward their point of view in groups
11. Provide tutorial questions to stimulate tutorial discussions

were introduced over 13 weeks. The techniques employed are shown in Table 1.

During the 13 week semester in semester 2 2005, the educational lecturer who is the partner in this project attended every lecture and tutorial and participated as a co-teacher. To further enhance learning, a genetic learning website was set up in which weekly exercises such as crosswords, matching, fill-in-blanks exercises were provided for individual study. At the end of the experimental period, a post-experimental questionnaire was also administered to each student. This questionnaire consisted of questions related to the design of the unit in general, the 11 techniques listed in Table 1 and several related open-ended questions.

QUANTITATIVE MEASURES AND STATISTICAL ANALYSIS

To evaluate the effectiveness of the above-described techniques, the achievement of the 2005 (experimental) cohort was compared to earlier student cohorts (2001-2004) who studied *Genetics* under a traditional transmissive mode of delivery. Comparative performance on common assessment items over the 2001-2005 periods was investigated (descriptive statistics and unpaired Student t-test), and analysis on the progression of the 2001-2004 versus 2005 students to higher final grades (P / CR / CR / DI) (Chi-square $\chi 2$) was investigated. Comparison of student performance in *Genetics* (final mark from 100) with their overall degree achievement, as reflected by degree Grade Point Average (GPA), was the subject of correlation (r) analysis for each year cohort who studied *Genetics* under a traditional approach (2001-2004) or the language-centred approach (2005). Students were analysed individually and then assigned to grade/year cohorts for correlation analysis {Zhang, 2006 #804}.

At the end of the semester, a post experimental questionnaire was conducted and students were asked to evaluate the range of techniques used in the experiment. Interestingly while students found the provision of lecture materials online very helpful towards their study (with techniques 5, 6, 11 in Table 1 ranked first, second and fifth place), only 62% of the students found the online exercises on the Genetic learning web page useful. Other higher ranked (8, 10, 1, 7 and 9 in Table 1) involved the use of group work to revise previously learned concepts in tutorials and readings; to apply new techniques such as using the structure of scientific articles to guide reading and to discuss and obtain feedback through discussion in the

Table 2. Results of pre-and post- genetic inventory tests in 2008 for students enrolled in second year Genetics at The University of Canberra

Test	Average	Std Dev	Highest/Lowest (out of 70 marks)
Pre-test	39.74	23.55	55/5
Post-test	48.35	9.40	61.7/16.7

learning community. Most students viewed the learning through group discussions as effective. In contrast, language learning specific techniques such as breaking down words to aid memory and using warm up activities were only considered by some students as effective means of learning.

The answers to the open-ended questions in the post-experimental questionnaire affirmed the positive impact of these techniques on students. Many students felt:

- That the breakdown skills used to read journal articles were useful in every other subject;
- That building upon genetics in previous units helped them in completing other subjects such as Pathobiology;
- They develop skills in understanding scientific papers (using language exercises);
- They became much more dependent on finding answers for themselves rather than relying on being told all the time.

Since the introduction of language intervention into the second year undergraduate genetics subject at the University of Canberra in 2005, subsequent teaching developments on learning enhancement by students studying genetics, statistical analysis methods have continued to be applied to assess the impact of language learning interventions on student performance.

In 2008, scientific language competence was surveyed on-line at the beginning of semester, and again in the final week of semester (identical survey). The language surveys comprised two parts; a gap-fill (Cloze) exercise and a multi-answer multiple-choice survey that also asked students how confident they were in their answers. In addition to this direct survey of language competence, a 12-question genetics concept inventory was also run at the beginning and at the very end of semester. The 12 question genetics concept inventory was a subset of 38 questions from Elrod's genetics concept inventory (GenCI) (Elrod, 2007) developed and published in 2007 at California Polytechnic State University. These 12 questions chosen for implementation are all multiple-choice questions.

During the first semester of 2008, 51 students took the selected 12 items from the objective GenCI test devised by Elrod (2007) as both pre and post UCGenCI tests. The results of the pre- and post-tests are shown in Table 2 below.

Table 3 shows the results for the questions that were answered with greater than 70% correct answers. These indicate concept areas where a solid majority of students demonstrated understanding. There were 4 out of 12 questions in this category and these questions tended to be about reproductive genetic concepts. This confirms one of the conclusions from Elrod's paper that students tended to have better understanding of reproductive genetic concepts (Elrod, 2007).

Table 4 shows the results for the questions that were answered with a moderate degree of correctness in the pretest (between 48-70%) based on pre-test ranges in 2008. These indicate concept areas where only between 48-70% of the students' demonstrated understanding. There were 5 out of 12 questions in this category. This also confirms with data analyzed in Elrod's paper. The concept areas in Tables 4 and 5 tend toward the more

Table 3. Questions answered with a high degree of correctness (greater than 70% correct) in the pretest in 2008

UC items	Corresponding items in GENCI (Elrod)	Content	P value	% Correct	D-value
6	23	If a skin cell starts with 24 chromosomes, what number of chromosomes will it have at the end of cell division?	0.69	68.63	0.35
11	11	Sex in humans (and other mammals) is determined by:	0.76	76.47	0.65
7	Changed.	If a human epithelial cell contains 46 chromosomes, how many chromosomes are present in the human gametes (egg or sperm) after sexual maturity?	0.88	88.24	0.35
9	7	In which of the following cell types within your body are genes found?	0.92	92.16	0.29

Table 4. Questions answered with a moderate degree of correctness (pre-test between 48-70%) in pre-test, 2008

UC items	Corresponding items in GENCI (Elrod)	Content	P value	% Correct	D-value
2	9	What are chromosomes made of?	0.52	52.27	0.41
10	33	What is meant by "genetic code"?	0.49	54.24	0.41
4	12	When chromosomes are drawn as an X structure, each of the lines in the X represent:	0.59	58.82	0.41
12	35	Which of the following molecules are the products of transcription?	0.61	60.78	0.65
8	13	If an organism is diploid, and has 16 chromosomes, how many sets of homologous chromosomes does it possess?	0.63	62.75	0.65

Table 5. Questions answered with a low degree of correctness (pre-test less than 48%) in 2008

UC items	Corresponding items in GENCI (Elrod)	Content	P value	% Correct	D-value
1	16	What is an allele?	0.29	29.41	0.24
3	10	Where are chromosomes located within eukaryotic cells of animal origin?	0.34	34.37	0.25
5	14	Which of the following is true of homologous chromosomes?	0.35	35.29	0.41

abstract as well as the molecular. The p-value and D-value measure similarity to Elrod's findings.

Table 5 shows the results for the questions that were answered with a low degree of correctness (less than 48%) based on pre-test ranges. These indicate concept areas where less than half of the students demonstrated understanding. There were 3 out of 12 questions in this category. This also confirms with data in Elrod's paper except for question 3 in UCGenCI's test which corresponds to Question 10 in Elrod's paper. At UC, students seemed to be confused about the location of chromosomes within eukaryotic cells. The UC data seems to agree with Elrod's conclusion that students tend to have less understanding in more abstract as well as molecular areas of genetics (Elrod, 2007).

Post-test UCGenCI Scores

Table 6 shows the post-test GENCI scores which demonstrated that students made a statistically

Table 6. Group gains between pre-post-tests for each item on a genetic concept inventory for undergraduate students studying a genetics unit at The University of Canberra

Items	Pre-test % Correct	Post-test % Correct	Gains %	Statistical significance with p=0.05 level
1	29.41	47.06	17.65	0.05*
2	52.27	61.43	9.16	0.04*
3	34.37	37.65	3.27	0.6
4	58.82	50.98	-7.84	0.35
5	35.29	64.78	29.41	0.00**
6	68.63	86.27	17.65	0.01**
7	88.24	90.20	1.91	0.74
8	62.75	56.86	-5.88	0.54
9	92.16	96.08	3.92	0.42
10	49.02	47.06	-1.96	0.77
11	76.47	86.27	9.80	0.10
12	60.78	70.59	9.80	0.17

Key: * suggests that the difference in gain for the item in the GENCI (UC) is statistically significant at p=0.05 level.
**: suggests that the difference in gain for the item in the GENCI (UC) is statistically significant at p=0.01 level

significant increase in their understanding of items 1, 2, 4 and 5, but there is a reduction in understanding in items 4, 8 and 10. For the rest of the items, the gains were not statistically significant in size.

With the UCGenCI test, we also would like to compare student's final exam results with the UCGenCI post-test data. In order to do this, a series of statistical tests to evaluate the reliability and validity of the UCGenCI test as an assessment instrument for this cohort of students were conducted. The attributes we tested were item difficulty, item discrimination and reliability (Ding, Chabay, Sherwood, & Beichner, 2006).

STATISTICAL EVALUATION OF UCGENCI

Item Difficulty Index

The item difficulty index 'P' is a measure of the difficulty of a single test question (see Figure 1). It is calculated by taking the ratio of the number (N_1) of correct responses on the question to the total number (N) of students who attempted the question:

$$P = \frac{N_1}{N}$$

This difficulty index P might more meaningfully have been called the "easiness index," since it is simply the proportion of correct responses on a particular question. The greater the P value is, the higher the percentage of respondents giving the correct answer and the easier this item is for this population. The range for the difficulty index P value is [0, 1]. If the P value is 0, then no one can answer the question correctly; on the other hand, if the P value is 1, then everyone can correctly answer this question. Under most circumstances, such extremes should be avoided in a test. A noteworthy aspect of the difficulty index is that the P value depends on the particular population taking the test. As an example, consider the first question in this test. Of the 51 UC students (who did the post-test), 29.41% of students answered

Table 7. Difficulty index (P value) and item discrimination index (D-value) for the pre and post-tests (new corrected) in 2008

UC items	Corresponding items in GENCI (Elrod)	Content	P value pre test	P value post-test	D-value pre test	D-value post-test
1	16	What is an allele?	0.29	0.47	0.24	0.53
2	9	What are chromosomes made of?	0.52	0.61	0.41	0.25
3	10	Where are chromosomes located within eukaryotic cells of animal origin?	0.34	0.38	0.25	0.21
4	12	When chromosomes are drawn as an X structure, each of the lines in the X represent:	0.59	0.51	0.41	0.53
5	14	Which of the following is true of homologous chromosomes?	0.35	0.65	0.41	0.29
6	23	If a skin cell starts with 24 chromosomes, what number of chromosomes will it have at the end of cell division?	0.69	0.86	0.35	0.29
7	Changed.	If a human epithelial cell contains 46 chromosomes, how many chromosomes are present in the human gametes (egg or sperm) after sexual maturity?	0.88	0.90	0.35	0.18
8	13	If an organism is diploid, and has 16 chromosomes, how many sets of homologous chromosomes does it possess?	0.63	0.57	0.65	0.18.
9	7	In which of the following cell types within your body are genes found?	0.92	0.96	0.29	-0.12
10	33	What is meant by "genetic code"?	0.49	0.47	0.41	0.18
11	11	Sex in humans (and other mammals) is determined by:	0.76	0.86	0.65	0.35
12	35	Which of the following molecules are the products of transcription?	0.61	0.71	0.65	0.47
		average	**0.59**	**0.66**	**0.42**	**0.28**

correctly, so the difficulty index for the first question is 0.29 (Table 7).

In evaluating this test, we chose a widely adopted criterion that requires the difficulty index value to be between 0.3 and 0.9 (Doran, 1980), a range which included the optimum value of 0.5. The averaged difficulty index 'P' is 0.59 for the pre test and 0.66 for the post-test in Table 7, which also fall into the criterion range [0.3, 0.9].

Item Discrimination Index

The item discrimination index (D) is a measure of the discriminatory power of each item in a test. In other words, it measures the extent to which a single test item distinguishes students who know the material well from those who do not. On a test item with a high discrimination index, students with more robust knowledge will usually answer correctly, while students whose understanding is weaker will usually get the item wrong. In contrast, a flawed test question might lead more thoughtful students to give answers that are judged wrong, while students who think less deeply give a correct answer. If a test contains many items with high discrimination indices, the test itself can be useful in separating strong students from weak students in a specific test domain.

In order to calculate the item discrimination index (D), we divide the whole sample of students into three different groups of equal size, a high group (H), a middle group and a low group (L), based on whether an individual total score is higher or lower than the median total score of the entire sample. For a specific test item, one counts the number of correct responses in both

Figure 1. Item difficulty indices for each item in the UCGenCI pre-and post tests in 2008

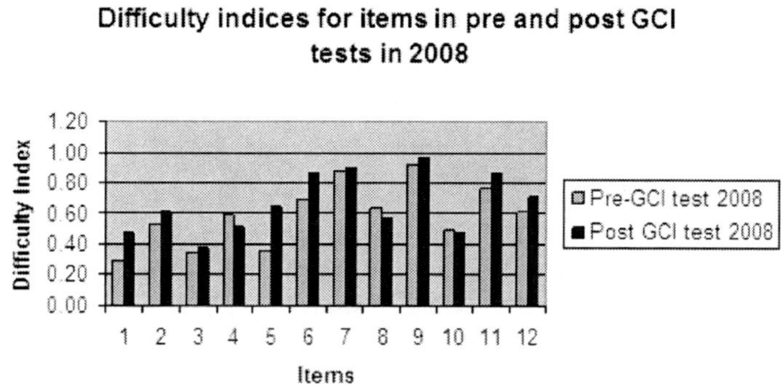

H and L groups: namely, N_h and N_l If the total number of students who take the test is N, then the discrimination index (D) of this item can be calculated as

$$D = \frac{Nh(top33.3\%) - Nl(bottom33.3\%)}{N/3}$$

The possible range for the item discrimination index D is [−1, +1], where +1 is the best value and −1 is the worst value. In the extreme ideal case all students in the high group get the item correct and all students in the low group get it wrong, giving a discrimination index D of +1. In the worst case the situation is reversed: everyone in the low group answers correctly, and everyone in the high group gets it wrong. In this case the discrimination index D will be −1. These extreme cases are unlikely, but it is important to eliminate any items with negative discrimination indices. An item is typically considered to provide good discrimination if $D \geq 0.3$ (Doran, 1980). Items with a discrimination index lower than 0.3 but greater than 0 are not necessarily bad, but a majority of the items in a test should have relatively high discrimination index values to ensure that the test is capable of distinguishing between strong and weak mastery of the material. The average $\bar{D}$ value for the items in the pre-test was 0.48 which satisfies the criterion that $\bar{D} \geq 0.3$ (see Figure 2). However, the average $\bar{D}$ value for the items in the post-test was 0.28 that is slightly below the criterion that $\bar{D} \geq 0.3$. This means that the higher the discrimination index for an item, the better that item is for discriminating between good students from bad students.

For instance, in the tests we administered at UC (Table 7), in the pre-test, only items 3 and 9 have values below the criterion level. They are 0.25 and 0.29 respectively. However, in the post-test, items 1, 4, 11 and 12 remained as good discriminators while item 8's discriminatory power is reduced. Item 9 (Item 7 in Elrod's paper) in the UC test has a negative index. This suggests that it is not a good discriminator of strong students from weak students.

What do the Scores for the Poor Discriminators Reflect?

In interpreting the data in the evaluation, we need to bear in mind that this cohort of students was second year Genetics students who had already completed the introduction unit in Biology. If we look at the Tables 5 and 6 together, the items with good discriminatory power (1, 5, 6, 11 and 12 except 4) also were items with which have reasonably large post-pre test gains. For example, item 1 has a gain of 17.65%. However, in the Pre-test, most students found this item difficult with only 29.41% getting it correct. In the post-test, slightly less than 50% of the students got this item

Figure 2. Item discrimination index for items in the UCGenCI pre-test in 2008

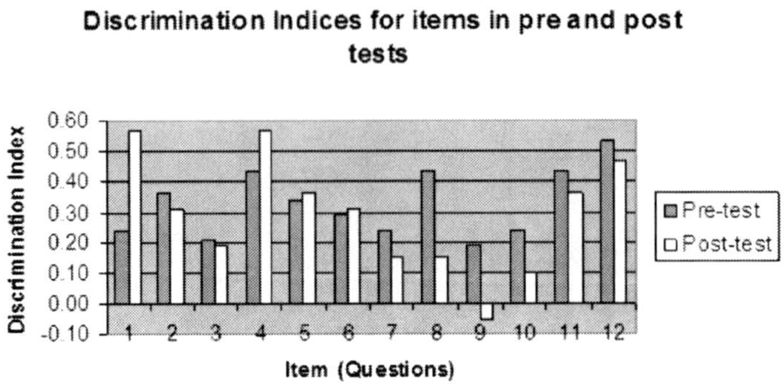

correct. Thus with a high discrimination index of 0.53, it can be predicted that the better students were able to correctly answer this item. The same reasoning can be used to explain the discriminatory power of item 4.

The large decreases in the pre and post-tests' D-values of items 5, 6, 11 and 12 (see Table 6) corresponds to the large increases in gains. For instance, item 5 has moved from being a difficult item (%correctness < 48%) to being a medium level difficulty item (%correctness between 48-70%). Items 11 and 12, however, were relatively easy in both pre and post-tests.

The difficulty of items 8 and 10 remained the same in both pre and post-tests. This means both good and poor students found these two items difficult even after instruction. Question 8 (Elrod's 13) deals with the nature of homologous chromosomes. In the pre-test the percentage of correctly answering this question was 62.75% and yet in the post-test, the percentage went down to 56.86%. Elrod's preliminary finding also stated that students have problems with the nature of homologous chromosomes. Clearly after 3 semesters of instruction in Biology and genetics, the nature of homologous chromosomes (as well as chromatids) is still not well understood within the UC cohort.

Elrod (2007) designed items 8 (Elrod's 13) and 5 (Elrod's 14) as a pair. In the UC GenCI post-test, the fact that students improved their gains significantly in item 5 and yet lost gains in item 8 only confirms that they did not understand the nature of homologous chromosomes.

Point Biserial Coefficient

The point biserial coefficient, sometimes referred to as the reliability index for each item is a measure of consistency of a single test item with the whole test. It reflects the correlation between students' scores on an individual item and their scores on the entire test, and is basically a form of the correlation coefficient. The point biserial coefficient has a possible range of [−1, +1]. If an item is highly positively correlated with the whole test, then students with high total scores are more likely to answer the item correctly than are students with low total scores. A negative value indicates that students with low overall scores were the most likely to get a particular item correct and is an indication that the particular test item is probably defective. For a detailed discussion on how to calculate the point biserial coefficient, see Ding et al paper on evaluating another concept inventory (Ding, et al., 2006). To calculate the point biserial coefficient for an item, one needs to calculate the correlation coefficient between the item scores and total scores. A student's score on one item is a dichotomous variable which can have only two values: 1 _correct_ or 0 _wrong_. Scores for the whole test usually can be viewed

Figure 3. Point biserial coefficients for the 12 items in the pre and post UC GENCI tests in 2008 for 51 students

Items	Pretest	Posttest
Q1	0.03	0.49
Q2	0.67	0.4
Q3	0.48	0.26
Q4	0.43	0.26
Q5	0.45	0.4
Q6	0.4	0.34
Q7	0.67	0.13
Q8	0.53	0.39
Q9	0.63	-0.04
Q10	0.31	0.15
Q11	0.59	0.54
Q12	0.5	0.51
average	0.47	0.32

as continuous _if the test has a relatively large number of items—say, _20_.

The criterion widely adopted (Kline, 1986) for measuring the "consistency" or "reliability" of a test item is ≥ 0.2. Items with point biserial coefficient lower than 0.2 can still remain in a test, but there should be few such items. Figure 3 lists the point biserial coefficient for the 12 items in the UC GENCI pre and post-tests in 2008 for a small cohort of 51 students.

The pre-test results reviewed that only item 1 has a point biserial coefficient of less than ≥ 0.2. All the other items have satisfactory coefficients indicating that almost all items in the test are reliable and consistent.

The post-test results, however, reviewed that items 7, 9 and 10 are a bit problematic in that they have below 0.2 level of the coefficients. Item 9 is of particular concern as it has a negative coefficient. This confirms the negative discrimination index for this item in Table 7.

Ferguson's Delta

Ferguson's delta Ferguson's delta is another whole-test statistic. It measures the discriminatory power of an entire test by investigating how broadly the total scores of a sample are distributed over the possible range. If a test is designed and employed to discriminate among students, one would like to see a broad distribution of total scores. If a test has Ferguson's delta greater than 0.9, the test is considered to offer good discrimination (Kline, 1986). Ferguson's delta for the UC GenCI pre test is 0.9 and for the post-test is 0.86.

CONCLUSION

In conclusion, the reliability and discriminatory power of the UC GenCI was evaluated by four statistical tests, two of them focus on individual items and two of which focus on the test as a whole. The results are, which are summarized in Table 9, indicate that the UC GenCI is a reliable test with adequate discriminatory power (with the exception of the post-test) to assess student learning in the unit Genetics.

Table 9. Evaluation of the UC GenCI using four statistical tests

Test statistic	Possible values	Desired values	UCGENCI Pre-test	UCGENCI Post-test
Item difficulty index P	[0, 1]	≥ 0.3	0.59	0.66
Item discrimination index D	[-1, 1]	≥ 0.3	0.42	0.28
Point biserial coefficient	**[-1, 1]**	≥ 0.2	**0.47**	**0.32**
Ferguson's delta	**[0, 1]**	**0.9**	**0.9**	**0.86**

Implications need to be discussed, with reference to the wider literature.

REFERENCES

Baldwin, J. A., Ebert-May, D., & Burns, D. J. (1999). The development of a college biology self-efficacy instrument for nonmajors. Science Education, 83(4), 397-408. doi: 10.1002/(SICI)1098-237X(199907)83:4<397::AID-SCE1>3.0.CO;2-#

Ding, L., Chabay, R., Sherwood, B., & Beichner, R. (2006). Evaluating an electricity and magnetism assessment tool: Brief electricity and magnetism assessment. *Physics Education Research, 2*(1).

Doran, D. (1980). *Basic measurement and evaluation of science instruction*. Washington, DC: NSTA.

Ellem, G. K., & McLaughlin, E. A. (2005). Tales from the coalface: From tragedy to triumph in a blended learning approach to the teaching of 1st year biology. Paper presented at the 2005 National UniServe Science Symposium: Blended Learning in Science Teaching and Learning, Sydney, NSW. Retrieved from http:// sydney.edu.au/

Elrod, S. (2007). *Genetics concept inventory*. Retrieved from http:// bioliteracy.net/ Readings/ papersSubmittedPDF/ Elrod.pdf

Fagen, A. P. (2003). *Assessing and enhancing the introductory science course in physics and biology: Peer instruction, classroom demonstrations, and genetic vocabulary*. Cambridge, Massachusetts: PhD, Harvard University.

Gardner, P. L., & Australian Science Education Project. (1972). *Words in science: An investigation of non-technical vocabulary difficulties amongst form I, II, III, & IV science students in Victoria, a report*. Melbourne, Australia: Australian Science Education Project.

Gardner, P. L. (1977). Logical connectives in science: A summary of the findings. *Research in Science Education*, *7*(1), 9–24. doi:10.1007/BF02643108

Garvin-Doxas, K., & Klymkowsky, M. W. (2008). Understanding randomness and it impact on student learning: Lessons learning from building the biology concept inventory (bci). *CBE Life Sciences Education*, *7*(2), 227–233. doi:10.1187/cbe.07-08-0063

Hewson, W. P., Robert Tabachnick, B., Zeichner, K. M., Blomker, K. B., Meyer, H., & Lemberger, J. (1999). Educating prospective teachers of biology: Introduction and research methods. *Science Education*, *83*(3), 247–273. doi:10.1002/(SICI)1098-237X(199905)83:3<247::AID-SCE1>3.0.CO;2-C

Hohenshell, L. M., & Hand, B. (2006). Writing to learn strategies in secondary school cell biology: A mixed method study. *International Journal of Science Education*, *28*(2/3), 261. doi:10.1080/09500690500336965

Khodor, J., Halme, D. G., & Walker, G. C. (2004). A hierarchical biology concept framework: A tool for course design. *Cell Biology Education*, *3*(Summer), 111–121. doi:10.1187/cbe.03-10-0014

Kline, P. (1986). *A handbook of test construction: Introduction to psychometric design*. London, UK: Methuen.

Klymkowsky, M. W., & Garvin-Doxas, K. (2008). Recognizing student misconceptions through ed's tools and the biology concept inventory. *PLoS Biology*, *6*(1), e3. .doi:10.1371/journal.pbio.0060003

Lawson, A., E. (2002). Sound and faulty arguments generated by preservice biology teachers when testing hypotheses involving unobservable entities (Vol. 39, pp. 237-52).

Lidbury, B., & Zhang, F. (2008). Comprehension of scientific language as a strategy to enhance learning and engagement for molecular biology students. *Australian Biochemist*, *39*(3), 10–13.

McDaniel, C. N., Lister, B. C., Hanna, M. H., & Roy, H. (2007). Increased learning observed in redesigned introductory biology course that employed web-enhanced, interactive pedagogy. *CBE Life Sciences Education*, *6*(3), 243–249. .doi:10.1187/cbe.07-01-0003

Michael, J. (2007). Conceptual assessment in the biological sciences: A national science foundation-sponsored workshop. *Advances in Physiology Education*, *31*(4), 389–391. .doi:10.1152/advan.00047.2007

Michael, J., McFarland, J., & Wright, A. (2008). The second conceptual assessment in the biological sciences workshop. *Advances in Physiology Education*, *32*(3), 248–251. .doi:10.1152/advan.90122.2008

Miller, V., Gregg, G., & Kelly, B. (2000). Peer assisted study sessions (pass) in a first year biology teaching program-strategies developed by undergraduate pass leaders. Paper presented at the Effective Teaching and Learning at University Conference, Duchesne College, The University of Queensland. Retrieved from http:// www.tedi.uq.edu.au/ conferences/

Newport, E. (1991). Contrasting conceptions of the critical period for language . In Carey, S., & Gelman, R. (Eds.), *The epigenesis of mind: Essays on biology and cognition*. Hillsdale, N.J.: Erlbaum.

Odom, A. L., & Barrow, L. H. (1995). Development and application of a two-tier diagnostic test measuring college biology students' understanding of diffusion and osmosis after a course of instruction. *Journal of Research in Science Teaching*, *32*(1), 45–61. doi:10.1002/tea.3660320106

Odom, A. L., & Kelly, P. V. (2001). Integrating concept mapping and the learning cycle to teach diffusion and osmosis concepts to high school biology students. *Science Education*, *85*(6), 615–635. .doi:10.1002/sce.1029

Pearsall, N. R., Skipper, J. E. J., & Mintzes, J. J. (1997). Knowledge restructuring in the life sciences: A longitudinal study of conceptual change in biology. Science Education, 81(2), 193-215.

Piattelli-Palmarini, M. (1989). Evolution, selection and cognition: From "learning" to parameter setting in biology and in the study of language. *Cognition*, *31*(1), 1–44. doi:10.1016/0010-0277(89)90016-4

Pickersgill, S., & Lock, R. (1991). Student understanding of selected non-technical words in science. *Research in Science & Technological Education*, *9*(1), 71–79. doi:10.1080/0263514910090107

Roy, H. (2003). Studio vs interactive lecture demonstration–effects on student learning. Bioscene . *Journal of College Biology Teaching*, *29*(1), 3–6. Retrieved from http:// acube.org/ pdf/bioscene/ v29-1.pdf.

Shepherd, G. M. (1994). *Neurobiology* (3rd ed.). Oxford: Oxford University Press.

Smith, M. K., Wood, B. W., & Knight, J. K. (2008). The genetics concept assessment: A new concept inventory for gauging student understanding of genetics. *CBE Life Sciences Education*, *7*, 422–430. doi:10.1187/cbe.08-08-0045

Wallace, J., & Mintzes, J. (1990). The concept map as a research tool: Explaining conceptual change in biology. *Journal of Research in Science Teaching*, *27*(10), 1033–1052. doi:10.1002/tea.3660271010

Wandersee, J. H. (1988). The terminology problem in biology education: A reconnaissance. *The American Biology Teacher*, *50*(2), 97–100. Retrieved from http:// www.jstor.org/.

Wellington, J. J., & Osborne, J. (2001). *Language and literacy in science education*. Buckingham, UK & Philadelphia, PA: Open University.</>

ADDITIONAL READING

Winter, B., Breitenstein, C., Mooren, F. C., Voelker, K., Fobker, M., & Lechtermann, A. (2007). High impact running improves learning. *Neurobiology of Learning and Memory*, *87*(4), 597–609. doi:10.1016/j.nlm.2006.11.003

Zhang, F., Lidbury, B., Schulte, J., Bridgeman, A., Yates, B., & Rodger, J. (2009). A cross-disciplinary approach to language support for first year students in the science disciplines. Paper presented at the HERDSA 2009, Charles Darwin University, Darwin.

Zhang, F., & Lidbury, B. A. (2006). It's all foreign to me: Learning through the language of genetics and molecular biology. Paper presented at the 2006 National UniServe Science Symposium: Assessment in Science Teaching and Learning, The University of Sydney, Sydney. Retrieved from http:// sydney.edu.au/

KEY TERMS AND DEFINITIONS

Ferguson's Delta: Ferguson's delta is another whole-test statistic. It measures the discriminatory power of an entire test by investigating how broadly the total scores of a sample are distributed over the possible range. If a test is designed and employed to discriminate among students, one would like to see a broad distribution of total scores. I If a test has Ferguson's delta greater than 0.9, the test is considered to offer good discrimination.(Kline, 1986).

The Item Difficulty Index (P): A measure of the difficulty of a single test question.

The Item Discrimination Index (D): A measure of the discriminatory power of each item in a test. In other words, it measures the extent to which a single test item distinguishes students who know the material well from those who do not. On a test item with a high discrimination index, students with more robust knowledge will usually answer correctly, while students whose understanding is weaker will usually get the item wrong. In contrast, a flawed test question might lead more thoughtful students to give answers that are judged wrong, while students who think less deeply give a correct answer. If a test contains many items with high discrimination indices, the test itself can be useful in separating strong students from weak students in a specific test domain.

The Point Biserial Coefficient: Sometimes referred to as the reliability index for each item is a measure of consistency of a single test item with the whole test. It reflects the correlation between students' scores on an individual item and their scores on the entire test, and is basically a form of the correlation coefficient. The point biserial coefficient has a possible range of [−1, +1]. If an item is highly positively correlated with the whole test, then students with high total scores are more likely to answer the item correctly than are students with low total scores. A negative value indicates that students with low overall scores were the most likely to get a particular item correct and is an indication that the particular test item is probably defective.

Chapter 6
Language Support for First Year Human Physiology and Biology

Karen Mate
University of Newcastle, Australia

John Rodger
University of Newcastle, Australia

Brett Lidbury
Australian National University, Australia

ABSTRACT

This chapter examines student performance in Human Physiology and Biology in response to a number of interactive language-focused and career-oriented interventions that were integrated into both lecture and tutorial formats. The study took place at the University of Newcastle, Australia and involved around 1000 students on two campuses. Although it was not possible to demonstrate a clear overall improvement in student performance as a result of these interventions, the activities were viewed positively by the majority of students. Arguably, the major impact of the language and other interventions was in the level of student motivation and engagement with the course topic. Contemporary university students are highly focused on the relevance of a course to their chosen career path, suggesting that language-based teaching and learning strategies would be most effective if placed within a vocation-specific context.

INTRODUCTION

Learning the language of science is a major obstacle to many students in first year human physiology and biology. It is particularly difficult for those students who have not studied science subjects previously or recently, and consequently struggle to recognize and differentiate concepts. These issues are exacerbated by the diversity of the student population in large first year courses in relation to background knowledge, their degree program and career path.

Students from the allied health programs (occupational therapy, speech pathology, nutrition/dietetics and podiatry) must complete first year human physiology as part of their program requirements for accreditation, but consistently perform poorly due largely to their lack of background in

DOI: 10.4018/978-1-61350-062-0.ch006

science (McCleary et al, 1999). It is obvious to those who teach human physiology that a basic understanding of the biochemical processes, cells and tissues that contribute to the functioning of human organ systems is essential in any clinical context. Many students, however, fail to recognize the relevance of physiological and biochemical processes that operate to maintain and integrate the specific organ systems that contribute to a healthy human body. Students often become overwhelmed with course content and question its relevance with comments such as "Why do I need to know all this? I just want to be a podiatrist/ speech pathologist/occupational therapist etc".

The situation for students undertaking first year biology, and student attitudes and engagement with particular subject matter in biology is similarly diverse. Students are enrolled in a variety of degree programs many of which require first year biology (e.g. Psychology, BSc Biology major, Biotechnology, Environmental Science and Management and Environmental Engineering). The unit which is the focus of the current chapter provides an introduction and overview of the animal kingdom and the principles of animal function. Inevitably this introduces the student to a very large array of new and often confusing technical terms and animal names in addition to the special uses and meanings certain 'normal' English words take on in biology (e.g. model, control, significant). As with the health programs, questions of relevance to the students' specific program of study are often expressed, "Do I really need to know these funny names, etc?"

The aim of this study was to develop teaching methods and tools that increased students' ability to see the interconnectedness between language and biological problems and situations. Students were also encouraged them to develop their own techniques to deal with language issues in biology, based on the resources provided. Some of the language-oriented techniques implemented were breaking down long words by identifying prefixes and suffixes,

1. exploring the roots and origin of words,
2. matching exercises with scientific terms and definitions,
3. flashcards for vocabulary revision and
4. role playing for students to practise conveying complex scientific information to their peers and the public.

Strategies were also developed to increase the motivation for embracing the language and concepts introduced. Vocation-specific clinical examples were presented as a "Disease of the Week" package in the human physiology course to enable students to put basic physiological information in a broader context with relevance to their future career. A series of regular online assessments were also introduced to provide consistent and immediate feedback to students on their progress. In the case of the animal function unit, examples were provided throughout the lecture material of how the particular animal group or its function related to:

1. Everyday life examples (e.g. Why the life style of some animals makes them dangerous parasites, or why certain stages in development are so critical for producing a normal baby);
2. Biodiversity and the environment (e.g. Kangaroo breeding and life in the Australian arid zone), or
3. How contemporary biology was engaging with evolutionary biology and taking it beyond anatomy and fossils (e.g. The major changes in understanding of animal evolution and relationships due to molecular biology).

METHODS EMPLOYED TO SUPPORT LEARNING IN HUMAN PHYSIOLOGY

Human Physiology is a first year undergraduate subject taught over two semesters. It is arbitrarily split into two parts (*Part 1* and *Part 2*) that run consecutively each year. Each part is delivered concurrently at two locations (Callaghan and Ourimbah, Newcastle, New South Wales (NSW), Australia) and utilize the same assessment items and criteria. Callaghan is the main campus of the University of Newcastle with and enrolment of approximately 17,000 full time student equivalents (FTSE). The Ourimbah campus is located on the central coast of NSW, approximately 100km south of the main campus. It is a multi sector campus offering university, TAFE and community college programs. There are approximately 3,000 FTSE studying at the University of Newcastle Ourimbah campus, many of whom study part-time.

Disease of the Week

Modules targeting a particular disease or condition that related to course content were introduced in 2009 via the subject Learning Management System (LMS; i.e. *Blackboard*). Each "Disease of the Week" module included a "user-friendly" introduction that provided a summary of the disease or condition, the relevance of related lecture content and the role of allied health professionals in treatment with vocation-specific references. The introduction was written in language aimed at an educated lay audience, avoiding unnecessary jargon and providing clear explanations of any technical terms. Where possible, the scientific papers that were cited in the introduction were also posted on the subject LMS to encourage students to take the next step in furthering their physiology-related language skills by reading on topics of particular interest to the student. These articles also introduced students to peer-reviewed scientific articles as a reliable source of information and as the major source of communication for new biomedical and health-related research for those working in the field.

Topics for "Disease of the Week" were generally chosen to be relatively common conditions that students may have heard of previously, with a variety of treatment modalities relevant to the study programs of the student cohorts. For example, the role of intermediate filaments and the cytoskeleton in maintaining cell shape, stabilizing the position of the organelles within the cell and stabilizing the position of the cell with respect to other cells is demonstrated by the module on *Epidermolysis bullosa (EB)*. A mutation in the gene coding for keratin proteins in intermediate fibres of superficial skin layers can result in blistering of the skin in response to even minor trauma. Individuals with *EB* may require treatment from:

- **Dieticians:** who have an understanding of the manifestations of EB and its associated nutritional problems to develop a strategy to ensure nutritional status (Birge, 1995; Bello et al, 2003).
- **Physiotherapists and Occupational Therapists:** to help prevent contractures (i.e. abnormal shortening of muscle tissue) and to maintain muscle tone through the use of exercise and special splints (Bello et al., 2003)
- **Podiatrists:** who can play a role in minimizing the blistering associated with EB by ensuring appropriate and well-fitting footwear that provides both mobility and protection (Kahn et al, 2002).
- **Speech Pathologists:** who may assist in management of communication problems in instances where oral scarring has occurred (Bello et al, 2003).

In this example it was also pointed out that biomedical students in particular may be interested in the brief account of how EB was discovered to be due to a keratin defect and the contradictory results that were reported during this process (Fuchs, 1996).

Other modules included Calcium Homeostasis and Osteoporosis, Mitochondrial Disease, Cystic Fibrosis, Multiple Sclerosis, Neurotoxins, and

Spinal Cord Injury. The information presented in "Disease of the Week" was not examined.

Tutorial/PASS Activities

Learning support was provided at the Callaghan campus via a Peer Assisted Study Session (PASS) for 1 hr/week. Participation in PASS was based on students' self selection and was voluntary. Student leaders of PASS have achieved either a Distinction or High Distinction in human physiology during the past 1-3 years. Student leaders are trained in active learning strategies and are provided with sample resources including "Mix and Match" exercises, "Fill in the Blanks", crosswords and concept mapping exercises (see examples below). Similar activities are used in a more traditional tutorial style session at the Ourimbah campus. These sessions have not been changed substantially during the course of this study.

The "Mix and Match" (see Table 1) and "Fill in the Blanks" (see Box 1) exercises were primarily aimed at improving language comprehension of topics presented in lectures during the previous week. The activity was done in paper format in small groups during the PASS or tutorial, and answers were then shared with the class. During the latter part of the semester, crosswords were also used during PASS and tutorials to reinforce some of the more complex language and concepts associated with lectures on the nervous system. All of these activities covered key language and concepts that had to be mastered in order to pass the course, but the activities themselves did not form part of the assessment.

Online Assessments

Regular online assessments were introduced during 2009 via the subject's LMS. Since many students are using an online LMS for the first time during this subject there was one practice quiz followed by four compulsory online assessments. Each online assessment consisted of 10 multiple choice questions that tested basic understanding of the language and concepts that had been presented during the previous 2-3 weeks. Students received feedback immediately after submission of the test including their score and the correct answer for any question that they had answered incorrectly. Each part is delivered concurrently at two locations (Callaghan and Ourimbah) and utilizes the same assessment items and criteria.

During 2008, *Human Physiology Part 1* and *Part 2* consisted of lectures with learning support provided by PASS or tutorial sessions, depending upon the campus location. 473 and 389 students participated in the unit on the Callaghan campus in semester 1 and 2 in 2009. On Ourimbah campus, 174 and 180 students participated in the unit in both semesters in 2009. Assessment consisted of two written mid-semester exams (each contributing 15% of final mark) and a final written exam (70% of final mark). Feedback to students on their progress during the semester was limited to the scores obtained on the written mid-semester exams. Although individual feedback on mid-semester exams was offered by appointment, very few students took advantage of this opportunity. During 2009, four online assessments (each contributing 5% of final mark) were introduced to both *Part 1* and *Part 2* of *Human Physiology*, in addition to a written mid-semester exam (worth 20%) and a final written exam (worth 60%). This was intended to be formative assessment to keep them on track. The "Disease of the Week" modules were incorporated into *Human Physiology Part 1* only.

IMPACT ON STUDENT PERFORMANCE IN HUMAN PHYSIOLOGY

Student data from Callaghan and Ourimbah was examined separately, as the cohorts differ in several aspects and historically perform very differently.

Table 1. Match the subject with its building blocks

Subject ("To be built")	"Building Blocks"
1. Life (Organic Complexity)	A. Systems (eg muscle, respiratory)
2. Atoms (Stuff of the Universe)	B. Elements(eg carbon, hydrogen)
3. "DNA" / "RNA"(Design & Function)	C. Phosphate, Sugar & Nitrogen Base
4. Proteins (Form & Function)	D. Atoms
5. Lipids (Oils, Fats & Waxes)	E. Phospholipids & Proteins
6. Matter (Solids, Liquids & Gases)	F. Nucleotides
7. Organism (Worm, Moss, Tree, You)	G. Cells
8. Nucleotides (Adenine, Cytosine, ...)	H. Glycerol and Fatty acids
9. Compounds (H_2O; $C_6H_{12}O_6$; CO_2)	I. Sugar molecules
10. Cell Membranes (Life's border)	J. ADP's; P's & ENERGY!
11. Sugar (Life's chemical energy)	K. Protons, Neutrons & Electrons
12. Systems (ex. digestive, skeletal)	L. Amino acids
13. Polysaccharides (eg starch, glycogen)	M. Organs (eg heart, kidney)
14. ATPs (Energy currency of life)	N. CO_2 & H_2O

Box 1. Fill in the blank

Fill in the gaps. Then summarize the information by drawing a concept map. Compare the concept maps produced by different groups.

There are two types of nucleic acids: (1) ___ which is ______ stranded and found primarily in the ______ and (2) ___ which ______ stranded and found primarily in the ________. The two strands of DNA are together by _______ bonds between complementary ___________.

Nucleic acids are polymers made up of ________, that are linked together by ________ bonds. Each nucleotide consists of three components (1) a _______-containing base that can be either a ______ or a ________; (2) a __ carbon sugar; (3) a _______ group that links the nucleotides. The sugar is RNA is _________, and in DNA is _______.

A summary of student performance in terms of grade distribution i.e. percentage and number of *High Distinction* (HD), *Distinction* (D), *Credit* (C), *Pass* (P), *Fail* (F) and *Withdrawn* (W) for *Human Physiology (Part 1)* at both campuses is shown in Table 2.

Statistical analysis using Chi-square (χ^2) indicated a significant improvement (Callaghan, $p<0.001$; Ourimbah, $p<0.0001$) in student performance in 2009 compared to 2008 at both Callaghan and Ourimbah. Whilst the overall failure rate remained stable or increased slightly, there were a higher percentage of students at both campuses that achieved a D or HD grade. There were also a greater percentage of students that withdrew from the subject 2009 although students' reasons for withdrawing from a course might be complex and might not be closely associated with a unit's teaching arrangement.

Despite the shift in distribution of grades there was no significant change in the mean final mark at Callaghan (Table 3; t-test, $p=0.09$). There was however a significant increase in the mean score achieved by students at Ourimbah in 2009 (Table 2; t-test, $p<0.05$), largely due to a significant increase in the continuous assessment component of the course.

The online assessments could be completed by students at any location within a restricted time frame of availability, and as such can be regarded

Table 2. Distribution of final student grades in undergraduate human physiology (Part I) at Callaghan and Ourimbah campuses of University of Newcastle in 2008 and 2009

	High Distinction	Distinction	Credit	Pass	Fail	Withdrawn
Callaghan						
2008 (473)	7.0 (33)	18.4 (87)	24.7 (117)	36.8 (174)	13.1 (62)	2.5 (12)
2009 (389)	10.8 (42)	19.4 (76)	23.1 (90)	30.08 (117)	16.4 (64)	6.9 (27)
Ourimbah						
2008 (174)	4.0 (7)	6.9 (12)	21.3 (37)	43.7 (76)	24.1 (42)	4.0 (7)
2009 (180)	4.4 (8)	15.0 (27)	23.3 (42)	31.67 (57)	25.5 (46)	7.8 (14)

Final grade categories: *High Distinction* (85-100%); *Distinction* (75-84%); *Credit* (65-74%); *Pass* (50-64%); *Fail* (≤49%). To pass overall, a minimum of 45% was required in the end of semester formal examination.

Table 3. Mean scores obtained for assessment items and final overall total in undergraduate human physiology (Part I) at Callaghan and Ourimbah campuses of University of Newcastle in 2008 and 2009. Assessment consisted of two written mid-semester exams in 2008; and four online tests and one written mid-semester exam in 2009

	Callaghan		Ourimbah	
	Assessment	Final	Assessment	Final
2008	60.5±0.7	63.3± 0.7	51.6± 1.3	56.4±1.2
2009	65.3±0.7*	65.1 ± 0.8	60.4± 1.1*	60.4±1.2*

* significantly different from previous year (t-test, p <0.05)

as an open-book examination. Not surprisingly, it was found that the written mid-semester exam (completed under supervision) was more highly correlated with final grade (Callaghan, r=0.78; Ourimbah, r=0.72) than the online assessments (Callaghan, r=0.60; Ourimbah, r=0.65). The marked improvement in both assessment tasks and final grades at Ourimbah in response to regular online assessments may be partly due to increased peer-assisted learning and interaction. The student cohorts at Ourimbah are generally smaller in number and form a more cohesive group than students at Callaghan, which appeared to be enhanced further by the introduction of regular language-oriented learning tasks and assessments.

Impact of "Disease of the Week" on Failure Rates of Program Cohorts

Historically, within any one year student performance is similar in the two (arbitrarily split) parts of *Human Physiology*. This similarity is demonstrated by the 2008 failure rate data for the various program cohorts (Table 4), which do not differ significantly between *Part 1* and *Part 2* (χ^2, p=0.97).

Following the introduction of online assessments as outlined above (in both *Part 1* and *Part 2*) and the "Disease of the Week" modules (in *Part 1* only) in 2009, there was a significant difference in student performance in the two parts

Table 4. Failure rates for student cohorts in undergraduate Human Physiology (Part 1 and Part 2) at the University of Newcastle in 2008 and 2009

Cohort	2008				2009			
	Part 1		Part 2		Part 1*		Part 2*	
	n	%F	n	%F	N	%F	n	%F
Physiotherapy	106	1.9	106	0.9	67	7.5	62	3.2
Biomed	82	7.3	72	9.7	87	9.2	82	13.4
Occupational Therapy	95	15.8	97	17.5	68	23.5	63	47.6
Nutrition & Dietetics	65	16.9	52	17.3	62	8.1	69	15.9
Podiatry	80	17.5	66	16.6	63	20.6	52	13.5
Exercise/Sport Science	60	21.7	53	17.0	62	21.0	53	18.9
Speech Pathology	94	22.3	60	16.7	93	23.7	62	27.4
Food science/Human Nutrition	41	24.4	35	25.7	53	30.2	47	38.3
Science	34	32.4	28	32.1	22	22.7	13	38.5
All Students	666	15.6	577	14.7	610	18.0	522	23.55

n = number of students
%F = percentage of students that failed course

of the subject (χ^2, $p<0.0001$). The failure rates of students from Occupational Therapy, Nutrition & Dietetics, Food Science & Human Nutrition and Speech Pathology programs were much lower in *Human Physiology Part 1* compared to *Part 2,* suggesting that the "Disease of the Week" modules found their target audience and made a significant impact on their performance. It was envisaged that students from Podiatry and Sport Science & Exercise programs may also benefit from the clinical contexts provided for the course content, however these students went against the trend and performed better in *Part 2* of the course. These modules were made available online and therefore, the number of times these modules were accessed by students from the various program cohorts would have provided useful information to interpret these differences, but was not collected on this occasion.

The differences found in effectiveness of "Disease of the Week" between cohorts may well be related to differences in their motivation to learn basic physiology and/or their learning style preferences. A study with non-science majors found that overall motivation to learn science could be measured over five domains: *intrinsic motivation and personal relevance*, *self-efficacy and assessment anxiety*, *self-determination*, *career motivation* and *grade motivation* (Glynn et al, 2009). A belief that course content was relevant to their future career was a dominant motivating factor for both non-science (Glynn et al, 2009) and science majors (Walker, 2010). An increased perception of the relevance of basic physiology to their future career and may be a factor in the improved performance of students from Occupational Therapy, Nutrition & Dietetics, Food Science & Human Nutrition and Speech Pathology. Student learning can also be classified according to the sensory modality that is preferred: *visual*, *auditory*, *reading-writing* or *kinaesthetic*. Students may have either a strong single preference ("unimodal") or multiple preferences ("multimodal") for learning. Previous studies have indicated differences in learning preferences related to gender (Dobson, 2009) and career choice (Breckler et al, 2009). The "Disease of the Week" intervention was primarily a *reading-writing* activity and was

Table 5. Mean student satisfaction rating achieved for undergraduate human physiology (Part I) at Callaghan and Ourimbah campuses of University of Newcastle in 2008 and 2009

Course Area	Student Satisfaction			
	Callaghan		Ourimbah	
	2008	2009	2008	2009
1. Improved knowledge	4.2	4.0	3.6	4.3
2. Assessment Items	3.7	3.8	3.1	4.2
3. Overall Quality	3.5	3.5	2.4	4.3

probably most effective in those students with a strong *reading-writing* learning preference.

Impact on Student Satisfaction

At the Ourimbah campus there was a significant increase in student satisfaction reported for *Human Physiology Part 1*, with respect to overall knowledge gained, relevance and effectiveness of assessment items, and overall course quality (Table 5). The same changes introduced concurrently at Callaghan resulted in no change in mean satisfaction ratings. Questions were phrased as follows:

- As a result of studying this course, I have improved my knowledge of the topics/material covered.
- The assessment items in this course supported my learning.
- Overall, I am satisfied with the quality of the course.

Student Responses: scored on a scale of 1-5 (1=*Strongly Disagree*; 2=*Disagree*; 3=*Neutral*; 4=*Agree*; 5=*Strongly Agree*).

It can be reported anecdotally however, that the number of complaints and negative comments posted on the subject LMS Discussion Board declined markedly in 2009 compared to 2008 especially on the Ourimbah campus. The total number of posts on the Discussion also declined from 43.5 posts per week in 2008 to 30 posts per week in 2009. Several unsolicited positive comments were posted by students on the Discussion Board during 2009 in relation to Disease of the Week:

I'm really liking that you can apply the information that we are learning to a real life example. I've been reading the articles and finding them very interesting. Thank you.

I've been finding them very useful for applying them to review questions as examples that I've been doing, along also in my summaries.

Although mean satisfaction rating at Callaghan remained steady, chi-square analysis indicated a significant polarization of student opinion on whether they had increased their knowledge of the area, and satisfaction with the overall quality of the course. For the 2009 cohort more students responded at either the positive or negative end of the scale, whereas the 2008 students generally responded neutrally. A study of genetics students (Lidbury, in press, Chapter 4) presented evidence that innovative instructional styles can actually provide an additional disincentive to lower level students, which may also be the case here. The impact might be clearer if a longitudinal study of this group of students over a number of years. The pattern of student response to assessment items was consistent over the two years of the study.

Table 6. Biology 1002 students at The University of Newcastle in semester two, 2008

Terms	'Yes' response for confidence in %	100% understanding	Delusion index	Ranking (easiest=1, hardest=10)
Significant	44.44	33.33	1.33	9
Equation	72.22	38.89	0.54	8
Research	77.78	44.44	0.57	7
Equilibrium	66.67	50.00	0.75	6
Concentration	77.78	61.10	0.79	5
Power	44.44	66.67	1.50	4
Graph	88.89	77.78	0.88	3
Experiment	66.67	83.30	1.25	2
System	72.22	88.89	1.23	1

Note: at The University of Newcastle, one item (significant) was repeated and the item on 'model' was not included. Therefore the table shows the data for nine items rather than ten.

METHODS EMPLOYED TO SUPPORT LEARNING IN BIOLOGY

At the University of Newcastle in 2008 and 2009 there were 209 and 250 students respectively in the unit: Biol 1002: Organisms to Ecosystems. Biology 1002 is taught as four modules: Ecological and evolutionary processes; Principles of Development; Animals – Who are they and how do they live; Green machines (Plant physiology). During 2008, lecturers taught the subject matter as they had previously and students were assessed in the normal fashion. The language interventions outlined below were implemented during 2009 in the "Animals – who are they and how do they live" module only, with the three other modules in the course serving as an additional control. Assessment during 2008 and 2009 consisted of four online MCQ quizzes (worth 10% each) and a formal written exam (worth 60%) with multiple choice and short answer questions. Short answer questions required students to fill in missing words, list key words, draw and label simple diagrams or write brief 2-3 sentence paragraphs.

In 2009, in the implementation phase of the project in semester two, VotApedia questions (Appendix B) were used during the lectures as well as online revision exercises and tutorial activities (Appendix C). These took place over a period of three weeks out of a total of 13 weeks for the unit.

In semester two, 2008 in the subject Biology 1002, students were asked for their consent to participate in the project through the completion of Test 1 and 2 of the language surveys (Appendix A). However, only 18 students participated (Table 6). In semester two 2009, in order to increase the number of students participating, the Tests 1 and 2 of the language surveys were built in as part of the out of class non-assessable assessment items. At the end of the second non-assessable item, a number of demographic questions were asked as well as the implementation of a Biology Self-Efficacy questionnaire which was based on Baldwin, Ebert-May and Burn's Biology Self-efficacy instrument (Baldwin, Ebert-May, & Burns, 1999). In the last item of the out of class survey, they were asked to tick 'yes' or 'no' on their participation. By building test 1 and 2 items into the non-assessable assessment item, the participation rate for Test 1 and 2 of the language surveys rose to 30.4%.

During semester two, 2009, the language strategies were implemented for three weeks over four hours per week. In the second two-hour block of each week, the language exercises (Appendix D) were administered in about half an hour each week. This amounts to one and half an hour in

Table 7. Biology 1002 students at the University of Newcastle in semester two, 2009

Terms	'Yes' response for confidence in %	100% understanding	Delusion Index	Ranking (easiest=1, hardest=10)
Research	80	5	16	9
Equilibrium	47	11	4.27	8
Equation	73	15	4.87	7
System	73	24	3.04	6
Model	34	26	1.31	5
Experiment	83	40	2.08	4
Concentration	76	43	1.77	3
Graph	83	48	1.73	2
Control	87	74	1.18	1

total. These tutorial exercises were also provided to students before each week's tutorial. These exercises provided students opportunities to classify, link and reflect on the linkage of words and concepts (Table 7).

The language difficulty surveys were also distributed to first year students at the University of Newcastle and second year students at the University of Canberra in both 2008 and 2009. Since both these universities were involved in the discipline of biology, the same language difficulty questionnaire appropriate to biology was distributed.

Interactive Lecture Format

Traditional lectures are essentially a one way communication. First year students in particular, are reluctant to ask or answer questions in the lectures. To increase two way communication we used a variant on the 'clicker' style of student interaction tool during lectures. Short survey questions were integrated into each lecture at the completion of each significant unit of content (3-4 per hour) through the VotApedia audience response system (http://www.urvoting.com) to gauge student understanding of lecture content and to engage the students in use of the language and discussion of the material just covered. The emphasis was not on correct answers, but discussing why an answer or answers were correct and why an alternative was incorrect or 'less' correct. The goal was discussion, and reflection, not just assessment. There was no specialist infrastructure required for the use of VotApedia and the lecturer was able to create, edit and assess their own questions using the site (Figure 1a). Students responded to each question via their mobile phone (at no cost), and were able to view the results immediately thus facilitating discussion (Figure 1b).

In-Lecture Tutorial Activities

Formal tutorials are not feasible in our system, but small group discussions (4-6 students) were used to complete tutorial activities such as the linking of terms and concepts within the Lecture Theatre environment. Over the course of several tutorial exercises, these activities progressed from relatively simple activities involving a single concept to the use of extensive and complex word lists associated with animal diversity (see Table 8 and Box 2).

The incremental increase in difficulty of the tutorial exercises enabled student confidence with the language to increase and encouraged students to put their points of view forward within their small groups. Reporters from each group then

Figure 1. Screen shots of the VotApedia audience response system used during lectures in the "animals" module of biology (a) before voting and (b) after voting

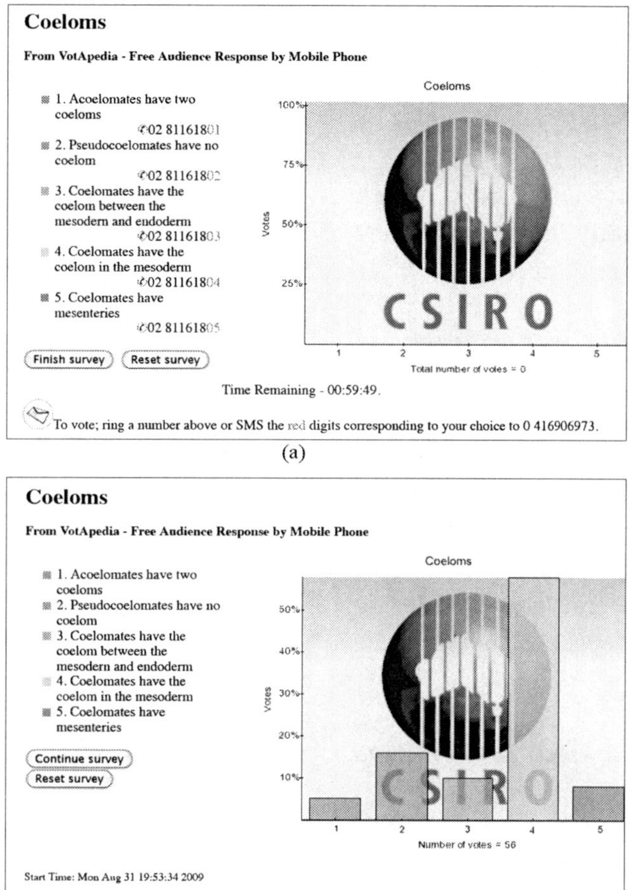

(b)

Table 8. Example of one of the more complex "in-lecture" tutorials used in the animals biology module

Animal Phyla	Animals in the Phylum Class or Common Name					Body Features Seen in the Phylum*					
Arthropoda											
Nematoda											
Chordata											

Task 1: From the word list below, fill in the rows opposite the Phyla with one word that fits in the above table.

Clue: Some Phyla are very diverse and not all Classes in the Phylum will share all the same body features.

Word List: Hydra (Greek - many headed god), Medusa (Greek – snake hair god), Coral, Anthozoa (Class), Anemone, Mouth only, Nerve Net, Nervous system, Excretory system, Turbellaria (Class), Mammalia (Class), Hair, Feathers, Flatworm, Mesoderm, Acoelomate, Nematocyst, Circulatory system, Anus, Respiratory system, Cephalization, Asexual - Sexual life cycle, Trematoda (Class), Amphibia (Class), Oviparity, Reptile, Exoskeleton, Podia, Tagmata (Tagmatisation), Jointed limbs, Terrestrial members, Marine/aquatic members, Mouth parts, Wings, Tracheal respiration, Euclitellata (Class), Parasites, Endothermy, Viviparity, Cephalochordata, Insecta (Class), Crustacea (Class), Cestoda (Class), Segmentation, Hydrostatic skeleton, Skin/surface respiration, Gills, Jelly Fish, Blue Bottle, Polychaeta (Class), Lungs, Amniote egg, Milk, *Caenorhabditis*

Box 2.

Task 2: For each of the tissue types below, indicate the germ layer(s) (i.e. ectoderm, mesoderm, endoderm) that it is derived from during embryonic development?
Tissue Types: Epithelium, Skin, Gut Lining, Lung Lining, Vaginal Lining, Muscle, Nerve Chord, Blood, Bone

contributed to whole class discussions, which provided another opportunity for the use of new language and terms to be reinforced.

Online Resources and Assessments

Practice quizzes were provided online, at weekly intervals throughout the semester for all four modules of the course. These quizzes consisted of MCQs as used previously, supplemented with mix and match exercises and gap-fill (Cloze) exercises to reinforce new language terms presented during the past week. The following is an example of the gap-fill exercises used:

Diabetes is a disease affecting more people around the world as societies become more affluent. A ________ of the disease is possible using mice that have been chemically-induced to be diabetic, or have been genetically selected to become diabetic over time. ____________ into this disease requires the conduct of laboratory __________ using the mouse _________. Diabetes is a complex disease, requiring the consideration of physiological ____________ that include sugar metabolism, water regulation via electrolyte ______________ in blood, and the blood acid/base ___________. The acid/base balance, or pH, can be described by an _________ showing the production of bicarbonate ions via the enzyme carbonic anhydrase, which helps buffer the blood pH around 7.4. Once the __________ are completed, the data can be represented in either tables or ___________. Analysis is also required to decide whether the experimental results are __________ or not.

***Words To Consider:** Research, Experiment(s), System(s), Concentration, Equation(s), Significant, Model(s), Equilibrium, Graph(s)*

There was a formative online quiz that consisted of between 40 and 60 multiple choice questions each week. Students received feedback immediately after submission of each practice quiz including their score and the correct answer for any question that they had answered incorrectly.

IMPACT OF CHANGES ON STUDENT LEARNING IN BIOLOGY

The language interventions outlined above were implemented during 2009 in the "Animals – who are they and how do they live" module only, with the three other modules in the course serving as an additional control.

Practice and Assessment Quizzes

An average of 61.2% of students completed the practice quiz each week, ranging from a low of 46.2% in week 5 of semester to peaks of 66-67% in weeks 1, 10, 11 and 12. The average score obtained in the practice quizzes ranged from 26-49% in 2009.

In 2009 students scored significantly higher in the online assessment for the "Animals" module that incorporated the language intervention than they did in the other three Biology modules (Table 7). The score obtained for "Animals" module was significantly higher in 2009 compared to 2008 (*t-test*, $p<0.01$), which was also the case for the "Ecological and Evolutionary Processes" module.

Table 9. Mean scores obtained in the online assessment for each of the biology modules during 2008 and 2009. The language intervention was incorporated into the "animals" module in 2009 only

Module	Online Assessment Quiz Score (% ± s.d.)	
	2008	2009
Ecological and evolutionary processes	66.6 ± 15.4	71.4± 14.3
Principles of Development	66.0 ± 16.7	55.4 ± 17.0
Animals: Who are they; how do they live	75.4 ± 20.0*	84.2 ± 18.3*
Green machines	83.2 ± 13.7*	78.7± 15.1

* Significantly different from other modules (t-test, $p<0.001$).

Table 10. Comparison of 2008 and 2009 biology 1002 cohorts at the University of Newcastle

Study Year	2008 n = 221				2009 n = 226			
Cohort	Language		Control		Language		Control	
Grade	PASS	FAIL	PASS	FAIL	PASS	FAIL	PASS	FAIL
Number	28	193	88	133	63	163	131	95
% Total Students	13%	87%	40%	60%	28%	72%	58%	42%
Range of Marks (%)	50 – 94	0 – 48	50 - 87	19 - 49	50-100	0 - 49	50 - 82	4.4 -49
Mean Mark ± SD (%)	61.7 ± 10.1	15.6 ± 12.1	63.8 ± 8.3	37.5 ± 8.0	64.3 ± 12.3	24.6 ± 12.8	60.5 ± 8.2	39.5 ± 8.4

The "Language" intervention was incorporated into the "Animals" module of Biol 1002 in 2009 only. The other three modules, where no language intervention was introduced were used as the "Control". Grades and marks presented are for the short answer component of the final exam only.

Performance in the online assessment for the other two modules declined in 2009, compared to 2008 (Table 9).

Performance in Final Exam

The following data assessed the impact of the interventions on performance in the short answer component of the exam only because it was the section that most critically depended on language and vocabulary (see Table 10). Analyses showed a significant improvement for the language group between 2008 and 2009 (χ^2, $p<0.01$), but this was also found for the control modules, indicating a general improvement across the 2009 group, not an improvement only associated with the language interventions.

A closer look at the marks obtained by students reveals some interesting comparisons. The mean score of students with a fail grade increased significantly in 2009 for the language group (24.6 *cf.* 15.6; *t-test*, $p<0.001$) but remained steady over both years in the control group (39.5 *cf.* 37.5; *t-test*, $p>0.05$). Conversely, the mean score of students that passed did not differ significantly between years for the language cohort, but increased significantly in 2009 for the controls (*t-test*, $p=0.004$).

CONCLUSION

A major limitation of both studies is that they are based on only two years of data. In both the Health and Science student cohorts there were significant variability between years, and where programs were identified, even between program cohorts from year to year. With this level of noise in the data it is not surprising that evidence of the impact of teaching interventions was not always clear and in some cases performance was apparently worse. The study of genetics students (Lidbury, chapter 4) is based on five years of data and even with this greater amount of information a simple improvement in grades was not evident. The major impact of the language and other interventions was in the level of student motivation and engagement with the course topic.

The engagement and motivation of the different career program cohorts for the Health students is arguably an issue for future study. Similar subcohort analysis is possible for the biology students but would require changes to the study design. Anecdotal evidence would suggest contemporary university students are highly focused on the relevance of a course to their chosen career path. First year basic science courses have always found it challenging to engage the interest of students for which that discipline is not seen as core to the student's career goals. Learning tools such as; 'Disease of the Week' with subcohort specific examples, or illustrative examples of the relevance of lecture material to everyday life are sensible strategies to address this problem.

Similarly although it is difficult to provide empirical evidence of positive impacts of increased two way communication in the lecture theatre context, It is another effective strategy to promote student motivation and engagement. Anecdotal evidence from unsolicited student comments is that activities like the use of 'VotApedia' questions and discussion sessions and tutorial exercises with discussion are viewed positively by students and improve their experience in the lecture environment.

Future study should focus more on the student experience and their career aspirations and how these factors interact with the specific course or unit. Teaching interventions should arguably be assessed in terms of these outcome areas rather than in improvements in pass rates and grades.

REFERENCES

Bello, Y. M., Falabella, A. F., & Schachner, L. A. (2003). Management of epidermolysis bullosa in infants and children. *Clinics in Dermatology*, *21*, 278–282. doi:10.1016/S0738-081X(03)00050-6

Birge, K. (1995). Nutrition management of patients with epidermolysis bullosa. *Journal of the American Dietetic Association*, *95*, 575–579. doi:10.1016/S0002-8223(95)00157-3

Breckler, J., Joun, D., & Ngo, H. (2009). Learning styles of physiology students interested in health professions. *Advances in Physiology Education*, *33*, 30–36. doi:10.1152/advan.90118.2008

Dobson, J. L. (2009). Learning style preferences and course performance in an undergraduate physiology class. *Advances in Physiology Education*, *33*, 308–314. doi:10.1152/advan.00048.2009

Fuchs, E. (1996). The cytoskeleton and disease: Genetic disorders of intermediate filaments. *Annual Review of Genetics*, *30*, 197–231. doi:10.1146/annurev.genet.30.1.197

Glynn, S. M., Taasoobshirazi, G., & Brickman, P. (2009). Science motivation questionnaire: Construct validation with nonscience majors. *Journal of Research in Science Teaching*, *46*, 127–146. doi:10.1002/tea.20267

Khan, T., Atherton, D., & Denyer, J. (2002). Epidermolysis bullosa and the podiatrist's role. *British Society of Podiatry Conference Abstracts, 2002*, 121.

Martini, F. H., & Nath, J. L. (2009). *Fundamentals of anatomy & physiology* (8th ed.). New York, NY: Pearson.

McCleary, V. L., Aasen, G., & Slotnick, H. B. (1999). Predictors of success in undergraduate human physiology. *Advances in Physiology Education*, *22*, 119–126.

Walker, R. (2010). Motivating science undergraduates: Ideas and interventions. *International Journal of Innovation in Science and Mathematics Education*, *18*, 1–13.

ADDITIONAL READING

Baldwin, J. A., Ebert-May, D., & Burns, D. J. (1999). The development of a college biology self-efficacy instrument for nonmajors. *Science Education*, *83*(4), 397–408.

Ellem, G. K., & McLaughlin, E. A. (2005). Tales from the coalface: From tragedy to triumph in a blended learning approach to the teaching of 1st year biology. Paper presented at the 2005 National UniServe Science Symposium: Blended Learning in Science Teaching and Learning, Sydney, NSW. Retrieved from http://sydney.edu.au/

Fagen, A. P. (2003). *Assessing and enhancing the introductory science course in physics and biology: Peer instruction, classroom demonstrations, and genetic vocabulary*. Cambridge, Massachusetts: PhD, Harvard University.

Garvin-Doxas, K., & Klymkowsky, M. W. (2008). Understanding randomness and it impact on student learning: Lessons learning from building the biology concept inventory (bci). *CBE Life Sciences Education*, *7*(2), 227–233. doi:10.1187/cbe.07-08-0063

Hewson, W. P., Robert Tabachnick, B., Zeichner, K. M., Blomker, K. B., Meyer, H., & Lemberger, J. (1999). Educating prospective teachers of biology: Introduction and research methods. *Science Education*, *83*(3), 247–273. doi:10.1002/(SICI)1098-237X(199905)83:3<247::AID-SCE1>3.0.CO;2-C

Hohenshell, L. M., & Hand, B. (2006). Writing to learn strategies in secondary school cell biology: A mixed method study. *International Journal of Science Education*, *28*(2/3), 261. doi:10.1080/09500690500336965

Khodor, J., Halme, D. G., & Walker, G. C. (2004). A hierarchical biology concept framework: A tool for course design. *Cell Biology Education*, *3*(Summer), 111–121. doi:10.1187/cbe.03-10-0014

Klymkowsky, M. W., & Garvin-Doxas, K. (2008). Recognizing student misconceptions through ed's tools and the biology concept inventory. *PLoS Biology*, *6*(1), e3. .doi:10.1371/journal.pbio.0060003

Lawson, A., E. (2002). Sound and faulty arguments generated by preservice biology teachers when testing hypotheses involving unobservable entities (Vol. 39, pp. 237-52).

Lidbury, B., & Zhang, F. (2008). Comprehension of scientific language as a strategy to enhance learning and engagement for molecular biology students. *Australian Biochemist*, *39*(3), 10–13.

McDaniel, C. N., Lister, B. C., Hanna, M. H., & Roy, H. (2007). Increased learning observed in redesigned introductory biology course that employed web-enhanced, interactive pedagogy. *CBE Life Sciences Education*, *6*(3), 243–249. .doi:10.1187/cbe.07-01-0003

Michael, J. (2007). Conceptual assessment in the biological sciences: A national science foundation-sponsored workshop. *Advances in Physiology Education*, *31*(4), 389–391. .doi:10.1152/advan.00047.2007

Michael, J., McFarland, J., & Wright, A. (2008). The second conceptual assessment in the biological sciences workshop. *Advances in Physiology Education*, *32*(3), 248–251. .doi:10.1152/advan.90122.2008

Miller, V., Gregg, G., & Kelly, B. (2000). Peer assisted study sessions (pass) in a first year biology teaching program-strategies developed by undergraduate pass leaders. Paper presented at the Effective Teaching and Learning at University Conference, Duchesne College, the University of Queensland. Retrieved from http://www.tedi.uq.edu.au /conferences/

Newport, E. (1991). Contrasting conceptions of the critical period for language . In Carey, S., & Gelman, R. (Eds.), *The epigenesis of mind: Essays on biology and cognition*. Hillsdale, N.J.: Erlbaum.

Odom, A. L., & Barrow, L. H. (1995). Development and application of a two-tier diagnostic test measuring college biology students' understanding of diffusion and osmosis after a course of instruction. *Journal of Research in Science Teaching*, *32*(1), 45–61. doi:10.1002/tea.3660320106

Odom, A. L., & Kelly, P. V. (2001). Integrating concept mapping and the learning cycle to teach diffusion and osmosis concepts to high school biology students. *Science Education*, *85*(6), 615–635. .doi:10.1002/sce.1029

Pearsall, N. R., Skipper, J. E. J., & Mintzes, J. J. (1997). Knowledge restructuring in the life sciences: A longitudinal study of conceptual change in biology. *Science Education*, *81*(2), 193–215. doi:10.1002/(SICI)1098-237X(199704)81:2<193::AID-SCE5>3.0.CO;2-A

Piattelli-Palmarini, M. (1989). Evolution, selection and cognition: From "learning" to parameter setting in biology and in the study of language. *Cognition*, *31*(1), 1–44. doi:10.1016/0010-0277(89)90016-4

Roy, H. (2003). Studio vs interactive lecture demonstration–effects on student learning. Bioscene . *Journal of College Biology Teaching*, *29*(1), 3–6. Retrieved from http://acube.org/ pdf/bioscene/v29-1.pdf.

Shepherd, G. M. (1994). *Neurobiology* (3rd ed.). Oxford: Oxford University Press.

Smith, M. K., Wood, B. W., & Knight, J. K. (2008). The genetics concept assessment: A new concept inventory for gauging student understanding of genetics. *CBE Life Sciences Education*, *7*, 422–430. doi:10.1187/cbe.08-08-0045

Wallace, J., & Mintzes, J. (1990). The concept map as a research tool: Explaining conceptual change in biology. *Journal of Research in Science Teaching*, *27*(10), 1033–1052. doi:10.1002/tea.3660271010

Wandersee, J. H. (1988). The terminology problem in biology education: A reconnaissance. *The American Biology Teacher*, *50*(2), 97–100. Retrieved from http://www.jstor.org/.

Winter, B., Breitenstein, C., Mooren, F. C., Voelker, K., Fobker, M., & Lechtermann, A. (2007). High impact running improves learning. *Neurobiology of Learning and Memory*, *87*(4), 597–609. doi:10.1016/j.nlm.2006.11.003

Zhang, F., Lidbury, B., Schulte, J., Bridgeman, A., Yates, B., & Rodger, J. (2009). A cross-disciplinary approach to language support for first year students in the science disciplines. Paper presented at the HERDSA 2009, Charles Darwin University, Darwin.

Zhang, F., & Lidbury, B. A. (2006). It's all foreign to me: Learning through the language of genetics and molecular biology. Paper presented at the 2006 National UniServe Science Symposium: Assessment in Science Teaching and Learning, The University of Sydney, Sydney. Retrieved from http://sydney.edu.au/

KEY TERMS AND DEFINITIONS

Chi-square (χ^2): The chi-square (chi, the Greek letter pronounced "kye") statistic is a nonparametric statistical technique used to determine if a distribution of observed frequencies differs from the theoretical expected frequencies. Chi-square statistics use nominal (categorical) or ordinal level data, thus instead of using means and variances, this test uses frequencies.

Intrinsic motivation: refers to motivation that comes from inside an individual rather than from any external or outside rewards, such as money or grades.

Self-efficacy: has been defined in a variety of ways: as the belief that one is capable of performing in a certain manner to attain certain goals, as a person's belief about their capabilities to produce designated levels of performance that exercise influence over events that affect their lives.

VotApedia: an audience response system that does not require issuing clickers or need specialist infrastructure. Known users (users who have signed on and have obtained permission from Ken.Taylor@csiro.au to create surveys) can create surveys and edit the pages on this site. VotApedia is intended for co-located audiences but also works with remote audiences. Surveys are in a multiple choice question (MCQ) form. (http://www.urvoting.com/)

APPENDIX A: TEST 1 AND 2 OF THE LANGUAGE DIFFICULTY SURVEYS USED AT THE UNIVERSITY OF NEWCASTLE IN SEMESTER 2, 2009

Test 1

- **Question 1.** Tick the definitions that best fit the statement containing the key word (there can be more than one correct answer). MODEL: *Experimental models can be ideas or actual things.*
  A: a microscope
  B: an anatomical model
  C: an experimental procedure
  D: a mouse made susceptible to cancer
  E: a blood specimen

Opinion scale:

- **Question 2.** Are you confident with your answer to the above question?
  - No: I do not understand the meaning of this word.
  - Not sure: I have some idea of the meaning of this word.
  - Yes: I understand the meaning of this word.

- **Question 3.** Tick the definitions that best fit the statement containing the key word (there can be more than one correct answer). SYSTEM: *A system is many things that work together.*
  A: The heart, veins and arteries
  B: a motor
  C: an electron
  D: a protein
  E: an event

Opinion scale:

- **Question 4.** Are you confident with your answer to the above question?
  - No: I do not understand the meaning of this word.
  - Not sure: I have some idea of the meaning of this word.
  - Yes: I understand the meaning of this word.

- **Question 5.** Tick the definitions that best fit the statement containing the key word (there can be more than one correct answer). EQUILIBRIUM: *Over 5 years, the number of people leaving town was identical to the number of people arriving, leading to population equilibrium.*
  A: The addition of chemical A to cells lead to an equal amount of oxygen production by these cells over 5 days of measurement.
  B: The temperature of the pond water regularly changed by 10 degrees Celsius over a 100 day period of measurement.
  C: the release of gas into the corner of the room soon spreads to fill the entire room.
  D: The concentration of sodium (NA) in your blood changes when you drink a lot of water.

Opinion scale:

- **Question 6.** Are you confident with your answer to the above question?
  - No: I do not understand the meaning of this word.
  - Not sure: I have some idea of the meaning of this word.
  - Yes: I understand the meaning of this word.

- **Question 7.** Tick the definitions that best fit the statement containing the key word (there can be more than one correct answer). EQUATION: *The tutor helped the student balance the equation.*
  A: A statement that two things are the same.
  B: A description of a reaction in terms of chemical symbols
  C: A mathematical tool for calculating molecular mass.
  D: A problem to solve to find the answer
  E: A representation of a chemical reaction

Opinion scale:

- **Question 8.** Are you confident with your answer to the above question?
  - No: I do not understand the meaning of this word.
  - Not sure: I have some idea of the meaning of this word.
  - Yes: I understand the meaning of this word.

- **Question 9.** Tick the definitions that best fit the statement containing the key word (there can be more than one correct answer). RESEARCH: *We carry out research to find out the answers to scientific questions.*
  A: Observing the results of a series of chemical reactions
  B: Testing a hypothesis by experimentation
  C: Answering an exam question
  D: Looking for information in the library or on-line
  E: Copying information out of a textbook

Opinion scale:

- **Question 10.** Are you confident with your answer to the above question?
  - No: I do not understand the meaning of this word.
  - Not sure: I have some idea of the meaning of this word.
  - Yes: I understand the meaning of this word.

- **Question 11.** Tick the definitions that best fit the statement containing the key word (there can be more than one correct answer). GRAPH: *The temperature was measured at 5 minute intervals and plotted on a graph.*
  A: An XY plot of temperature versus time.
  B: A table of numbers representing temperature and time.
  C: A diagram expressing the connection between temperature and time.
  D: A picture of the experiment
  E: A piece of paper with lines on it.

Opinion scale:

- **Question 12.** Are you confident with your answer to the above question?
  - No: I do not understand the meaning of this word.
  - Not sure: I have some idea of the meaning of this word.
  - Yes: I understand the meaning of this word.

- **Question 13.** Tick the definitions that best fit the statement containing the key word (there can be more than one correct answer). CONTROL: An experiment can usually only be valid if it has a control or controls.
  A: The results are carefully measured.
  B: A trained observer is in control of the experiment
  C: The same measurements are made on treated and non-treated groups.
  D: Experimental animals are treated in an ethical way.
  E: Experimental animals are not allowed to escape from their enclosure.

Opinion scale:

- **Question 14.** Are you confident with your answer to the above question?
  - No: I do not understand the meaning of this word.
  - Not sure: I have some idea of the meaning of this word.
  - Yes: I understand the meaning of this word.

Test 2

Fill in the Blank (10 points)

- **Question 1.** Diabetes is a disease affecting more people around the world as societies become more affluent. A ________ of the disease is possible using mice that have been chemically-induced to be diabetic, or have been genetically selected to become diabetic over time.
  WORDS TO CONSIDER: *Research*, *Experiment(s)*, *System(s)*, *Concentration*, *Equation(s)*, *Significant*, *Model(s)*, *Equilibrium*, *Graph(s)*
  - Answer: *Model(s)*

Fill in Multiple Blanks (10 points)

- **Question 2.** (*This paragraph continues on from the sentence in the previous question*) **[a]** into this disease requires the conduct of laboratory **[b]** using the mouse **[c]**. Diabetes is a complex disease, requiring the consideration of physiological **[d]** that includes sugar metabolism, water regulation via electrolyte **[e]** in blood, and the blood acid/base **[f]**. The acid/base balance, or pH, can be described by an **[g]** showing the production of bicarbonate ions via the enzyme carbonic anhydrase, which helps buffer the blood pH around 7.4. Once the **[h]** are completed, the data can be represented in either tables or **[i]**. Analysis is also required to decide whether the experimental results are **[j]**, or not.
  WORDS TO CONSIDER: *Research*, *Experiment(s)*, *System(s)*, *Concentration*, *Equation(s)*, *Significant*, *Model(s)*, *Equilibrium*, *Graph(s)*

- Answer for **[a]**: *Research*
- Answer for **[b]**: *Experiment(s)*
- Answer for **[c]**: *Model(s)*
- Answer for **[d]**: *System(s)*
- Answer for **[e]**: *Concentration*
- Answer for **[f]**: *Equilibrium*
- Answer for **[g]**: *Equation(s)*
- Answer for **[h]**: *Experiment(s)*
- Answer for **[i]**: *Graph(s)*
- Answer for **[j]**: *Significant*

APPENDIX B: VOTAPEDIA QUESTIONS USED IN BIOLOGY 1002 AT THE UNIVERSITY OF NEWCASTLE

Biology clicker/votapedia questions on Animal Form and Function, written by Professor John Rodger, john.rodger@newcastle.edu.au

Some questions require multiple answers and the ticked items [✓] are the answers for the questions.

1. Monotremes don't produce milk the egg is self sufficient
   - Marsupials favour lactation over the placenta to support the growth phase of development [✓]
   - Marsupials don't have a placenta the embryo is supported only by milk
   - Marsupials lay eggs and the embryo hatches and climbs into the pouch
   - Placental mammals favour the placenta over lactation to support the growth phase of development [✓]
2. Place these membrane structures in order from the embryo outwards
   - Amnion – Yolk Sac – Allantois – Chorion [✓]
   - Allantois – Yolk Sac – Amnion – Chorion
   - Amnion – Allantois – Yolk Sac – Chorion [✓]
   - Amnion – Chorion – Yolk Sac – Shell
3. Place these Structures in their Order of Appearance in Vertebrate Evolution
   - Jaws – Limbs – Tail – Notochord – Dorsal hollow nerve cord
   - Jaws – Vertebrae – Bone – Lungs – Limbs – Amniote egg [✓]
   - Pharyngeal gill slits – Lungs – Notochord – Lactation
   - Ray fins – Gills – Tail – Feathers – Milk
4. Place Deuterostome Groups in Correct Order
   - Chordates – Vertebrates – Crainiates – tetrapods – Amphibians
   - Amniote – Vertebrate – Mammal – Bird – Craniates
   - Sarcopterygii – Actinopterygii – Vertebrates – Mammals
   - Chordates – Crainiates – Vertebrates – Tetrapods – Amniotes [✓]
5. Deuterostome chordates
   - Have aquatic larvae with gill slits
   - Share a common developmental process [✓]
   - Have a notochord and hollow nerve cord [✓]
   - Form the mouth first and have radial cleavage
6. Phylum Arthropoda, the arthropods
   - Arthropods are protostomes [✓]
   - Arthropods have smooth growth and an exoskeleton
   - Insects have a non-segmented body plan and mouth parts
   - Insects use gills to respire
   - Class Insecta and Class Crustacea are arthropods [✓]

7. Phylum Mollusca, the molluscs
   - Molluscs grow by ecdysis
   - Many molluscs us a radula to scrape food off surfaces [✓]
   - The mollusc body plan is made up of segments
   - slugs and oysters are mollusks [✓]
   - Bivalvia refers to the two valves in the mollusc heart
8. Phylum Annelida, the annelid worms
   - All annelids live underground
   - All annelids are segmented [✓]
   - Annelids have one heart
   - Annelids have circular and longitudinal muscle [✓]
   - Annelids are often parasites
9. The simplest animals
   - have a reduced coelom
   - are bilaterally symmetric
   - have a mouth only [✓]
   - have 3 body layers – triploblast
   - have 2 body layers – diploblast [✓]
10. Flatworms
    - Cestoda are free living flatworms
    - Platyhelminthes means flatworm [✓]
    - Turbellaria are free living flatworms [✓]
    - Flatworms are pseudocoleomate
    - Flatworms lack a circulatory system [✓]
11. Evolution of animals
    - Phylogeny is now based on structure of fossils and living forms
    - Molecular sequence data suggests that acoelomates gave rise to pseudocoelomates then coelomates
    - Sequence data suggests that the pseudocoelomate condition arose at least twice from coelomate ancestors [✓]
    - Nematodes are closely related to arthropods [✓]
    - All worms are more closely related to one another than other phyla
12. Sexual or asexual reproduction
    - Meiosis is a feature of sexual reproduction [✓]
    - Meiosis is a feature of asexual reproduction
    - Mitosis is a feature of sexual reproduction
    - Mitosis is a feature of asexual reproduction [✓]
    - Meiosis results in a doubling of chromosome number

13. Osmosis
    - Osmosis describes molecules moving from a site of high concentration to a site of low concentration
    - Diffusion is the process by which molecules move due to current in the water
    - Osmosis requires a selectively permeable membrane that water cannot cross
    - Osmosis requires a selectively permeable membrane that some solutes cannot cross [✓]
    - Water moving across a selectively permeable membrane to equalise water concentration is called osmosis [✓]
14. Respiration
    - Surface area to volume ratio means that very small animals can 'breath' through their 'skin' [✓]
    - Large animals cannot control body temperature in the cold as easily as small animals
    - Water is a good source of oxygen if not stagnant
    - Fish use counter current flow in their gills to more effectively remove carbon dioxide
    - Counter current flow is only important in marine mammals
    - Very thin animals can 'breath' through their 'skin' without gills or lungs [✓]
15. Digestion
    - Animals with the simplest form of digestion system have a mouth and anus
    - Birds use teeth to grind their food
    - Parasites often have reduced or no digestive organs [✓]
    - Mammals have specialised teeth to eat very different diets [✓]
    - Nematodes have only a mouth and no anus
16. Coeloms
    - Acoelomates have two coeloms
    - Pseudocoelomates have no coelom
    - Coelomates have the coelom between the mesodern and endoderm
    - Coelomates have the coelom in the mesoderm [✓]
    - Coelomates have mesenteries [✓]
17. Epithelia:
    - A type of tissue [✓]
    - A cellular lining [✓]
    - Part of the mesoderm
    - Connect other tissues together
    - Regulate movement of materials [✓]
18. What is an animal?
    - Dog
    - Bird
    - Fly
    - Worm
    - All of the above [✓]

19. Cost of Nitrogen Excretion
    - All animals must excrete the waste products of the breakdown of nitrogen containing molecules like proteins and nucleic acids because they are toxic. Access to water and the cost of producing less toxic versions of the nitrogen containing waste are the two key issues for animals.
    - Ammonia is produced in terrestrial animals
    - Urea is produced by all marine or aquatic animals
    - Ammonia, Urea, Uric acid, Guanine is the ascending order of energy cost√
    - All animals produce Urea
    - Guanine, Ammonia, Urea, Uric acid is the decreasing order of water cost

APPENDIX C: TUTORIAL LANGUAGE EXERCISES

1. **Activity 1:** body plan
   Instruction to instructors:
   Aim: to practice the names of different cells in the body plan of different types of animals.
   Language practiced:
   Procedure: Divide the class into even numbered groups. Within each 2-group, students group A gets the Task sheet. One student gets one task sheet. Each student in group B gets a definition sheet. The task of the students in group A is to walk around and talk to group B students and get the names of all the things they are looking for. You need to photocopy enough for all students. (see Tables 11 and 12)

Table 11. Task sheet

Find the:
groups of similar cells that are organized into structural units.
Find the name of the organism that have cells that cover the surface
Name of the animal group that has two types of tissue layers.
Name of the animal group that has three types of tissue layers
Find the name of the inner layer in two buds animals.
Find the name of the outer layer in two buds animals
What is the name of the layer in three buds animal which is between the outer and inner layer?

Table 12. Definition sheet

All animals have these groups of similar cells that are organised into the tightly integrated structural and functional units. My name is 'tissue'.
Sponges have a layer of tightly joined such cells that covers the surface. I am epithelium.
These animals have embryos that have two types of tissue layers. (diploblast)
These animals have embryos that have three types of tissue lay-ers. (triploblast)
Embryonic tissues are organized in such layers. (germ layers)
This is the name of inner layers in "two buds" animals. (ectoderm)
This is the name of outer layers in "two buds" animals. (endoderm)
In "three buds" animals, there is another tissue layer between the outer and inner skin. (mesoderm)

2. Why is epithelium a particular important tissue in achieving homeostasis?
   Hints for answering this question:
   - Step 1: fill in the blanks in the drawing (below)
   - Step 2: fill in the blanks in the sentence: 'Epi' from Greek means '_____ ' or '______'.
   - Step 3: what is epithelium (singular)/epithelia (plural)?
   - Step 4: Where are they found in an animal's body?
   - Step 5: Give two examples of epithelia
   - Step 6: use the two examples to illustrate what the epithelia do to maintain homeostasis.

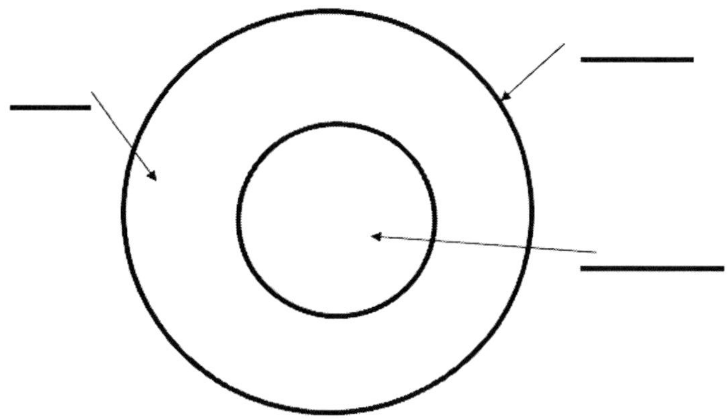

Tissue Layers

3. Explain the difference between a diploblast and a triploblast. Why was the evolution of a third embryonic tissue layer important?
   Hints for answering this question:
   - 'Di' means _________, 'tri' means ________.
   - The third layer is known as '_______'.
   - What does the third layer enable the organism to do?
   - What does the third layer develop into?
   - What does this further development allow the organism to do?
4. Explain why cephalisation was important.
   Hints for answering this question:
   - Step 1: what is cephalisation?
   - Step 2: Give two examples of cephalisation.
   - Step 3: use the two examples to illustrate what cephalisation enables the animal to do better.
5. Provide students with a phylogeny of animals, then ask them to fill in the blanks with the following words: *Lophotrochozoa, protostoma, ecdysozoa, deuterostoma, bilateria, animalia, deuterostome development, protostome development, triploblasty, bilateral symmetry, multicellularity, annelida, mollusca, nematode, arthropoda, echinodermata, chordata*

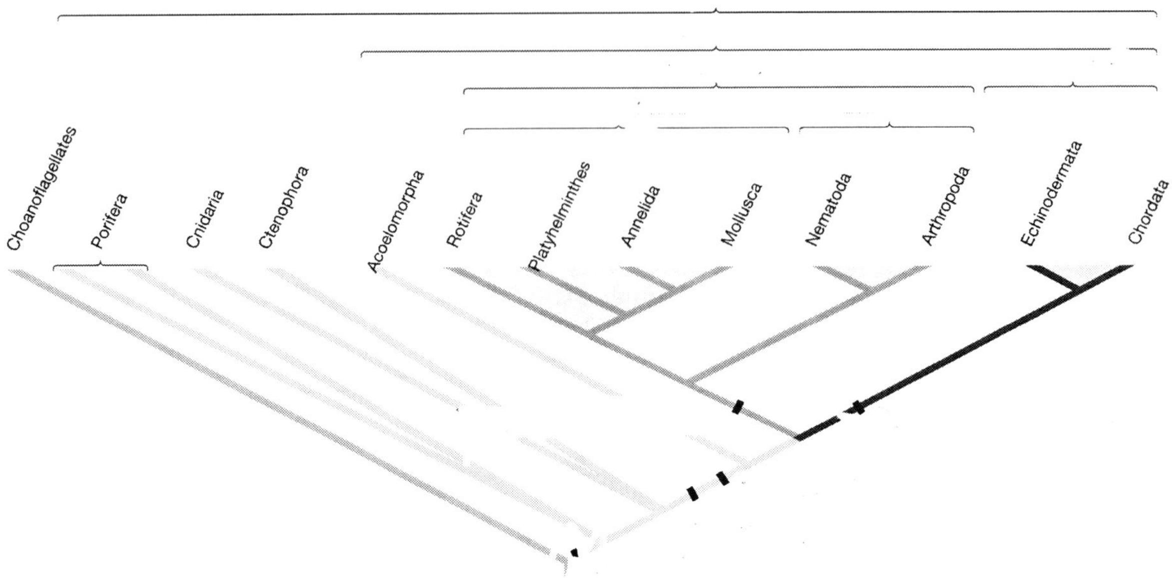

Aim: what is the relationship between the 'ma' in 'protostoma' and 'me' in protostome development?

6. Complete the following table on diversification of animals: **Instructions:** Provide each student with a table without the answers and get them in groups 3-4 to discuss this for five minutes, then let them report back their solutions. (see Tables 13 & 14)

Table 13.

Examples of animals	Types of feeders	environment	Shapes/mouthparts	Function of shapes	Movement
Worms, fish					
Hummingbirds, ticks, mosquitoes, bees					
Humans, dogs, eagles					
Clams					

Table 14. Answers

Examples of animals	Types of feeders	environment	Shapes/mouthparts	Function of shapes	Movement
Worms, fish	*Deposit feeder*	*Surrounded by substrate such as soil, water*	*tubelike*	*Take in substrate from front, waste out behind.*	*Hydroskeleton?*
Hummingbirds, ticks, mosquitoes, bees	*Fluid feeders*	*Trees, flowers*	*Long tubular mouthparts*	*Piercing or reaching into food source such as nectar by suction.*	
Humans, dogs, eagles	*Food-mass feeders*		*Different types of teeth*	*Puncturing prey*	
Clams	*Suspension feeders*				

7. Compare and contrast the types of mouthparts and body shapes you would expect to find in herbivorous insects that such plant fluids from stems, bore through stems, bite leaves, and such nectar from flowers. **Instructions:** Provide each student with a table without the answers and get them in groups 3-4 to discuss this for five minutes, then let them report back their solutions. (see Tables 15 & 16)

Table 15.

Types of insects	mouthparts	Uses of mouthparts	Body shapes	Uses of body shapes
Insects that such fluids, e.g. mosquito				
Insects that bore into stems				
Insects that bite leaves				
Insect that suck nectar				

Table 16. Answer

Types of insects	mouthparts	Uses of mouthparts	Body shapes	Uses of body shapes
Insects that such fluids, e.g. mosquito	*Sharp and point syringe like (proboscis)*	*Pierce woody stems and suck out the fluid.*	*Have Legs or wings*	*Cling to the stems of the tree*
Insects that bore into stems	*Mouthparts that can bite and chew and a strong proboscis*	*To bite and chew stems to get food*	*Smooth/streamlined outer surface and appendages*	*Enable easy movement through the stem*
Insects that bite leaves	*Mouthparts that bite and chew*		*Legs and wings*	*Cling to the leaves while eating*
Insect that suck nectar	*Long proboscis that can be inserted into the flowers*	*To such nectar from flowers*	*Eyes to sense the flower and wings*	*Wings to get to the flowers.*

8. Explain how a hydrostatic skeleton works. Compare the types of movement that are possible with a hydrostatic skeleton to the types of movement observed in animals that lack a hydrostatic skeleton. Hints for answering this question:
   - Step 1: Fill in the blanks for the drawing (below)
   - Step 2: Fill in the blanks in the sentence: 'Hydro-' from Greek means '_____ ' or '______'. 'Static' means ________. In this case, the water is not moving.
   - Step 3: By what means is the movement possible in this animal?
   - Step 4: What is the 'coelom'?
   - Step 5: Where is the coelom found in an animal's body?
   - Step 6: At what stage of phylogeny did the coelom first appear?
   - Step 7: Use the two examples to illustrate how the hydrostatic skeleton functions to enable movement. 1.____________________________ 2.___________________________
   - Step 8: what happens if muscles on both sides stiffen? Or one side relax and one side stiffen?

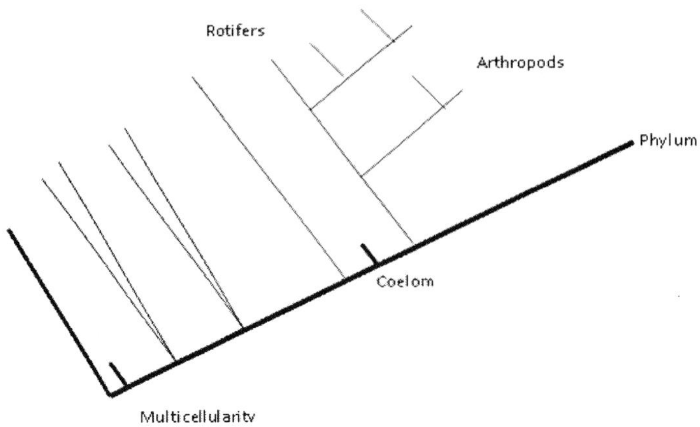

Body Cavities

9. Suppose you are walking along an ocean beach at low tide and find an animal that is unlike any you have ever seen before. How would you go about determining how the animal feeds and how it moves? How would you go about determining the major features of its body plan? Hints for consideration: Construction of animal body plan from external observation, dissection and embryo culture. (see Table 17)

Table 17. Construct a phylogenetic tree from the following body plan observations summarised below, and answer the questions embedded in the exercise

Phylogenetic Characteristic	Mode of Observation								
	External Examination			Dissection of Organism			Embryo Culture		
Bilateral Symmetry + Cephalisation	Yes	No	NA	Yes	No	NA	Yes	No	NA
	✓					✓			✓
Mouth & Anus	Yes	No	NA	Yes	No	NA	Yes	No	NA
	✓					✓			✓
Mouth first? (________stome)							✓		
Anus first? (________stome)								✓	
Body Cavity	Yes	No	NA	Yes	No	NA	Yes	No	NA
			✓	✓					✓
Fluid-Filled Body Cavity has Muscles & Blood Vessels on both sides	Yes	No	NA	Yes	No	NA	Yes	No	NA
			✓		✓				✓
Number of Tissue Layers? (2=______blastic OR 3=_______blastic)?	Two	Three	NA	Two	Three	NA	Two	Three	NA
			✓	✓					✓
Spiral or Radial Cleavage	Spiral	Radial	NA	Spiral	Radial	NA	Spiral	Radial	NA
			✓			✓		✓	
Language & Definitions	(1) Which "Phylogenetic Characteristic" above can be described as a *coelom*? (2) With *diploblasty* there was the evolution of the ______derm and __n____derm. With *triploblasty*, there was the addition of a ___e_____derm. (3) *Ceph-* refers to what part of the body?								

NA = Not Applicable (i.e. mode of observation cannot detect this phylogenetic characteristic).

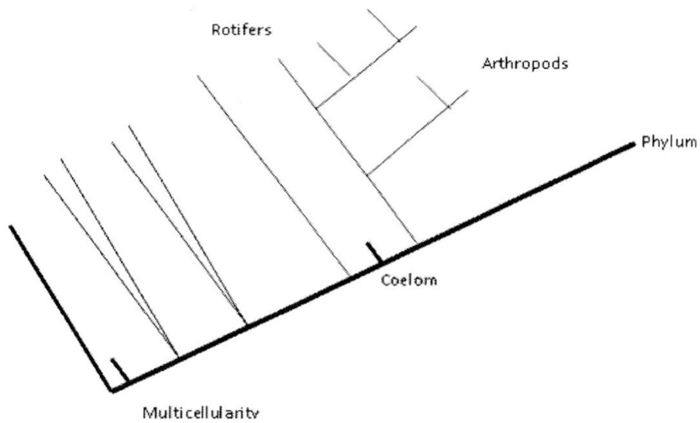

In [Table 18], tick the appropriate boxes to construct your own representative animal at the level of Phylum. Be aware that the observations recorded in the table cannot always be detected by external examination, dissection or embryo culture, so you need to be aware of what technique is appropriate for the particular phylogenetic observation.

Table 18. + Student phylogenetic construction of their own animal

Phylogenetic Characteristic	**Mode of Observation**								
	External Examination			**Dissection of Organism**			**Embryo Culture**		
Bilateral Symmetry + Cephalisation	Yes	No	NA	Yes	No	NA	Yes	No	NA
Mouth & Anus	Yes	No	NA	Yes	No	NA	Yes	No	NA
Mouth first? (________stome)									
Anus first? (________stome)									
Body Cavity	Yes	No	NA	Yes	No	NA	Yes	No	NA
Fluid-Filled Body Cavity has Muscles & Blood Vessels on both sides	Yes	No	NA	Yes	No	NA	Yes	No	NA
Number of Tissue Layers? (2=______blastic OR 3=_______blastic)?	Two	Three	NA	Two	Three	NA	Two	Three	NA
Spiral or Radial Cleavage	Spiral	Radial	NA	Spiral	Radial	NA	Spiral	Radial	NA
Language & Definitions (List your own here)	(1) (2) (3)								

Chapter 7
The Benefits of Teaching Students the Language of Physics

Jurgen Schulte
University of Technology, Sydney, Australia

ABSTRACT

First year physics courses for non-physics majors are among the most difficult course to cope with for science and engineering students alike. Not only are students confronted with physics specific concepts and mathematics applied in unfamiliar environment, students also have to learn the specific lingo of physics and keep it separate from common language and language used in their own field of major. Anecdotally, we know that students have to master the language of their respective field of study in order to master the field. We investigated the link between language of physics used in our first year physics lectures, understanding of related physics concept, and student performance after a language focused intervention. We found that a language conscious approach in first year physics not only improved the performance of students of non-English background but also benefited the performance of students of native English speaking background.

INTRODUCTION

The first year physics course is one that many science, engineering and medical science students will have to master in their career at university.

DOI: 10.4018/978-1-61350-062-0.ch007

While some may have been exposed to the language, others will have had little or no experience of physics or its language. Many non-science majors find first year physics quite daunting, if not intimidating. Of the many factors contributing to this experience are two that stand out quite prominently, students believes that physics is difficult; and difficulties trying to learn physics at a pace that presumes full command of the language of science and physics.

Research into the problem of enabling students to better acquire scientific vocabulary suggests that 'technical' words make up only a small percentage of vocabulary in scientific texts. They therefore pose fewer difficulties than vocabulary which is used in everyday English as well as in a science context. With words that are peculiar to physics, many words have both a scientific and everyday meaning such as 'work', 'energy' or 'power' which can cause confusion for learners. As students bring the everyday association of the words with them, these same words, when introduced in a physics class, can cause confusion. Itz-Ortiz, Rebello, Zollman and Rodriguez-Achach (2003) suggested that a weak version of the Sapir and Whorf hypothesis might be dictated by the language habits of our community predisposing certain choices of interpretation of words. This would cause confusion in students with scientific terms. Researchers have studied semantics in physics (Touger, 1991; Williams, 1999).

In a recent study on the perception of high school students about physics, Angell, Guttersrud, Henriksen, Isnes (2004) found that students find physics difficult because they are confronted with many different representations of physics concepts simultaneously and have to quickly make transformations among them: formulas, graphs, conceptual explanation, experiments. As Redish (1994) pointed out in his implications for cognitive studies for teaching physics, "Physics requires the ability to use algebra and geometry and to go from the specific to the general and back. This makes learning physics particularly difficult for many students" (p. 13). In addition to that standard English enhanced with physics lingo (an English dialect spoken by physicists) presents considerable difficulties similar to the confusion encountered when confronted with a totally foreign language (Stoddard, et. al., 2002; Touger, 2000; Novemsky, 2004).

The challenge facing students with little background in physics is addressed in our course design for non-science students, i.e., students are generally required to take an introductory or foundation mathematics class prior or parallel to their first physics class. Hence, opportunities to practice algebra, formulae, geometry and graphs are provided to assist students with employing those skills a physics context. Conceptual explanations and experiments rely heavily on the use of common day language within a physics context, often assumes a quite different and distinct meaning.

Many students find it difficult to solve physics based problems although they have a fairly good understanding of how to employ the symbolic representation of physics (formulas, geometry, graphs), the physical (mathematically symbolic) modeling of a physics problem. One student in Redish's study (1994)put it eloquently "it is as if physics were a collection of equations on fallen leaves." Mehmet (2009, p. 8) in a study about expectations and beliefs about physics and physics learning that students "tend to regard learning physics as a kind of memorization of separate pieces of information and take what is given by instructors without evaluation."

The difficulty that students have is in making the link between the collections of leaves and associating each leaf with a particular variety of tree and a particular branch where it came from. Our hypothesis for the study presented here is that a good knowledge of the specific meaning of words used in a physics context will lead students to an improved comprehension of physics concepts and establish a more insightful link between the mathematical modeling of aspects of physics and underlying concepts.

LANGUAGE IN PHYSICS

Our observation and testing pool for this study was the Physical Modeling course at the University of Technology, Sydney. Physical Modeling is a one semester first year physics course reserved for engineering students. The course is offered twice each year. The enrollment in a typical autumn semester is about 420 students and in typical spring semester 200 students. In the Physical Modeling course, the teaching of physics concept and problem solving strategies is predominantly done in lectures while the laboratory classes provide for more opportunities to practically apply physics concepts to selected situations in an inquiry based environment. It is well document in literature (Gijbels, Dochy, Bossche, & Segers, 2005) that complex problems such as encountered in first year physics require students to work together and rely on each other to solve them. This poses some problems in large lecture and tutorial classes (we consider classes to be large when there are more than 150 students in one room) where the interaction and communication among students as well as between students and lecturer can be only very limited. Hence, students rely even more on lecturers' communication skills and spoken words, i.e. the language used in the lecture context.

Here is where in addition to trying to understand how to apply physics concepts to mathematically model a problem posed by nature, students in Physical Modeling also have to cope with the specific language used in physics. That is, language that is related to technical terms ("new words"), words that have a different meaning or connotation in physics compared to their use in daily life, and words that students from non-English speaking background may interpret differently compared to their native English speaking peers. Here, we consider a student coming from a non-English speaking background as one where English is not the dominant language spoken at home. In this study, we focus on the use and understanding of technical terms and the words that are common in a physics context, but which have a different meaning in daily life. A purely physics technical term, for instance, would be "specific heat capacity"; a common daily language word with entirely different meaning in physics would be "work."

Our observation over years has been that our first year engineering students generally learned how to sort the "collection of leaves" with equations on them and use them to solve a simple problem. However, we noticed that when problems dealt with more than just one physics concept, students were put off the track and found it difficult to find the correct sets of leaves and to put them in correct order to solve the problem. Obviously, students had little difficulty in applying the equations laid in front of them to solve a simple problem. However, they had difficulties in understanding individual concepts in physics and, consequently, to identify the relevant physics concepts in given problems.

A typical textbook example problem that involves two physics concepts is a stainless steel nail of certain mass that is driven into a block of wood by means of a hammer of certain mass and strike velocity. The student is asked to find the rise of the nail's temperature after the nail has come to rest. This is a problem which is easy to visualize and quite practical to engineering inclined students. The problem involves the concept of energy and related energy conservation and the concept of thermal heat and related heat capacity. Here, all "mechanical" kinetic energy that the nail acquires is turned into thermal energy which in turn causes the temperature of the nail to rise. Kinetic energy as one type of energy is traditional covered in the context of mechanics, i.e., potential and kinetic energy of moving solid objects whereas thermal energy and rise of temperature is generally treated the context of mixing of liquids, melting or vaporization. While words of the common daily language are allowed to cruise through a multitude of contexts, the use of similar words used in physics is often not given the same privilege. This is where we see some students trying to find a single concept (and formula) that can be applied to an inherently multiple concept problem. In the above case, students usually try to make Newton's second law work because the hammer "forces" the nail into the wood. As one of our students put it, "I can't seem to get my formulas set to get past a certain point."

Surely, we have explained the concepts to them in lectures, provided beautiful and meticulously written lecture notes, and had selected a well tested textbook. And yet, something seemed to have not been communicated well enough. Here begins our journey with the support of a linguist and foreign language teacher and a grant from the Australian Learning and Teaching Council.

From a foreign language teacher's point of view, it was heartening to see that the three-language problem (home language, instructional language, science language) that exists for most of our science language learners is acknowledged as parallel to that of English language learning. From a linguist's opinion, this also applies to foreign language learning, in which learners need to move across many discourse communities of their family, school, science and foreign language cultures. However, as pointed out by Yore & Treagust (2006, p. 297): "There has been little consideration between science education and language arts reforms…because it is difficult to convince science majors in teacher education programs that language is an essential part of doing science." This study thus represents one of the first attempts to link practices frequently used in language teaching to the teaching of science by providing examples of how and what to include in science curricula. These examples have been chosen for their simplicity and for the fact that they can be easily constructed with or without the help of technology. In other words, this project is an exemplification that even if one is not a linguist, one can still incorporate language-focused practices in one's teaching.

LANGUAGE AND SCIENCE EDUCATION

Scientists became interested in how to teach science as far back as the 1980s. Initial attempts had been to describe and understand what it is that we would like students to learn. This is evident from titles such as 'Scientific approaches to science education' or 'Learning to think like a physicist: A review of research-based instructional strategies' (Yore, et al., 2006). However, as Jay L Lemke pointed out (Lemke, 1998) such research constitutes a cognitive model of science education which cannot tell us enough how to become better teachers of science as 'it lacks the necessary vocabulary to tell us just what we must lead students to do in order to learn to reason and act scientifically' (p. 1).

In order to transfer to students what they must do in learning to reason and act scientifically, the teacher is the role model 'who can model for students how scientists talk and write and diagram and calculate, how scientists plan and observe and record, how we represent and analyze data, how we formulate hypotheses and conclusions, how we connect theories, models, and data, how we relate our work and result to those of other researchers.' (Lemke, 1998, p. 1). In addition, teachers do not model these abilities by simply telling students a list of concepts they need to 'understand'. They participate in social semiotic dialogues (Lemke, 1998) in which teachers and students become engaged together in scientific sense-making, and scientific doing. These dialogues are carried out in languages of science which integrate multiple languages of representation such as visual representation, mathematical symbolism and languages of experimental operations. We would argue that these should be seen to include the meta-language of learning such as checking, problem solving, reflecting and so on, to learn how to synthesize, monitor, critique and reflect on the information gathered in the process of learning. Furthermore, acquiring scientific terms is, of course, a basic part of scientific literacy education which would include the ability to read and write a science text which is, in turn, a dominant genre used by practicing scientists (Yore, Hand, Prain, 2002).

In order not to fall into the trap of 'activitymania' (Yore & Treagust, 2006), we place teachers in the centre of learning by asking them to explicitly model (a verb) how to test reality 'by checking, monitoring, coordinating, and controlling deliberate attempts to execute learning activity (Koch, 2001). This is why, in the lectures and tutorials involved in this project, interventions were designed to increase student to student and student to staff discussions in activities which promote sense-making rather than just completing calculations.

CHANGED TEACHING PRACTICES REGARDING THE USE OF LANGUAGE

This project also documents the change process the lecturers went through in order to build language-focused and inquiry-based practices into their day to day teaching.

Gunstone (1994) argued that metacognition is central to constructivist perspectives of learning. This idea is reaffirmed and embraced by Yore and Treagust (2006). Metacognition was first used by Flavell in 1976 (Flavell, 1976). He describes it in these words:

Metacognition refers to one's knowledge concerning one's own cognitive processes or anything related to them, e.g., the learning-relevant properties of information or data. For example, I am engaging in metacognition if I notice that I am having more trouble learning A than B; if it strikes me that I should double check C before accepting it as fact. J. H. Flavell (1976, p. 232).

A number of strategies were deployed. These are outlined in Table 1, which includes the research that supported the use of these strategies.

BUILDING ON EXISTING KNOWLEDGE OF PEDAGOGICAL ADVANCES

Research into why students did so poorly in tertiary level physics courses has been going on since 1985. The 'common sense theory' of the physical world, which the students brought with them into the classroom, had long been blamed as one of the major causes for instructional failure in introductory physics (Halloun & Hestenes, 1985; Hestenes & Wells, 1992; Hestenes, Wells, & Swackhamer, 1992; Jacobs, 1989; Touger, 1991). Work has been carried out on two fronts: (1) to establish diagnostic instruments to diagnose students misconceptions such as concept inventories; and (2) to explore different ways of organizing learning activities in order for students to learn.

Since the publication of the 'Mechanical baseline Test' (Hestenes & Wells, 1992) and the 'Force Concept Inventory' (FCI) (Hestenes et al., 1992) appeared in 1992 in the Journal *Physics*, a large number of concept inventories have been developed and validated in different areas of science. These are the Brief Electricity Magnetism Assessment (BEMA) (Pollock, 2008) and the Conceptual Survey of Electricity and Magnetism (CSEM) (Maloney, O'Kuma, Hieggelke, & Van Heuvelen, 2001), the Chemistry Concept Inventory (CCI) (S. Krause, Birk, Bauer, Jenkins, & Pavelich, 2004), Genetics Concept Inventory (S. Elrod, 2007) and Biology Concept Inventory (BCI) (Garvin-Doxas & Klymkowsky, 2008). While these inventories are useful, they have essentially been established as diagnostic instruments to diagnose students' misconceptions or conceptual understanding. Data from such diagnostic tools, while informing educators what the problem areas are, do not offer advice on how these problem areas can be addressed in teaching.

In the meantime, many researchers have also invested a great deal of energy into investigating 'how' problem areas identified by the diagnostic tools cited above can be taught better or how

Table 1. Strategies implemented in the project and the research that supported them

Strategies	Meta-cognitive skills practiced	Research supporting use
1. Small group work in tutorials using guided questions	Learning to use oral language to express and explain scientific ideas	(Kempa & Ayob, 1995; Ritchie, 2001)
2. Students are provided with a list of terms and, through the process of group work, place these terms in relation	Physical concept mapping, exploring relationships between each term	(Roth & Roychoudhury, 1994; Wellington & Osborne, 2001)
3 Giving students opportunities to put forward their points of view in groups	Creating a supportive atmosphere for idea exploration and debate	(Chin & Brown, 2002)
4. Using online language exercises such as crosswords, gap-fill (cloze) exercises and simplified scientific readings	Explicit instruction, practice, applying concepts	No research discovered in science education
5. Providing stimulus questions for lecture and tutorial materials on WebCT* thus encouraging students to prepare before the lecture	Preparation or reflection, just in time learning, online feedback	(Zhang & Lidbury, 2006)
6. Breaking down long words to aid memory by identifying prefixes and suffixes, and exploring the roots and origin of words	Explicit instruction on how to acquire new vocabulary and how to see patterns in the roots and origins of words	(Wellington & Osborne, 2001)
7. Using warm up activities such as matching scientific terms to definitions for revision purposes	Categorizing and systematizing terms	(Richardson & Zhang, 2008; Richardson, Zhang, & Lidbury, 2008)
8. Using of flashcards for vocabulary revision, creating a glossary	Metalinguistic maintenance	(Zhang & Lidbury, 2006)

*Now called Blackboard LMS

learning activities can be organized better to enable better learner outcomes in science education.

In physics, many studies in designing instructional sequences in teaching concepts such as force, motion and Newton's Third Law have been carried out (Alonzo & Steedle, 2008; Halloun, 1998; Savinainen, Scott, & Viiri, 2005). Analogies have also been used extensively to remediate misconceptions in physics since the early 1990s. (Brown, 1992; Dupin & Johsua, 1989) .

Interactive Engagement strategies have been shown to enhance the gain in learning. In an Interactive Engagement class, the lecturer sets up a demonstration and asks the students to predict and write down, with consultation among themselves, their prediction of how the demonstration will work out. Then the demonstration is carried out, the results discussed and the relevant theory presented. The central idea is to offer students opportunities to engage with each other, commit to a position, confront this with reality and then use the event to trigger reflection and foster understanding. Extensive research has been done using Interactive Engagement strategies and the finding is that the 'use of Interactive Engagement strategies can increase the effectiveness of conceptually difficult courses well beyond that obtained with traditional methods' (Hake, 2007). Professor Hake's claim is based on the achievement scores obtained using standardized tests such as the FCI (see above).

Harvard's physics Professor Eric Mazur implemented 'Peer Instruction' which also uses interactive-engagement strategies. Data obtained in classes using Peer Instruction in a variety of disciplines show that learning gains nearly triple. Most importantly, students not only perform better on a variety of conceptual assessments, but also improve their traditional problem-solving skills (Crouch & Mazur, 2001).

In recent years, Interactive Engagement and Peer Instruction, in various scientific disciplines, have been enhanced by the use of computer technology such as concept mapping tools, visualization tools (Wu, Krajcik, & Soloway, 2001) audience response devices such as clickers and other web-enhanced strategies(Martyn, 2007; McDaniel, Lister, Hanna, & Roy, 2007; Odom & Kelly, 2001).

In the late 1990s, studio classes replaced straight lecturing in a number of universities in the US. Studio teaching consisted of a mixture of student exercises, instructor coaching, and sometimes laboratory experiments. It drew inspiration from the idea of interactive learning and generally taking advantage of modern technology to deliver instructional materials (Cummings, Marx, Thornton, & Kuhl, 1999; Pipes & Wilson, 1996; Roy, 2003). While studio classes have shown their merit, they are by nature limited to small classes or small groups and would be impractical for the large classes we are dealing with in this our study.

Due to these factors, these units were delivered in the traditional transmissive mode at each institution prior to 2009. Despite these inhibiting factors, the project still intended to change the teaching of these subjects in the participating institutions to a more interactive mode of teaching by incorporating a number of interactive strategies implemented by the lecturers themselves. Studio teaching and radical changes in curriculum design were simply either too expensive to implement or too disruptive to other lecturers who co-teach with the participating lecturers. Detailed implementation will be reported below.

EDUCATIONAL CONTEXT AND INTERVENTIONS

Physical Modeling is a first year course at the University of Technology, Sydney, Australia, which has been designed specifically for the needs of engineering students. The course has been developed by physicists and is taught by physicists to engineering students for more than ten years (we call it "service teaching"). The teaching philosophy of the course can best described by problem based teaching and learning as defined in the works of Gijbels, Dochy, Bossche and Segers (2005). The key characteristic of our lectures and laboratory work is putting a complex problem to students to initiate the cycle of problem solving, where students should take on an active role in constructing knowledge and engaging in inquiry and problem-solving skills (Mehmet, 2009). There are many research-based reforms that help build students' conceptual knowledge. As in many first year physics classes we too apply a cognitive-conflict (Nussbaum & Novick, 1983) or elicit-confront-resolve teaching model (McDermott, 1991) in which students are confronted with a physics problems and are asked to make predictions prior to having learned about the underlying physics concept. In this process, conceptual misunderstandings become evident at early stage and become alerted to terminology used in physics and their divergence from similar words used in common daily language.

The "active role" part of the learning process is to a large extent left to our laboratory classes. Necessity requires us to have lectures as large as possible, i.e., in a typical autumn semester there would be a class of 420 engineering students in one lecture room and in spring about 230 students. In the following, we will be frequently using the terms autumn and spring semester and, hence, some further explanation may be useful for those who are not quite familiar with the southern hemisphere teaching terms (Australia). The autumn semester (March-June) is the first semester in the year which has the main intake of freshly graduated high school students; the spring semester (August-November) is the second semester in the calendar year with a typically smaller, "mid-year" intake of students. Physical Modeling is a one semester course taught in both in autumn and spring.

The mode of lectures in this course is teaching students in groups of up 420. If student numbers are large, a repeat class is time tabled with the remainder, usually about another 100 students.

Figure 1. Lecture slide with multiple-choice question aiming to test the understanding of the use of the term normal force

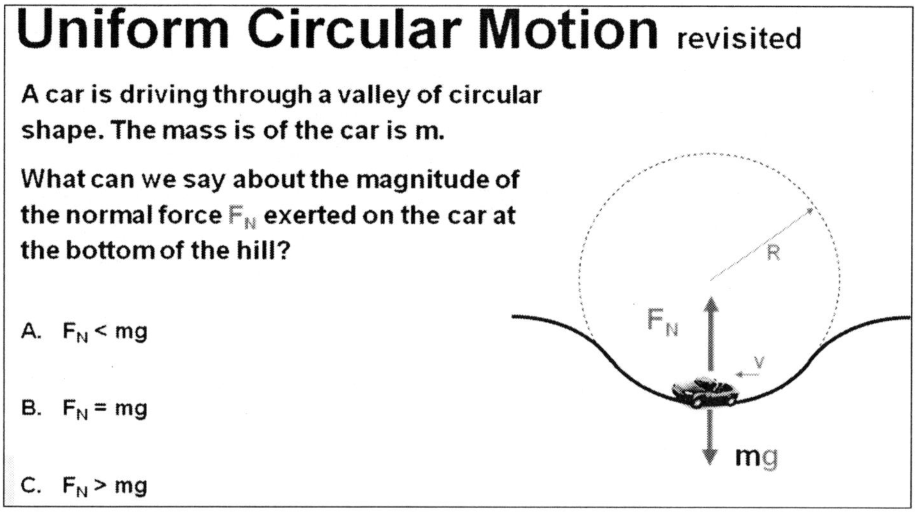

Lectures are dominated by power point slides presented in a paced way to allow for questions by students as well as lecturers as a form of feedback. Our lecture halls do not provide for preparation rooms for lecture demonstrations, lecture halls are shared with all disciplines present within the university. So, lecture demonstrations play a minor role, if at all, and demonstrations usually have to fit in a lecturer's briefcase. Under such circumstances, a two-way communication to be able to adjust to students' feedback is at times quite challenging.

In order to improve student engagement in lectures, we have been using RF clickers (eInstruction, 2010) capable of two-way communication and immediate processing and display of feedback/answer demographics. This gave us an opportunity to incorporate physics concept multiple-choice questions in lecture slides that emphasize on the use of our selected technical terms and test their understanding within the context. From the demographics of student answers the lecturer receives immediate feedback on how well students' understanding of a particular concept defining word is and how much of confusion with the word's common daily use remains. Figure 1 illustrates an example of a question put to students after the concepts of forces and circular motion were introduced. This particular question aims to find clarity about the term "normal force", i.e., the perpendicular component of the force exerted by a surface onto an object which is in contact with the surface. A frequent misconception among students is that the normal force is simply the weight of the object (F_N=mg; answer B), which is clearly illustrator by almost 50% of students opting for this answer (Figure 2). Other students mix the common language use of force and slack interpretation of their own experience to conclude the magnitude of normal force in this case is less than the weight of the car (F_N<mg; answer A). In this example, only 22% of students in this lecture correctly that apart from the weight of the car, which contribution and direction is indicated in the graph, there is also a contribution the normal force due to the rotational motion of the car. Some students did not know how to answer this question ("Invalid"), which we usually interpret as students who find it difficult to untangle their daily experience and the technical terms and definitions used in the lecture.

Figure 2. Audience response to the question in Figure 1

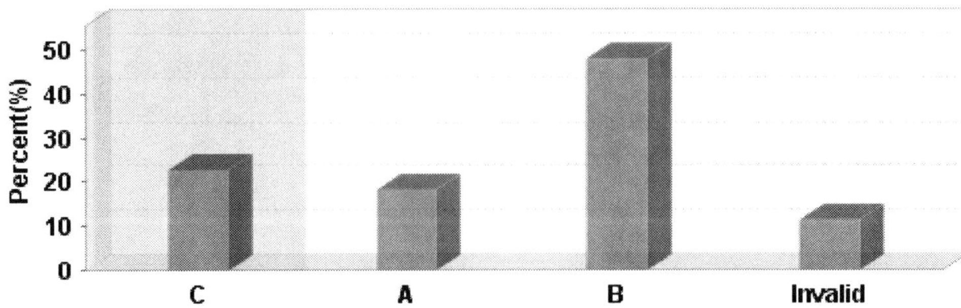

Depending on the outcome of the responses as illustrated above, particular features of the physics concept were then revisited and/or additional explanations regarding the use of the technical term in a physics context were given. At the end of a particular topic, a different set of multiple-choice questions was used to test the same concept and technical term. This so called follow up question consistently had a more positive response. The questionnaire slides were seamlessly integrated into the lecture flow following the line of thinking and discussion. There were between four and six such slides per hour of lecture.

The lecturer was complemented by small group student -to -student group discussions and then students to teacher discussion in biweekly full class size tutorials. Only one hour was available in these tutorials.

Students were presented with a number of quizzes online before each lecture. This protocol involves the implementation of technique 5 in Table 1. The physics lecturer created, implemented and collected data on a set of language specific online quizzes for the respective disciplines in 2008 and 2009. In 2008 and 2009, the physics assignments deployed through the WileyPLUS website (WileyPlus, 2010) consisted mainly of calculation types of questions with randomized parameters where answers comprise numerical values and units and are entered in a online answer form. Below is an example:

Problem: On a dry road, a car with good tires may be able to brake with a constant deceleration of 5.28 m/s^2. (a) How long does such a car, initially travelling at 24.4 m/s, take to stop? (b) How far does it travel in this time?

(a) *Number Units s*
(b) *Number Units*

Here, the student is expected to enter the correct answer number *4.62* and SI unit symbol *s* for solutions (a) in the provided form field spaces, and number *56.4* and unit SI symbol *m* for solution (b), respectively. Other students are presented with the identical problem statement, however, different problem parameters, e.g., deceleration of 3.12 m/s^2 and initial travelling speed of 21.3 m/s, which of course lead to different numerical solutions.

In order to get away from the assumption that if students can correctly do the calculations, then they have understood the subject matter, we also introduced a physics concept survey.

During the one hour tutorials, the lecturer also incorporated into small group activities the use of multiple choice and concept questions related to language use. These concept questions were created specifically to test students' understanding of particular concepts such as 'force' in physics and the use of 'force' in real life. For example:

Meaning of 'force'

Which one(s) of the following sentences containing 'force' have meanings that are close to the meaning of 'force' in Physics:

1. *I forced the box into the closet.*
2. *Jim was forcing the nut on the bolt.*
3. *I forced myself to go to class every day.*
4. *My parents forced me to go to university.*
5. *The force on the ball made it move.*
6. *The bomb exploded with great force.*
7. *I was hit by the force of the 18 wheeler.*
8. *She used a very forceful tone of voice.*
   Answer: (A) 1, 2, 4, 3 (B) 3, 4, 8 (C) 1, 2, 5 (D) 5, 6, 7

These sentences have two conceptually relevant things in common: (1) the word 'force' was used as a verb linked to an agency [or an assumed agency as in (5)] and every use contains a preposition such as 'into' or 'onto' or 'on' and another object. This makes the verb 'force' a transitive verb involving the interaction of two objects. This seems to loosely fit in with the common definition of force as a push or pull on an object. At the University of Technology, Sydney, the textbook used by this group of students at the time was Fundamentals of Physics (8th Edition) by Jearl Walker (Halliday, Resnick, & Walker, 2008). Unfortunately, the way it discusses 'force' on page 87 is a somewhat confusing. For instance, the sentence 'The force is said to *act* on the object to change its velocity.' (Italic is theirs). This gave the impression that somehow 'force' itself is an agency like a person causing the object to change its velocity. Such differences in interpretation were discussed in small group activities.

During semester one in 2009 at the University of Technology, Sydney, two calculation type class tests and a final examination were conducted. This enabled the results of these tests and examination to be compared with similar tests and examination used in semester one 2008. In addition to this, a Physics concept survey (Appendix A) was also administered. This survey combines 16 questions related to definitions of physics concepts such as 'force', 'momentum' and 25 questions on 'thermodynamics' taken from (Yeo & Zadnik, 2001). 269 students completed the survey. However, because this test was not administered in 2008, we did not attempt to make comparative interpretations.

DO STUDENTS REALLY HAVE A LANGUAGE PROBLEM?

Though we suspected that language problem might be the cause of failure to engage, we liked to collect evidence to prove from student performance that a genuine problem actually existed. As part of a larger study across first year courses in Chemistry, Physics and Biology, a set of technical terms common to the various science disciplines was selected for testing. Our study revealed that the common denominator of technical terms in first year science subjects was rather small, 10 with chemistry and biology, and 5 with physics. So, apart from these 5 common terms some disciplines involved in the study chose to include more terms or to apply the intervention to almost all technical terms used in class, or include all daily language words that have a different meaning in the context of their discipline. The physicist in this study chose the latter option, i.e., a wholesome application of the intervention. The study took place from autumn 2007 to autumn 2010.

In our reference class, in autumn 2007, we conducted a pre- and post evaluation of the understanding of selected technical words, which have slightly different meanings in and out of disciplinary context. The pre-evaluation was done in a multiple choice survey in week 2 of our 13 weeks semester where students would not have encountered the technical terms in the lecture yet. The post-evaluation was conducted in week 8 after all technical terms were covered and using the same survey questions. University

Table 2. Proportion of students studying Physical Modeling from non-English speaking background in one teaching year (there are two intakes of Physical Modeling students each year)

	A2007	S2008	A2008	S2009
Non-English background	67%	49%	46%	64%

Table 3. One of the multiple-choice multiple answer questions used in this study to discriminate complete and incomplete understanding of the word 'research'

Choice		Correct answer	Value
A	Observing the results of a series of chemical reactions	Yes	33.33%
B	Answering an exam question	No	–10%
C	Looking for information in the library or on-line	Yes	33.33%
D	Copying information out of a textbook	No	–10%
E	Testing a hypothesis	Yes	33.33%

policy required that all our surveys had to be done voluntarily for students. Not surprisingly, the end of semester survey in week 8 had only a limited overlap with students who also did the pre-evaluation survey in week 2.

In order to address the problem of understanding the technical language of physics in a first year physics class, we introduce an elicit-confront-resolve approach in lectures as described by McDermott (1991). We focused on those technical terms that in daily language may have a different meaning and those terms that we suspected a student of non-English speaking background might have some problems with. The fraction of students in our Physical Modeling course from non-English speaking background is fairly high compared to what one would expect in European or North American colleges with similar class sizes (Table 2).

In this project, we adopted a model similar to that of Jacobs (1989); we distributed three tests through our Learning Management System (LMS). In Test 1, we tested comprehension of five common terms (research, power, density, equilibrium, graph) (Zhang, Lidbury, Schulte, Bridgeman, Yates, et al., 2008). This approached allowed us to discriminate between complete and incomplete understanding and respective confidence levels.

For example, in the question related to 'research', the following is provided:

'We carry out research to find out the answers to scientific questions.' What is meant by the word 'research' in the sentence above? (There may be more than one correct answer).

Questions are then provided with five answers in Table 3.

Each correct answer is allocated 5 marks each. If a student chose the three correct answers, a score of 15/15 would be achieved. This indicates a complete understanding of the term in different contexts. Any score less than 15 suggest an incomplete understanding of the term. The second part following each question was a related confidence question which assesses how confident a student is of the choice(s) made, shown in Table 4 and Table 5.

From the 412 students who were asked to voluntarily participate in the tests we received a 9.7% response rate for Test 1 and 5.6% for Test 2, respectively. (Kerri-Lee Krause, et al., 2005)

Table 4. Multiple-choice question to determine the confidence level; a single answer response was enforced

Choice		Value
A	Yes, I understand the meaning of this word	100%
B	No, I do not understand the meaning of the word	-100%
C	I have some idea of the meaning of this word	0%

Table 5. Level of complete understanding of commonly used terms in first year physics

Terms	Complete understanding	Ranking (easy=1, hard=5)
Research	15.0%	5
Equilibrium	17.5%	4
Power	25.0%	3
Density	27.5%	2
Graph	42.5%	1

The relatively low response rate correlates with observations made by Krause et al. (Krause et al., 2005) who conducted a study of first year students experience in Australian universities. Of the 8,486 students surveyed, there were 2344 usable responses. Of the 2,344 usable responses the response rate for science and engineering students for a typical, newer metropolitan university was 8% and 5%, respectively.

The survey confirmed our suspicion, i.e., language was indeed a problem with physics students (Table 5). It was surprising that only 15% of the participating students achieved full understanding of the word "research". For a word which is used more specifically in physics, such as 'density', only 27.5% of the students achieved full understanding. A close examination of final exam papers also provided evidence that students were struggling with these terms and with understanding related physics concepts. In addition to the five terms which were used in all science discipline of this research project, for the intervention part we also paid close attention to the terms of velocity and speed used in conjunction with the concept of kinetics; impulse in conjunction with concept of momentum; mass, weight and force in conjunction with the concept of Newton forces; and work and energy in conjunction with the concept of thermal processes (Zhang, et al., 2009).

RESULTS

In the (1999)suggested that 'Part of the difficulty (of Physics), perhaps even a large part, lies in language and the way we use it in the practice of physics, and in the teaching of physics.....we simply do not spend enough time with our students for us to use the laboratory vernacular and expect them to assimilate it' (Williams, 1999, p. 671). The importance of precise use of language, particularly in introductory teaching of physics, has been emphasized by Arons (1990):

Since the words, to begin with, are metaphors, drawn from everyday speech, to which we give profoundly altered scientific meaning, only vaguely connected to the meaning in everyday speech, the students remain unaware of the alteration unless it is pointed to explicitly many times-not just once'.

The construction of the Physics Concept survey took the advice of Williams (1999). This gives the impression that somehow 'force' itself is an agency like a person causing the object to change its velocity'.

Students also seemed not to have understood the definition of 'mass' (Question 11). This is demonstrated by 3% of the students choosing (d: sensation one has when trying to kick a football) the correct answer whereas 50% of the students chose (b: density) as the correct answer. The writer of the textbook tried their best to clear up the confusion between the use of 'mass' in everyday language with the concept of 'mass' in physics by saying 'you can have a physical sensation of mass only when you try to accelerate a body, as in the kicking of a baseball or a bowling ball'(Halliday, et al., 2008, p. 91). However, since most students might not have read the textbook in detail, this useful discussion is likely to be missed totally.

Question 14 is a question on the definition of Newton's first law. However, the key to get the correct answer lies in the students' understanding of the words 'constant' and 'uniform'. On page 88 of the textbook, the writer writes 'we can conclude that a body will keep moving with constant velocity if no force acts on it'. However, from students' answers, only 28% chose (a) the correct answer, 18% chose (b), 4% chose (c) and 48% chose both (a) and (b). This means 48% of the student body thought 'constant' has the same meaning as 'uniform'.

Question 15 on the understanding of 'net force' tests students' precise understanding of the cause of an object's acceleration. The fact that 50% of the students chose (a) as the correct answer ((b) is the correct answer) suggests that students lacked the ability to use this concept precisely. Of course, as we know if the vector sum of the forces is zero, there will be no acceleration. Only when the vector sum of the forces is larger than 0 N, will there be acceleration.

Williams (1999) suggested that students often associate the terms 'action' and 'reaction' improperly in the physics context. This is because the normal definitions of action and reaction often suggest a 'temporal delay between action and reaction' (p.676). In fact, the physics definition of Newton's third law emphasizes 'the simultaneity of the forces or the symmetry of the force relationship'. This group of students did not fall into this trap with 68% of students choosing (a) as the correct answer. The following tables show the questions answered with a high and medium degree of correctness (Table 6 and Table 7 respectively).

The rest of the Physics Concept survey (Question 17-41) concerns concepts in Thermodynamics. Questions 17-41 were taken from (Yeo & Zadnik, 2001). Unfortunately, all students only achieved a low degree of correctness (less than 48%) with this part of the survey. This is much lower than the results reported in Yeo & Zadnik (2001). Because these surveys were completed anonymously, it was difficult to conduct statistical tests such as item discrimination tests to verify the appropriateness of this survey for this cohort of students in physics.

Final Examination Performances

In this study, we had the final exam performance as our baseline for improved performance, which was what we had planned to investigate at the beginning of our project. Distribution of marks is shown in Table 8.

Despite an increase in total student numbers of 23% from autumn 2008 to autumn 2009, students generally performed better in the kinetics, momentum and forces sections of the final exam by between 4.46% - 14%, and showed lower performance in the energy conservation related section (Table 5). The number of questions in the energy conservation section differed from 2008 to 2009. In that sense, a clear significant improvement in final exam performance was observed for only the forces section ($p=0.005$; $p<0.01$).

Absorption of language and correct use of its association with physics concepts involves a

Table 6. Questions answered with a high degree of correctness (greater than 70% correct) in semester one in 2008

UTS question	Concepts	Content	% Correct
2	Impulse	I equal the **change** in an object's **momentum**, i.e. the product of the total mass and the velocity of the centre of mass.	74
4	Momentum	Which one(s) of the following sentences containing 'momentum' have meanings that are close to the meaning of 'momentum' in Physics: 1.After their touchdown, the other team had the momentum. 2. The football player has a lot of momentum when he tacked his opponent. 3. Our team gained momentum in the game after intercepting the ball. 4. As the car rolled down the hill it gained momentum.	76
6	Normal force	This is the force that is acting along the normal (perpendicular) to the contact surface.	83
7	Static friction	These are forces that are acting parallel to the contact surface. This force exists when the surfaces are not moving relatively to each other.	74
8	Gravitational force	It is the force that the earth exerts on any object. It is directed towards the centre of the earth. Its magnitude is given by Newton's second law.	88
9	Centripetal acceleration	This is the acceleration that is due to change in direction, not speed (in uniform circular motion) and it points toward the centre. a=v2/R	88
10	Weight	This is a vector force with which Earth is pulling on an object with.	83
12	Force	This is the vector describing the interaction between two objects (pull or push). The unit of force is Newton, N.	86

Table 7. Questions answered with a medium degree of correctness (between 48%-70% correct) in semester one in 2008

UTS question	Concepts	Content	% Correct
1	Momentum	I am a vector quantity of a particle which is defined as product of the mass of the particle and its velocity. The SI unit for me is kg. m/s.	49
3	Impulse	Which one(s) of the following sentences containing 'impulse' have meanings that are close to the meaning of 'impulse' in Physics: **1.An impulse made her change her mind.** 2. My first impulse was to kick him. 3. In time of crisis we act on our impulses. 4. My sister is an impulsive shopper.	49
5	Force	Which one(s) of the following sentences containing 'force' have meanings that are close to the meaning of 'force' in Physics: **1.I forced the box into the closet. 2. Jim was forcing the nut on the bolt. 3. I forced myself to go to class everyday.** 4. My parents forced me to go to college. **5. The force on the ball made it move**. 6. The bomb exploded with great force. 7. I was hit by the force of the 18 wheeler. 8. She used a very forceful tone of voice.	36
11	Mass	This is a scalar quantity which describes how difficult it is to change an object's velocity (sluggishness or inertia of the object). Which one of the statements below describes the Physics definition of mass?	49
13	Newton's first law	An object cannot continue to move with the same speed and in the same direction. It will eventually stop.	65
14	Newton's first law	Every object ________ in its state of rest or _______ velocity in a _________ line, unless it is compelled to change that state by _____ force acting on it.	28
15	Net force	When will an object accelerate?	45
16	Newton's third law	Which of the following statements about Newton's 3rd law are correct? 1. For every action there is an equal and opposite reaction simultaneously. 2. For every action there is an equal and opposite reaction but a time delay is allowed. 3. Forces occur in the action-reaction pairs simultaneously.	68

Table 8. Distribution of marks in the final exam across different sections

		Percentage of full marks			
	Students	Kinetics	Momentum	Forces	Energy
2008	388	79.77	69.3	32.2	63
2009	478	83.33	75.1	46.3	53.5
change	23.19%	4.46%	8.37%	14.1%	-15.0%
p-value		0.57	0.32	0.005	0.07

more lateral learning process and an improved understanding of physics may show up in many different ways, not just final exam performance. This is also appreciated in the overall weighting of assessment where final exam is worth 45% and class tests, home work and laboratory reports 55%. So, we had a closer look at the overall semester performance of students, which included weekly problem solving assignments, weekly laboratory classes, a major laboratory report and two class tests. In analyzing the overall performance of students before and after the learning intervention, the use of RF clickers enabled us to discriminate between students who participated in the intervention ("opt in" students) and those who chose not to participate ("opt out"). The "opt in" students were those who came to the lectures, engaged in the dialogue during concept development, participated in the multiple-choice lecture questions and following revision of concepts and further clarification of technical terms used where needed. The "opt out" students did not attend lectures; instead they chose to rely on copies of the lecture slides without additional clarification of concepts and related technical terms.

Our baseline semester was autumn 2008 (A2008). In A2008, we used clickers, however, without language specific intervention. Overall, students who used clickers in lectures had better results in the final exam (Figure 3) with an average mark 7.5% higher (65.4%) compared to their peers (57.5%) that chose to use printed lecture slides only (Table 6). The semester failure rate for the opt-out students within their own group was 65% (Figure 4) compared to 12% for those who used clickers in class (there is a 98% or higher presence of clickers in lectures). The end of semester average mark was 60.3%.

In the following semester (spring 2008, S2008), where the language specific intervention was introduced, the average end of semester mark dropped to 46.6%, which was consistent with similar atypical poor performance of this class in other courses, indicating a cohort specific phenomenon rather than intervention specific one. Yet overall, opt-in students performed with an average end of semester mark of 52.4% (Figure 4) considerably better by a margin of 13.4% (Table 6) compared to their peers (39.1%) who were not subject to the intervention.

In the following semester, autumn 2009 (A2009) we could not use RF clickers in class and, hence, had no verifiable means to discriminate between opt-in and opt-out students. We continued with the intervention in A2009 by using the simple raising-of-hands method in class. In spring 2009 (S2009), RF clickers were available again and we continued with the collection of data. By then, we gained more experience with the routine application of the language intervention in class. Our opt-in students achieved an average end-of-semester mark of 56.9% (Table 6), which was a distinctive 17% higher compared to opt-out students who had an average of only 39.9% (Table 6). Not surprisingly, within their own group opt-out students had a failure rate of 72% compare to 21% of their opt-in peers (Figure 4). The overall, average end of semester mark was 49.3% (Figure

Figure 3. Overall student performance of "opt in" and "opt out" students as measured by the total end of semester mark

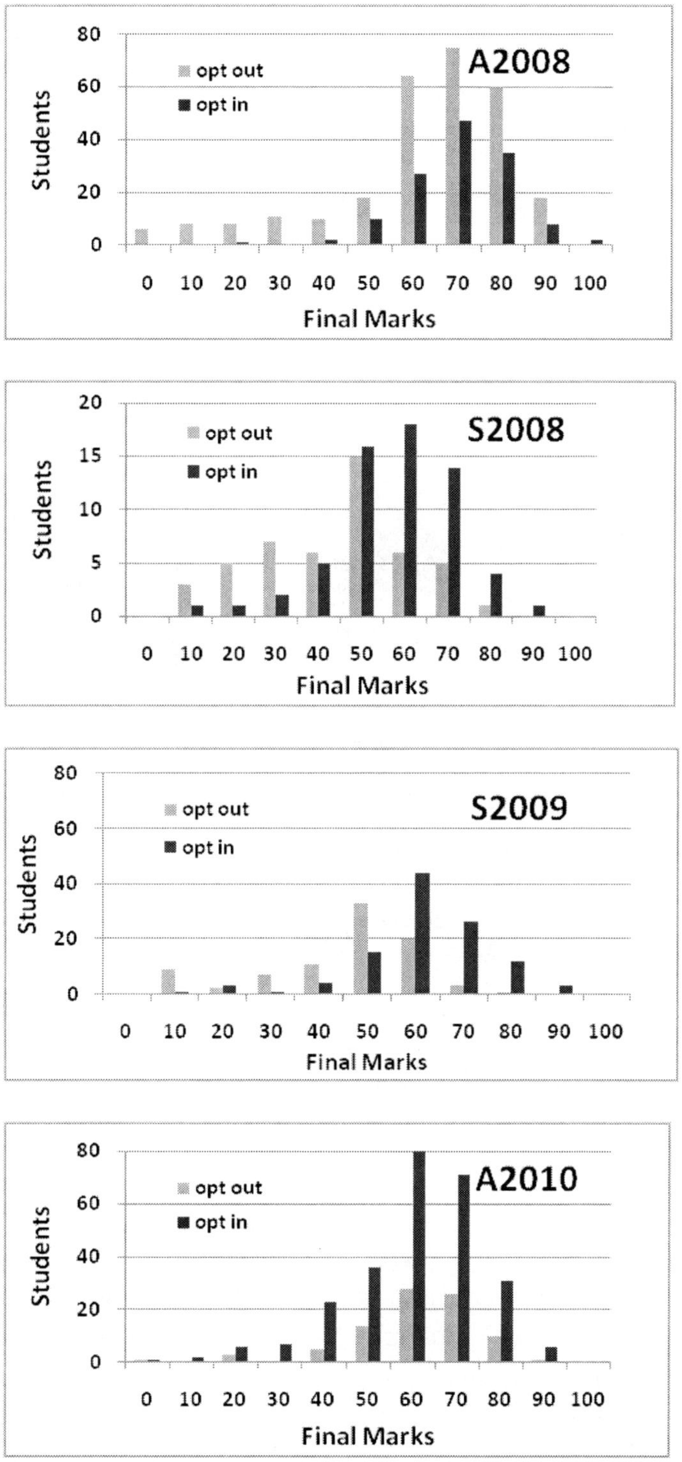

Figure 4. Failure rates within the group of "opt in" and "opt out" students, respectively

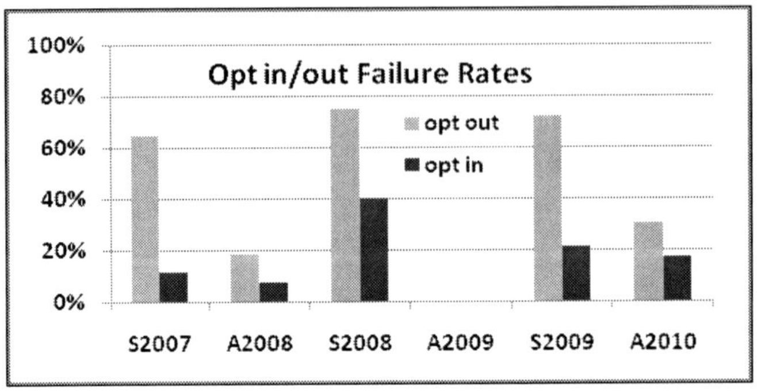

Figure 5. Average final marks of "opt in" and "opt out" students in comparison

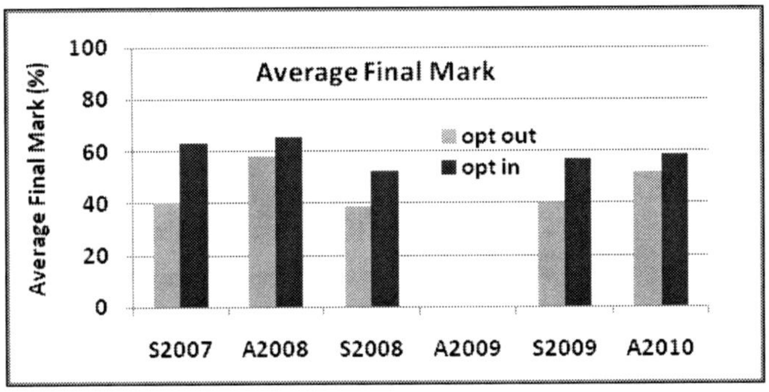

Table 9. Average final marks of intervention participating (opt-in) and non-participating (opt-out) students between over the period between 2008 and 2010

	Average final marks and respective standard deviations				
Student groups	**A2008**	**S2008**	**A2009**	**S2009**	**A2010**
opt -out	(57.9±20.1)%	(39.1±18.1)%	n/a	(39.9±17.0)%	(51.3±15.9)%
opt –in	(65.4±12.5)%	(52.4±14.4)%	n/a	(56.9±13.2)%	(58.4±12.9)%
% improve	+7.5%	+13.4%	n/a	+17.0%	+7.1%

5). Average final marks from 2008 to 2010 are shown in Table 9.

Looking at the student failure rates, it appears that having spent quality time on key words defining the physics context and concepts has proved to be a worthwhile effort. Naturally, with only a limited time available for face to face teaching in lectures, something has to be omitted to make space for the time spent on language clarification. In our case, we reduced the number of real world examples related to the topic covered as well as some example problems. Despite this, we found that our opt-in students, those who were exposed to the teaching intervention, achieved marks that were an average 15.2% higher compared to their opt-out peers.

We performed a what-if analysis of opt -out students to see how many students of the opt-out group might have passed the course (>50%) if they had participated in the teaching intervention.

Figure 6. Proportion of "opt out" students within 15% below a pass mark (50%)

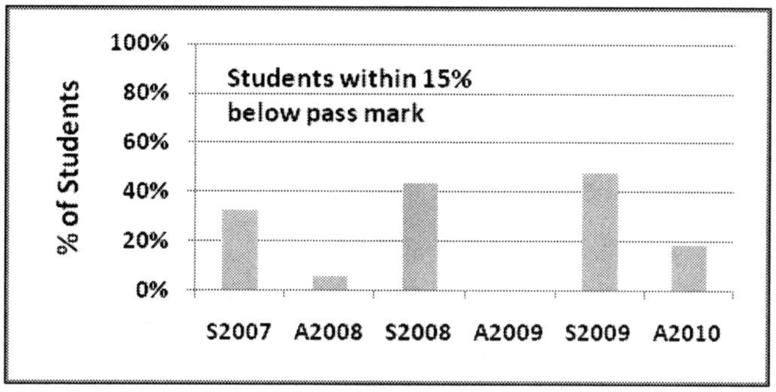

Table 10. Course failure rates within language groups

	S2007	A2008	S2008	A2009	S2009
English background	47%	10%	20%	17%	39%
Non-English background	24%	10%	56%	17%	34%

Their average end of semester mark would have indeed been an average of 15% higher than their original mark. Figure 6 shows a summary of the percentage of students that were within 15% below pass mark. We were surprised to see that in the first semester of the intervention, in S2008, about 44% of the opt-out students had been within the reach of a close pass if they had participated in the intervention. In S2009, where we had very large number of students from a non-English speaking background (68%), we found 48% of opt-out students within just 15% below pass mark.

On the basis of the picture that is emerging from the above student statistics one might be tempted to take the view that our fairly large proportion of students from a non-English speaking background may skew the data. A closer look at the failure rates in the context of students' language background reveals (Table 7) that apart from the atypical irregularity in S2008 there does not appear to be much evidence pointing toward a language background issue in terms of learning the language of physics (Jacobs, 1989). In fact, during the semester where we had the highest proportion of students from a non-English speaking background (Table 3, S2009/64%), we find that the proportion of students from an English speaking background who failed the subject (39%) is larger than that of their non-English speaking peers (34%), as shown in Table 10.

In 2008, semester one (A2008), the question 'I received constructive feedback when needed' on the Student feedback results (SFR) only received a rating of 2.70/5 and this lack of satisfaction was confirmed by the open ended questions section of the SFR which showed that 50% of students' complaints centered on how and where the tutorials were run. Students tended to see them as basically just another extension of the lecture. As one student put it:

The idea of a single tutorial for the whole subject in the lecture theatre was terrible. It basically turned into a tedious lecture on how to do questions which went too fast to grasp and didn't give any value since we weren't actually doing the questions. Proper tutorials are needed.

In 2009, semester one, the tutorial classes were renamed as 'workshops' and language strategies were consistently and extensively included. These interventions not only improved the rating on feedback to 3.56/5 but hardly any negative responses were received about the workshops although they were still conducted in a single full class lecture room. In fact one student suggested the following:

I reckon it will help the students if there were more workshops than lectures. To be successful in this subject, students must understand physics concepts inside out. In order for this to happen, problem solving workshop is the key. I felt that the workshops helped me more than lectures as it is more hands-on. What I suggest is that instead of having 3 hour lectures and 1 hour workshop a week, why not have 2 hour workshops and 2 hour lectures instead.

Clearly, although our constraints demanded that workshops were held in a full class lecture hall (conducted by a single lecturer), the break-up of students into small working groups within the large lecture class improved students' learning experience. Ideally, one would like to have smaller problem solving workshops or tutorials. However, such arrangements are resource intensive and sometimes cannot be arranged for all first year courses. Nevertheless, as a result of this study students have obviously identified our full class problem solving workshop arraignment as a preferred mode of learning within the given constraints.

FUTURE RESEARCH DIRECTIONS

We were pleased that the hunch that we had or our hypothesis, that if we are conscious about the dual problem in physics learning, i.e., understanding the special language used in physics and the intimately related physics concepts, our students will have a better understanding of physics they learn and, consequently, will be able to untangle physics problems more successfully. Nevertheless, one cannot help to feel that we have merely scratched the surface of problem that impacts all first year science service teaching courses. That is, students are required to learn concepts and skills of a technical area, which is not an immediate part of the core pool of knowledge of their major and does not lie within students' main interest. The question that arises is, how can we best teach non-major students the multitude of concepts and skills they are expected to learn within the limited time prescribed by their course schedule. What we have learned in this study is that if we take effort to untangle the language what we use in our first year physics lectures, all students show improved performance in dealing with physics problems. Is there more that we can learn from linguistic and language learning that can be embedded in our teaching of freshmen in science classes? Is it possible and feasible to restructure a first year science class in such a way that it a methodology of language learning as overarching pedagogy for all science skills and concepts to be learned?

CONCLUSION

Integrating language learning practices in first year physics has shown to have some merit, not only where there is a fairly large proportion of students coming from a non-English speaking background but appears that students from an English speaking background benefit from such intervention too. We identified words known as key words within a specific physics concepts and introduced opportunities for students to further contemplate about the meaning of these words within the context. Multiple choice questions emphasizing on key words were provided online as additional practice questions or occasionally embedded in online assignment. A more rigorous approach was taken in lectures where each 50 minutes lecture had a least two episodes of word and context related language exercises in form of simple multiple-choice questions. Using the immediately presented demographic of responses, each answer was analyzed for its correctness and use of language and was then placed into context of physics concept at hand. This approach provided an opportunity to discriminate between the common use of a key word and its specific meaning within the physics context. It also allowed addressing still prevailing misconceptions about the physics concept and common experience and communication. Once prevailing misconceptions were discovered, they were covered again through small group activities in the bi-weekly problem solving workshops. This approach was deemed to be very effective by students.

The technical means (clickers) that we had at our disposal allowed us to discriminate between students who participated in the learning intervention (opt-in) and those who were not exposed to the intervention (opt-out).We found that opt-in students benefitted most by performing better in the kinetics, momentum and forces part of the final exam compared to their opt-out peers. Since assessment of students covers many individual components (assignments, reports, tests, etc.) and each component requires a good understanding of the language of physics, we analyzed the overall performance of students in this course and found that opt-in students performed an average 15% better than their opt-out peers. The positive impact makes it worthwhile for us to continue with this approach in our lectures in future.

One of the key achievements of this study was that it succeeded in changing the teaching practices of the physics lecturer; not by imposing a set of 'best practices,' but by directly involving the lecturer in designing, testing and implementing practices that would be used in teaching. Materials created by the lecturer have been provided in this book for teaching to the general public who might be interested in science education. The study provides a sustainable model for professional development for academic staff.

It is worthwhile to note that although the encouraging result we received is noteworthy, our circumstances might be unique and results may not be easily generalized. Almost all science and technology inclined universities have large first year physics service teaching courses. Our cohort of students has a 40% and larger proportion of students from a non-English speaking background. Nevertheless, since the improvement in student performance we saw is quite remarkable, it would be interesting to see what impact the language intervention has on a similar course with more than 90% of students from a native English speaking background. After all, the language of physics is foreign to most non-physics majors.

ACKNOWLEDGMENT

This work has been supported by a grant (CG7-441) provided by the Australian Learning and Teaching Council. The author is in particular grateful to the supporting linguist in this project, Felicia Zhang (University of Canberra, Australia), who sit in my lectures pretending to be one of my students with little or no physics background, sieving through hundreds of pages of textbook and lecture notes and trying to get her head around the language used in physics lectures and the many assignment problems, and who, most importantly, gave advise on the implementation of the interventions we have introduced throughout the project. The author is also grateful to Les Kirkup for valuable comments to the manuscript.

REFERENCES

Angell, C., Guttersrud, O., Henriksen, E. K., & Isnes, A. (2004). Physics: Frightful, but fun- Pupils' and teachers' views of physics and physics teaching. *Science Education*, *88*, 683–706. doi:10.1002/sce.10141

Arons, A. B. (1990). *A guide to introductory physics teaching*. New York, NY: Wiley.

Brown, D. E. (1992). Using examples and analogies to remediate misconceptions in physics: Factors influencing conceptual change. *Journal of Research in Science Teaching*, *29*(1), 17–34. doi:10.1002/tea.3660290104

Chin, C., & Brown, D. E. (2002). Student-generated questions: A meaningful aspect of learning in science. *International Journal of Science Education*, *24*(5), 521–549. doi:10.1080/09500690110095249

Cummings, K., Marx, J., Thornton, R., & Kuhl, D. (1999). Evaluating innovation in studio physics. *American Journal of Physics*, *67*(S1), S38–S44. doi:10.1119/1.19078

Dupin, J. J., & Johsua, S. (1989). Analogies and modeling analogies in teaching: Some examples in basic electricity. *Science Education*, *73*(2), 207–224. doi:10.1002/sce.3730730207

Elrod, S. (2007). *Genetics concept inventory*. Retrieved from http://bioliteracy.net/Readings/papersSubmittedPDF/Elrod.pdf

Flavell, J. H. (1976). Metacognitive aspects of problem solving . In Resnick, L. B. (Ed.), *The nature of intelligence* (pp. 231–236). Hillsdale, NJ & New York, NY: Lawrence Erlbaum Associates.

Garvin-Doxas, K., & Klymkowsky, M. W. (2008). Understanding randomness and its impact on student learning: Lessons learning from building the biology concept inventory (BCI). *CBE Life Sciences Education*, *7*(2), 227–233. doi:10.1187/cbe.07-08-0063

Gijbels, D., Dochy, F., Van Den Bossche, P., & Segers, M. (2005). Effects of PBL: A meta-analysis from the angle of assessment. *Review of Educational Research*, *75*(1), 27–36. doi:10.3102/00346543075001027

Gunstone, R. F. (1994). The importance of specific science content in the enhancement of metacognition . In Fensham, P. J., Gunstone, R. F., & White, R. T. (Eds.), *The content of science: A constructivist approach to its teaching and learning*. London, UK: Routledge.

Halliday, D., Resnick, R., & Walker, J. (2008). *Fundamentals of physics* (8th ed.). Hoboken, NJ: John Wiley & Sons, Inc.

Halloun, I. A., & Hestenes, D. (1985). The initial knowledge state of college physics students. *American Journal of Physics*, *53*(11), 1043–1055. doi:10.1119/1.14030

Hestenes, D., & Wells, M. (1992). A mechanics baseline test. *The Physics Teacher*, *30*(3), 159–166. doi:10.1119/1.2343498

Hestenes, D., Wells, M., & Swackhamer, G. (1992). Force concept inventory. *The Physics Teacher*, *30*(3), 141–158. doi:10.1119/1.2343497

Itza-Ortiz, S. F., Rebello, N. S., Zollman, D. A., & Rodriguez-Achach, M. (2003). The vocabulary of introductory physics and its implications for learning physics. *The Physics Teacher*, *41*(6), 330–336. doi:10.1119/1.1607802

Jacobs, G. (1989). Word usage misconceptions among first-year university physics students. *International Journal of Science Education*, *11*(4), 395–399. doi:10.1080/0950069890110404

Kempa, R. F., & Ayob, A. (1995). Learning from group work in science. *International Journal of Science Education*, *17*(6), 743–754. doi:10.1080/0950069950170606

Koch, A. (2001). Training in metacognition and comprehension of physics texts. *Science Education*, *85*(6), 758–768. doi:10.1002/sce.1037

Krause, S., Birk, J., Bauer, R., Jenkins, B., & Pavelich, M. J. (2004). *Development, testing, and application of a chemistry concept inventory*. Paper presented at the 34th ASEE/IEEE Frontiers in Education Conference. from http://fie-conference.org/fie2004/

Krause, K.-L., Hartley, R., James, R., Mcinnis, C., & Centre for the Study of Higher Education, University of Melbourne (2005). *The first year experience in Australian universities: Findings from a decade of national studies*. Retrieved from http://www.dest.gov.au/

Lemke, J. L. (1998). *Teaching all the languages of science: Words, symbols, images, and actions*. Retrieved from http://academic.brooklyn.cuny.edu/education/jlemke/papers/barcelon.htm

Maloney, D. P., O'kuma, T. L., Hieggelke, C. J., & Van Heuvelen, A. (2001). Surveying students' conceptual knowledge of electricity and magnetism. *American Journal of Physics*, *69*(S1), S12–S23. doi:10.1119/1.1371296

Martyn, M. (2007). Clickers in the classroom: An active learning approach. *EDUCAUSE Quarterly*, *30*(2). Retrieved from http://www.educause.edu/.

Mcdaniel, C. N., Lister, B. C., Hanna, M. H., & Roy, H. (2007). Increased learning observed in redesigned introductory biology course that employed Web-enhanced, interactive pedagogy. *CBE Life Sciences Education*, *6*(3), 243–249. doi:10.1187/cbe.07-01-0003

Mcdermott, L. C. (1991). Millikan lecture 1990: What we teach and what is learned-Closing the gap. *American Journal of Physics*, *59*, 301–315. doi:10.1119/1.16539

Mehmet, S. (2009). Exploring university students' expectations and beliefs about physics and physics learning in problem based learning context. *Eurasia Journal of Mathematics . Science & Technology Education*, *5*(4), 321–333.

Nussbaum, J., & Novick, S. (1983). Alternative frameworks, conceptual conflict and accommodation: Toward a principled teaching strategy. *Instructional Science*, *11*(2), 183–200.

Odom, A. L., & Kelly, P. V. (2001). Integrating concept mapping and the learning cycle to teach diffusion and osmosis concepts to high school biology students. *Science Education*, *85*(6), 615–635. doi:10.1002/sce.1029

Pipes, R. B., & Wilson, J. M. (1996). A multimedia model for undergraduate education. *Technology in Society*, *18*(3), 387–401. doi:10.1016/0160-791X(96)00020-6

Pollock, S. J. (2008). *Comparing student learning with multiple research-based conceptual surveys: CSEM and BEMA*. Paper presented at the Physics Education Research Conference 2008. Retrieved from http://www.colorado.edu/

Redish, E. F. (1994). The implications of cognitive studies for teaching physics. *American Association of Physics Teachers*, *62*(6), 796–803.

Richardson, A., & Zhang, F. (2008). *Language support for statistics learners*. Paper presented at the OZCOTS 2008 - 6th Australian Conference on Teaching Statistics. Retrieved from http://sky.scitech.qut.edu.au/~macgilli/ozcots2008/

Richardson, A., Zhang, F., & Lidbury, B. A. (2008). *Activating multiple senses in learning Statistics*. Paper presented at the Assessment in Science Teaching and Learning Symposium. from http://sydney.edu.au/

Ritchie, S. M. (2001). Actions and discourses for transformative understanding in a middle school science class. *International Journal of Science Education*, *23*(3), 283–299. doi:10.1080/095006901750066529

Roth, W.-M., & Roychoudhury, A. (1994). Science discourse through collaborative concept mapping: New perspectives for the teacher. *International Journal of Science Education, 16*(4), 437–455. doi:10.1080/0950069940160405

Roy, H. (2003). Studio vs interactive lecture demonstration–Effects on student learning. *Bioscene: Journal of College Biology Teaching, 29*(1), 3-6. Retrieved from http://acube.org/pdf/bioscene/v29-1.pdf

Touger, J. S. (1991). When words fail us. *The Physics Teacher, 29*(2), 90–95. doi:10.1119/1.2343227

Wellington, J. J., & Osborne, J. (2001). *Language and literacy in science education.* Buckingham, UK & Philadelphia, PA: Open University.

Wileyplus. (2010). *Homepage.* Retrieved from http://www.wileyplus.com

Williams, H. T. (1999). Semantics in teaching introductory physics. *American Journal of Physics, 67*(8), 670–680. doi:10.1119/1.19351

Wu, H.-K., Krajcik, J. S., & Soloway, E. (2001). Promoting understanding of chemical representations: Students' use of a visualization tool in the classroom. *Journal of Research in Science Teaching, 38*(7), 821–842. doi:10.1002/tea.1033

Yeo, S., & Zadnik, M. (2001). Introductory thermal concept evaluation: Assessing students' understanding. *The Physics Teacher, 39*(8), 496–504. doi:10.1119/1.1424603

Yore, L. D., Florence, M. K., Pearson, T. W., & Weaver, A. J. (2006). Written discourse in scientific communities: A conversation with two scientists about their views of science, use of language, role of writing in doing science, and compatibility between their epistemic views and language. *International Journal of Science Education, 28*(2 & 3), 109–141. doi:10.1080/09500690500336601

Yore, L. D., & Treagust, D. F. (2006). Current realities and future possibilities: Language and science literacy-empowering research and informing instruction. *International Journal of Science Education, 28*(2 & 3), 291–314. doi:10.1080/09500690500336973

Zhang, F., Lidbury, B., Schulte, J., Bridgeman, A., Yates, B., & Rodger, J. (2008). *Language difficulties in first year science - An interim report.* Paper presented at the 2008 National UniServe Science Symposium: Assessment in Science Teaching and Learning. Retrieved from http://sydney.edu.au/

Zhang, F., Lidbury, B., Schulte, J., Bridgeman, A., Yates, B., & Rodger, J. (2009). *A cross-disciplinary approach to language support for first year students in the science disciplines.* Paper presented at the HERDSA 2009.

ADDITIONAL READING

Beichner, R. (1994). Testing student interpretation of kinematics graphs. *American Association of Physics Teachers, 62*(8), 750–762.

Bond, Z. S., & Small, L. H. (1983). Voicing, vowel, and stress mispronunciations in continuous speech. *Perception & Psychophysics, 34*(5), 470–474. doi:10.3758/BF03203063

Brookes, D. T., & Etkina, E. (2007). Using conceptual metaphor and functional grammar to explore how language used in physics affect student learning. *Physics Education Research, 3*(1), 1–16.

Brown, D. E. (1992). Using examples and analogies to remediate misconceptions in physics: Factors influencing conceptual change. *Journal of Research in Science Teaching, 29*(1), 17–34. .doi:10.1002/tea.3660290104

Chi, M. T. H., Feltovich, P. J., & Glaser, R. (1981). Categorization and representation of physics problems by experts and novices*. *Cognitive Science*, *5*(2), 121–152. .doi:10.1207/s15516709cog0502_2

Chi, M. T. H., Roy, M., & Hausmann, R. G. M. (2008). Observing tutorial dialogues collaboratively: Insights about human tutoring effectiveness from vicarious learning. *Cognitive Science*, *32*(2), 301–341. .doi:10.1080/03640210701863396

Ding, L., Chabay, R., Sherwood, B., & Beichner, R. (2006). Evaluating an electricity and magnetism assessment tool: Brief electricity and magnetism assessment. Physics Education Research, 2.

Fagen, A. P. (2003). *Assessing and enhancing the introductory science course in physics and biology: Peer instruction, classroom demonstrations, and genetic vocabulary*. Cambridge, Massachusetts: PhD, Harvard University.

Furió, C., Guisasola, J., Almudí, J. M., & Ceberio, M. (2003). Learning the electric field concept as oriented research activity. *Science Education*, *87*(5), 640–662. doi:10.1002/sce.10100

Genosko, G. (1994). *Baudrillard and signs: Signification ablaze*. London, New York: Routledge. doi:10.4324/9780203201145

Grimellini Tomasini, N., Pecori-Balandi, B., Pacca, J. L. A., & Villani, A. (1993). Understanding conservation laws in mechanics: Students' conceptual change in learning about collisions. *Science Education*, *77*(2), 169–189. doi:10.1002/sce.3730770206

Hake, R. (2007). Six lessons from the physics education reform effort. Latin-American Journal of Physics Education, 1(1), 24-31. Retrieved from http://www.journal.lapen.org.mx/

Hake, R. R. (1998). Interactive-engagement versus traditional methods: A six-thousand-student survey of mechanics test data for introductory physics courses. *American Journal of Physics*, *66*(1), 64–74. doi:10.1119/1.18809

Halloun, I. (1998). Schematic concepts for schematic models of the real world: The newtonian concept of force. Science Education, 82(2), 239-63. doi: 10.1002/(SICI)1098-237X(199804)82:2<239::AID-SCE7>3.0.CO;2-F

Halloun, I. A., & Hestenes, D. (1985). The initial knowledge state of college physics students. *American Journal of Physics*, *53*(11), 1043–1055. .doi:10.1119/1.14030

Halloun, I. A., & Hestenes, D. (1998). Interpreting vass dimensions and profiles for physics students Retrieved from http://modeling.asu.edu/R%26E/IHalloun/InterpVASS_SC&ED98.pdf

Heller, P., Keith, R., & Anderson, S. (1992). Teaching problem solving through cooperative grouping. Part 1: Group versus individual problem solving. American Journal of Physics, 69(7), (7), 627–36.

Jenaro Guisasola, J. M. A. J. L. Z. (2004). Difficulties in learning the introductory magnetic field theory in the first years of university. *Science Education*, *88*(3), 443–464. doi:10.1002/sce.10119

Kautz, C. H., Heron, P. R. L., Loverude, M. E., & McDermott, L. C. (2005). Student understanding of the ideal gas law, part 1: A macrosopic perspective. American Journal of Physics, 11(Novermber), 1055-63.

Kirkup, L., Scott, D., & Sharma, M. (2007). Teaching physics to non-physics majors: Models extant in Australian universities. Proceedings of the UniServe, Sydney.

Koch, A. (2001). Training in metacognition and comprehension of physics texts. *Science Education, 85*(6), 758–768. .doi:10.1002/sce.1037

Lee, L., & Nusbaum, H. C. (1993). Processing interactions between segmental and suprasegmental information in native speakers of english and mandarin chinese. *Perception & Psychophysics, 53*, 157–165. doi:10.3758/BF03211726

Maloney, D. P., O'Kuma, T. L., Hieggelke, C. J., & Van Heuvelen, A. (2001). Surveying students' conceptual knowledge of electricity and magnetism. *American Journal of Physics, 69*(S1), S12–S23. .doi:10.1119/1.1371296

Pollock, S. J. (2008). Comparing student learning with multiple research-based conceptual surveys: Csem and bema. Paper presented at the Physics Education Research Conference 2008, Edmonton, Alberta, Canada. Retrieved from http://www.colorado.edu/

Sabella, M, C. Henderson, C. Singh & American Institute of Physics (Eds.). (2009). Proceedings of the 2009 physics education research conference. Proceedings of the 2009 Physics Education Research Conference, Ann Arbor, MI. Melville, NY: American Institute of Physics.

Sabella, M., Henderson, C., Singh, C., & American Institute of Physics. (2009). 2009 physics education research conference: Ann arbor, mi, 29-30 July 2009. Melville, NY: American Institute of Physics.

Schulte, J. (2006). Delivering first year physics assignments with limited resources - an Australian three-centre study. Paper presented at the 2006 National UniServe Science Symposium: Assessment in Science Teaching and Learning, The University of Sydney, Sydney. Retrieved from http://sydney.edu.au/

Schulte, J., & Zhang, F. (in press). Physics concept surveys.

Touger, J. (2000). The role of language in learning physics: Beyond semantics. *American Journal of Physics, 68*(4), 306–307. doi:10.1119/1.19428

Touger, J. S. (1991). When words fail us. *The Physics Teacher, 29*(2), 90–95. doi:10.1119/1.2343227

Werker, J., & Logan, J. (1985). Cross-language evidence for three factors in speech perception. *Perception & Psychophysics, 37*, 35–44. doi:10.3758/BF03207136

Yeo, S., & Zadnik, M. (2001). Introductory thermal concept evaluation: Assessing students' understanding. *The Physics Teacher, 39*(8), 496–504. .doi:10.1119/1.1424603

Zhang, F., Lidbury, B., Schulte, J., Bridgeman, A., Yates, B., & Rodger, J. (2009). A cross-disciplinary approach to language support for first year students in the science disciplines. Paper presented at the HERDSA 2009, Charles Darwin University, Darwin.

KEY TERMS AND DEFINITIONS

Interactive Engagement (IE): strategies have been shown to enhance the gain in learning. In an IE class, the lecturer sets up a demonstration and asks the students to predict and write down, with consultation among themselves, their prediction of how the demonstration will work out.

Metacognition: refers to one's knowledge concerning one's own cognitive processes or anything related to them, e.g., the learning-relevant properties of information or data. For example, I am engaging in metacognition if I notice that I am having more trouble learning A than B; if it strikes me that I should double check C before accepting it as fact. (J. H. Flavell 1976, p. 232).

Peer Instruction: In 1991, Mazur began designing an instructional strategy for teaching called peer instruction. In 1997, he published a book called Peer Instruction: A User's Manual which provides details on this strategy. The seating arrangement plays an important role in the outcome of this method. For example, when low-performing students are seated in the front, their chance to do better increases. Meanwhile, the results of high-performing students who are seated in the back are not affected. In addition, when high-performing students are seated in the outer four corners of the classroom, the performance of the class as a whole increases.

Sapir-Whorf hypothesis: The linguistic relativity principle, or the Sapir–Whorf hypothesis is the idea that differences in the way languages encode cultural and cognitive categories affect the way people think, so that speakers of different languages will tend to think and behave differently depending on the language they use. The hypothesis is generally understood as having two different versions: (i) the *strong* version that says language determines thought and that linguistic categories limit and determine cognitive categories and (ii) the *weak* version that linguistic categories and usage influence thought and certain kinds of non-linguistic behavior. (http://www.wikipedia.com)

APPENDIX A: LANGUAGE DIFFICULTY SURVEY USED AT THE UNIVERSITY OF TECHNOLOGY, SYDNEY AUSTRALIA IN SEMESTER ONE 2008 AS A PRE-INTERVENTION ASSESSMENT

Physics Modeling Unit

Description: The following questions are part of a research project which is supported by the Australian Learning and Teaching Council. There are five universities participating in this research. The research project aims to improve our teaching in first year science subjects. Student participation in this project, i.e., answering the short questions, is voluntarily.

By answering the following short questions, you acknowledge that you have read information about this research given in the separate document on this folder.

Instructions: Answer the following set of five multiple choices multiple answer questions.

- **Question 1 (50 points).***The performance car has been tuned to have maximum power*. What is meant by the word "*power*" in the preceding sentence? (There may be more than one correct answer)
  A: The electricity produced by the car.
  B: The heat produced during combustion of fuel.
  C: The roaring sound from the exhausted pipe.
  D: The force exerted by the explosion on the piston.
  E: The force exerted by the engine on the wheels.
- **Question 2 (50 points).***Cold milk was poured and then stirred in a cup of hot coffee until it reached equilibrium*. What is meant by the word "*equilibrium*" in the preceding sentence? (There may be more than one correct answer)
  A: When the mixture shows a uniform blend of colour.
  B: When the mixture cools down.
  C: When the coffee and milk reach a state of complete mixture.
  D: When the temperature of both coffee and milk are the same.
  E: When no more stirring occurs.
- **Question 3 (50 points).***Jane carefully mixed 5 grams of sand with 100 mL of water in order to get the correct density for the mixture*. What is meant by the word "*density*" in the preceding sentence? (There may be more than one correct answer)
  A: The number of grams of sand and water per volume they occupy.
  B: The weight of sand that Jane added to the water.
  C: The softness of the mixture.
  D: The compactness of the sand after it has settled.
  E: The amount of something per unit volume
- **Question 4 (50 points).***To test scientific ideas, we conduct experiments*. What is meant by the word "*experiments*" in the preceding sentence? (There may be more than one correct answer)
  A: Observing animals or birds in the bush.
  B: Counting cells under a microscope.
  C: Growing bacteria on an agar plate.

D: Measuring the response of cells in tubes treated with or without a test substance.

E: Measuring how well some cells eat bacteria

- **Question 5 (50 points).***The temperature was measured at 5 minute intervals and plotted on a graph*. What is meant by the word "*graph*" in the preceding sentence? (There may be more than one correct answer)

  A: An XY plot of temperature versus time.

  B: A table of numbers representing temperature and time.

  C: A diagram expressing the connection between temperature and time.

  D: A picture of the experiment

  E: A piece of paper with lines on it.

APPENDIX B: PHYSICS CONCEPT SURVEY ADMINISTERED AT THE UNIVERSITY OF TECHNOLOGY, SYDNEY

1. **Definition 1:** I am a vector quantity of a particle which is defined as product of the mass of the particle and its velocity. The SI unit for me is kg. m/s.
   a. linear momentum
   b. momentum
   c. angular momentum
   d. both a and b
2. **Definition 2:** I equal the change in an object's momentum, i.e. the product of the total mass and the velocity of the center of mass.
   a. momentum
   b. acceleration
   c. impulse
3. **Definition 3:** Meaning of impulse. Which one(s) of the following sentences containing 'impulse' have meanings that are close to the meaning of 'impulse' in Physics: 1.An impulse made her change her mind. 2. My first impulse was to kick him. 3. In time of crisis we act on our impulses. 4. My sister is an impulsive shopper.
   a. 2, 3,
   b. 1
   c. 4
4. **Definition 4:** Meaning of momentum. (Taken from F. S. Itza-Ortiz, S. N. Rebello, D. Zollman, & M. Rodriguez-Achach, 2003). Which one(s) of the following sentences containing 'momentum' have meanings that are close to the meaning of 'momentum' in Physics: 1.After their touchdown, the other team had the momentum. 2. The football player has a lot of momentum when he tacked his opponent. 3. Our team gained momentum in the game after intercepting the ball. 4. As the car rolled down the hill it gained momentum.
   a. 3 and 4
   b. 1 and 2
5. **Definition 5:** Meaning of force. (Taken from F. S. Itza-Ortiz, et al., 2003, p. 42) Which one(s) of the following sentences containing 'force' have meanings that are close to the meaning of 'force' in Physics: 1.I forced the box into the closet. 2. Jim was forcing the nut on the bolt. 3. I forced myself to go to class everyday. 4. My parents forced me to go to college. 5. The force on the ball made it move. 6. The bomb exploded with great force. 7. I was hit by the force of the 18 wheeler. 8. She used a very forceful tone of voice.
   a. 1, 2, 4, 3,
   b. 3, 4, 8
   c. 1, 2, and 5
   d. 5, 6, 7
6. **Definition 6:** This is the force that is acting along the normal (perpendicular) to the contact surface.
   a. kinetic friction
   b. Normal force
   c. static friction

7. **Definition 7:** These are forces that are acting parallel to the contact surface. This force exists when the surfaces are not moving relatively to each other.
   a. kinetic friction
   b. static friction
   c. dynamic force
8. **Definition 8:** It is the force that the earth exerts on any object. It is directed towards the centre of the earth. Its magnitude is given by Newton's second law.
   a. gravitational force
   b. net force
   c. frictional force
9. **Definition 9:** This is the acceleration that is due to change in direction, not speed (in uniform circular motion) and it points toward the centre. a=v2/R
   a. gravitational force
   b. centripetal acceleration
   c. average acceleration
10. **Definition 10:** This is a vector force with which Earth is pulling on an object with.
    a. mass
    b. weight
    c. density
11. **Definition 11:** This is a scalar quantity which describes how difficult it is to change an object's velocity (sluggishness or inertia of the object). Which one of the statements below describes the physics definition of mass?
    a. weight
    b. density
    c. body size
    d. sensation one has when trying to kick a football
12. **Definition 12:** This is the vector describing the interaction between two objects (pull or push). The unit of force is Newton, N.
    a. Displacement
    b. Acceleration
    c. Force
13. **Definition 13:** Newton's first law. An object cannot continue to move with the same speed and in the same direction. It will eventually stop.
    a. True
    b. False
14. **Definition 14:** Newton's first law. Every object ________ in its state of rest or _______ velocity in a _________ line, unless it is compelled to change that state by _____ force acting on it.
    a. continues, constant, straight, net
    b. continues, uniform, straight, net
    c. stops, constant, straight,
    d. both a and b.

15. **Definition 15:** Net force. When will an object accelerate?
    a. When there is a net force on the object.
    b. When the net force is more than 0 N.
    c. When the net force on the object is zero N
16. **Definition 16:** Newton's 3rd law. Which of the following statements about Newton's 3rd law are correct? 1. For every action there is an equal and opposite reaction simultaneously. 2. For every action there is an equal and opposite reaction but a time delay is allowed. 3. Forces occur in the action-reaction pairs simultaneously.
    a. 1 and 3
    b. 2
    c. All the above

The following questions have been taken from (Yeo, S., & Zadnik, M., 2001).

17. **Thermal Concepts 1:** Ken takes six ice cubes from the freezer and puts four of them into a glass of water. He leaves two on the counter top. He stirs and stirs until the ice cubes are much smaller and have stopped melting. What is the most likely temperature of the water at this stage?
    a. -10°C
    b. 0°C
    c. 5°C
    d. 10°C
18. **Thermal Concepts 2:** What is the most likely temperature of ice cubes stored in a refrigerator's freezer compartment?
    a. -10°C
    b. 0°C
    c. 5°C
    d. It depends on the size of the ice cubes.
19. **Thermal Concepts 3:** The ice cubes Ken left on the counter have almost melted and are lying in a puddle of water. What is the most likely temperature of these smaller ice cubes?
    a. -10°C
    b. 0°C
    c. 5°C
    d. 10°C
20. **Thermal Concepts 4:** On the stove is a kettle full of water. The water has started to boil rapidly. The most likely temperature of the water is about:
    a. 88°C
    b. 98°C
    c. 110°C
    None of the above could be right.

21. **Thermal Concepts 5:** Five minutes later, the water in the kettle is still boiling. The most likely temperature of the water now is about:
    a. 88°C
    b. 98°C
    c. 110°C
    d. 120°C
22. **Thermal Concepts 6:** What do you think is the temperature of the steam above the boiling water in the kettle?
    a. 88°C
    b. 98 °C
    c. 110 °C
    d. 120 °C
23. **Thermal Concepts 7:** Lee takes two cups of water at 40°C and mixes them with one cup of water at 10°C. What is the most likely temperature of the mixture?
    a. 20°C
    b. 25°C
    c. 30°C
    d. 50°C
24. **Thermal Concepts 8:** Jim believes he must use boiling water to make a cup of tea. He tells his friends: 'I couldn't make tea if I was camping on a high mountain because water doesn't boil at high altitudes' Who do you agree with below?
    a. Joy says: 'Yes it does, but the boiling water is just not as hot as it is here'
    b. Tay says: 'That's not true. Water always boils at the same temperature'
    c. Lou says: 'The boiling point of the water decreases, but the water itself is still at 100 degrees'
    d. Mai says: 'I agree with Jim. The water never gets to its boiling point'
25. **Thermal Concepts 9:** Sam takes a can of cola and a plastic bottle of cola from the refrigerator, where they have been overnight. He quickly puts a thermometer in the cola in the can. The temperature is 7°C. What are the most likely temperatures of the plastic bottle and cola it holds?
    a. They are both less than 7°C.
    b. They are both equal to 7°C.
    c. They are both greater than 7°C.
    d. The cola is at 7° C but the bottle is greater than 7°C.
    e. It depends on the amount of cola and/or the size of the bottle.
26. **Thermal Concepts 10:** Pat believes her Dad cooks cakes on the top shelf inside the electric oven because it is hotter at the top than at the bottom. Which person do you think is right?
    a. Pam says that it's hotter at the top because heat rises.
    b. Sam says that it is hotter because metal trays concentrate the heat.
    c. Ray says it's hotter at the top because the hotter the air the less dense it is.
    d. Tim disagrees with them all and says that it's not possible to be hotter at the top.

27. **Thermal Concepts 11:** Ron reckons his mother cooks soup in a pressure cooker because it cooks faster than in a normal saucepan but he doesn't know why. [Pressure cookers have a sealed lid so that the pressure inside rises well above atmospheric pressure.] Which person do you most agree with below?
    a. Emi says: 'It's because the pressure causes water to boil above 100°C.'
    b. Col says: 'It's because the high pressure generates extra heat.'
    c. Fay says: 'It's because the steam is at a higher temperature than the boiling soup.'
    d. Tom says: 'It's because pressure cookers spread the heat more evenly through the food.'
28. **Thermal Concepts 12:** Dan simultaneously picks up two cartons of chocolate milk, a cold one from the refrigerator and a warm one that has been sitting on the counter top for some time. Why do you think the carton from the refrigerator feels colder than the one from the counter top? Compared with the warm carton, the cold carton?
    a. contains more cold.
    b. contains less heat.
    c. is a poorer heat conductor.
    d. conducts heat more rapidly from Dan's hand.
    e. conducts cold more rapidly to Dan's hand.
29. **Thermal Concepts 13:** Amy took two glass bottles containing water at 20°C and wrapped them in washcloths. One of the washcloths was wet and the other was dry. 20 minutes later, she measured the water temperature in each. The water in the bottle with the wet washcloth was 18°C, the water in the bottle with the dry washcloth was 22°C. The most likely room temperature during this experiment was:
    a. 26°C
    b. 21°C
    c. 20°C
    d. 18°C
30. **Thermal Concepts 14:** Kim takes a metal ruler and a wooden ruler from his pencil case. He announces that the metal one feels colder than the wooden one. What is your preferred explanation?
    a. Metal conducts energy away from his hand more rapidly than wood.
    b. Wood is a naturally warmer substance than metal.
    c. The wooden ruler contains more heat than the metal ruler.
    d. Metals are better heat radiators than wood.
    e. Cold flows more readily from a metal.
31. **Thermal Concepts 15:** Jan announces that she does not like sitting on the metal chairs in the room because "they are colder than the plastic ones." Who do you think is right below?
    a. Jim agrees and says: "They are colder because metal is naturally colder than plastic."
    b. Kip says: "They are not colder, they are at the same temperature."
    c. Lou says: "They are not colder, the metal ones just feel colder because they are heavier."
    d. Mai says: "They are colder because metal has less heat to lose than plastic."

32. **Thermal Concepts 16:** Mel is boiling water in a saucepan on the stove top. What do you think is in the bubbles that form in the boiling water? Mostly:
    a. Air
    b. Oxygen and hydrogen gas
    c. Water vapour
    d. There's nothing in the bubbles.
33. **Thermal Concepts 17:** Pam asks one group of friends: 'If I put 100 grams of ice at 0°C and 100 grams of water at 0°C into a freezer, which one will eventually lose the greatest amount of heat?' Which of her friends' opinion below do you most agree with?
    a. Cat says: 'The 100 grams of ice.'
    b. Ben says: 'The 100 grams of water.'
    c. Nic says: 'Neither because they both contain the same amount of heat.'
    d. Matt says: 'There's no answer, because ice doesn't contain any heat.'
    e. Jed says: 'There's no answer, because you can't get water at 0°C.'
34. **Thermal Concepts 18:** A few minutes later, Ned picks up the cola can and then tells everyone that the counter top underneath it feels colder than the rest of the counter.
    a. Jon says: 'The cold has been transferred from the cola to the counter.'
    b. Rob says: 'There is no energy left in the counter beneath the can.'
    c. Sue says: 'Some heat has been transferred from the counter to the cola.'
    d. Eli says: 'The can causes heat beneath the can to move away through the counter top.'
    Whose explanation do you think is best?
35. **Thermal Concepts 19:** After cooking some eggs in the boiling water, Mel cools the eggs by putting them into a bowl of cold water. Which of the following explains the cooling process?
    a. Temperature is transferred from the eggs to the water.
    b. Cold moves from the water into the eggs.
    c. Hot objects naturally cool down.
    d. Energy is transferred from the eggs to the water.
36. **Thermal Concepts 20:** Bev is reading a multiple-choice question from a textbook: 'Sweating cools you down because the sweat lying on your skin: Which answer would you tell her to select?
    a. 'wets the surface, and wet surfaces draw more heat out than dry surfaces.'
    b. 'drains heat from the pores and spreads it out over the surface of the skin.'
    c. 'is the same temperature as your skin but is evaporating and so is carrying heat away.'
    d. 'is slightly cooler than your skin because of evaporation and so heat is transferred from your skin to the sweat.'
37. **Thermal Concepts 21:** When Zack uses a bicycle pump to pump up his bike tires, he notices that the pump becomes quite hot. Which explanation below seems to be the best one?
    a. Energy has been transferred to the pump.
    b. Temperature has been transferred to the pump.
    c. Heat flows from his hands to the pump.
    d. The metal in the pump causes the temperature to rise.

38. **Thermal Concepts 22:** Why do we wear sweaters in cold weather?
    a. To keep cold out.
    b. To generate heat.
    c. To reduce heat loss.
    d. All three of the above reasons are correct
39. **Thermal Concepts 23:** Vic takes some Popsicles from the freezer, where he had placed them the day before, and tells everyone that the wooden sticks are at a higher temperature than the ice part. Which person do you most agree with?
    a. Deb says: 'You're right because the wooden sticks don't get as cold as ice does.'
    b. Ian says: 'You're right because ice contains more cold than wood does.'
    c. Ross says: 'You're wrong, they only feel different because the sticks contain more heat.'
    d. Ann says: 'I think they are at the same temperature because they are together.'
40. **Thermal Concepts 24:** Gay is describing a TV segment she saw the night before: 'I saw physicists make super-conductor magnets, which were at a temperature of 260°C.' Who do you think is right?
    a. Joe doubts this: 'You must have made a mistake. You can't have a temperature as low as that.'
    b. Kay disagrees: 'Yes you can. There's no limit on the lowest temperature.'
    c. Leo believes he is right: 'I think the magnet was near the lowest temperature possible.'
    d. Gay is not sure: 'I think super-conductors are good heat conductors so you can't cool them to such a low temperature.'
41. **Thermal Concepts 25:** Four students were discussing things they did as kids. The following conversation was heard: Ami: 'I used to wrap my dolls in blankets but could never understand why they didn't warm up.' Who do you agree with?
    a. Nick replied: 'It's because the blankets you used were probably poor insulators.'
    b. Lyn replied: 'It's because the blankets you used were probably poor conductors.'
    c. Jay replied: 'It's because the dolls were made of material which did not hold heat well.'
    d. Kev replied: 'It's because the dolls were made of material which took a long time to warm up.'
    e. Joy replied: 'You're all wrong.'

Chapter 8
Clickers in a First Statistics Course

Alice Richardson
University of Canberra, Australia

ABSTRACT

This chapter reports on a three-year project to study to improve student outcomes in an introductory statistics course. Collaboration with a language expert led to the introduction of a variety of student-centred activities during the course. Some were online such as matching terms to definitions and multiple choice quizzes. Some were face-to-face such as the introduction of a student response system (commonly known as clickers). Results show a small but significant increase in grades following the introduction of these two interventions. A meta-analysis of research into the use of clickers was also conducted. This shows a small but significant increase in exam marks for students who used clickers in class. A Statistics Concept Inventory (SCI) was also used to assess students' understanding of the course concepts. The usefulness of the SCI was partially supported, as many questions were better answered by more able students. The SCI was also used to assess retention from one semester to another. Again, the usefulness of the SCI is partially supported, in that students who remembered more of their introductory statistics study did better in a second statistics course.

INTRODUCTION

A first or introductory statistics course forms a compulsory part of many undergraduate degrees. This may be because of professional requirements e.g. psychology, or university-wide requirements for a wide range of first-year courses. Because of the large numbers of students taking these courses, the course has already been the subject of a large body of research in the last few decades. One of the most recent outcomes of this research has been the Guidelines for Assessment and Instruction in Sta-

DOI: 10.4018/978-1-61350-062-0.ch008

Table 1. Language teaching strategies implemented in introduction to statistics

1. Providing a list of terms and placing these terms in relation, in groups.
2. Using warm up activities such as matching terms to definitions for revision purposes.
3. Using online language exercises such as crosswords, gap-fill (Cloze) exercises.
4. Using of flashcards for vocabulary revision.
5. Providing stimulus questions for lecture and tutorial materials online, thus encouraging students to prepare before the lecture.

tistics Education (GAISE) report (Aliaga, Cobb, Cuff, Garfield, Gould, Lock, Moore, Rossman, Stephenson, Utts, Velleman and Witmer, 2005) setting out recommendations for the teaching of introductory statistics.

The first course in other scientific disciplines has also received attention. Zhang and Lidbury (2006) identified difficulties with language as contributing significantly to problems students experience in studying Genetics at the University of Canberra (UC). They identified ten strategies from the teaching of foreign languages, and implemented a selection of them in a Genetics classroom with much success (Lidbury & Zhang, 2008). (Zhang, Lidbury, Schulte, Bridgeman, Yates, et al., 2008) applied the same principles to first year Biology, Chemistry and Physics at a number of Australian universities. A selection of language teaching strategies was also implemented in Introduction to Statistics at the University of Canberra (UC) (Richardson and Zhang, 2008a, b). The strategies are given in Table 1. This paper reports on initial attempts to assess retention of knowledge from that revamped introductory statistics course to subsequent courses.

Other successful activities were found in Schaeffer, Gnanadesikan, Watkins & Witmer (1996) including the living boxplot, similar to the living histogram created and photographed by Joiner (1975).

Successful completion of a concept inventory is one of the possible ways of assessing whether learning outcomes of courses have been met. A concept inventory consists of a set of multiple choice questions that test understanding of key concepts in a discipline. The answers are carefully designed to include common misconceptions amongst the distractors. The first one was developed for forces in Physics (Hestenes et al. (1992)) and others soon followed in other science disciplines. These include more general Physics (Yeo & Zadnik, 2001), Biology (Khodor, Halme, & Walker, 2004; Michael, 2007; Michelle K Smith, Wood, & Knight, 2008) and Chemistry (Bowen & Bunce, 1997; S. Krause, et al., 2004; Michael, 2007). The Comprehensive Assessment of Outcomes in Statistics (CAOS) inventory consists of 40 questions (DelMas, Garfield, Ooms, and Chance et al., 2005) covering all the concepts instructors would like to see instilled by the end of a first statistics course.

The Statistics Concept Inventory (SCI) (Stone, Allen, Rhoads, Murphy, Shehab, and Saha, 2003) consists of 25 questions, concentrating on descriptive and inferential statistics with a few graphical and probability questions (Table 2). The topics appear in a haphazard order in the SCI, which is reproduced in full in Appendix A.

Our study will employ the SCI, with three questions omitted that pertain to topics not included in our course.

Clicker technologies were first used in educational settings in the 1960s (Judson and Sawada, 2002). Before that they had been used by businesses to collect data in meetings, and by government to collect and display votes in parliamentary settings, for example. Much academic literature makes reference to television programs that also use audience response systems, such as "Who wants to be a Millionnaire?." Lowery

Table 2. Distribution of topics in the statistics concept inventory

Topic Area	Number of questions
Inference	7
Probability	3 (2 omitted)
Descriptive	8 (1 omitted)
Graphical	4
Total	22 (3 omitted)

(2002) describes three phases in the development of clickers: hard-wired systems in the 1960s and 1970s; infra-red and radio frequency keypads in the 1980s onwards; and Web or mobile device-based systems from the 1990s onwards.

A variety of themes can be identified in the literatures on clickers. Some discuss the pros and cons of different types of technology (Lowery, 2002) or give advice on how to prepare for and use the technology in class (De Jong, Lane, Sharp & Kershaw, 2009). Some applications are at the school level (Penuel, Boscardin, Masyn & Crawford, 2007) but most are at the post-school level (Berry 2009; Czekanski & Roux, 2008; Crossgrove & Curran, 2008; Patterson, Kilpatrick and Woebkenberg, 2009; Snavely & Chan, 2008). Some report data on students' attitudes toward clickers (Beavers, 2010). Others record more quantitative learning outcomes such as exam results.

An array of different disciplines have tried clickers in classes, ranging from nursing (Berry, 2009; Patterson et al., 2009) and biology (Crossgrove & Curran, 2008) to environmental science (Czekanski & Roux, 2008) and finance (Snavely & Chan, 2008).

EDUCATIONAL SETTING

Introduction to Statistics at UC is a service unit that consisted of three hour lectures and one hour tutorial per week. Although no extraneous mathematics was introduced with the statistical concepts, the unit contained a significant amount of theory and formulae. There was some attempt to contextualize the learning however students often failed to acquire the Statistical language appropriate for application in their education and future professions. We recognized this as a key area for improvement and following a successful bid for institutional teaching and learning development funds, we set about revamping the way in which statistics was taught, in order to make the subject more successful for the students by utilizing some language learning strategies.

The project was concerned with redesigning the delivery of statistics to first year undergraduates and postgraduates with no prior Statistics study. In line with Garfield's (1995) recommendations we initially concentrated on what it was that we felt students needed to know and be able to do following the teaching sessions. The goals of the project were to:

- ensure that students were equipped with skills in interpreting data that would enhance their future performance in their chosen field of study;
- improve students' ability to use data to inform their practice in their chosen area of study, thus arriving at greater understanding of the underlying statistical principles;
- increase the amount of learner-centeredness in appreciating the role of statistics;
- develop online materials that allowed students flexible access, and more opportunities to interact with materials at their own pace.

This project concentrates on benefits that students perceive and display over one semester only. A future longitudinal study could address benefits over longer periods of time.

EXPERIMENT DESIGN: FACTORS

The project could be viewed as an experiment with the aim of examining whether language strategies and clickers have improved student learning in introductory Statistics. We also want to know whether students enjoy their experience using language strategies and educational technologies such as clickers.

Baseline: Online Support

In order to cater for different learning styles, lecture notes before the lecture and after the lecture with student questions and lecturer-written explanations, along with lecture recordings, were uploaded onto the learning management system at UC. This also included online quizzes (such as multiple choice, matching and cloze) created using *Hot Potatoes* or created directly within UC's Learning Management System, Moodle. Many questions focused on the use of language e.g. wording conclusions to hypothesis tests, use of words such as "stratify" and "outlier" and their variants.

Some were from the textbook (Moore, 2005) and some derived from activities (see for example Schaeffer et al (1996). Others grew from class discussions. From the written and recorded lecture notes created during each lectures, it would be possible to formalize their use and to create a bank of student questions that can be used as stimulus questions or quiz questions in future unit development.

Intervention: Clickers

As well as all the strategies described above, clickers were used on a weekly basis during semester. The questions were multiple choice questions, with generally just three choices, gathered from Moore (2010). There is no anecdotal evidence that students realised the source of the questions. If they had, they would have been able to prepare for the clicker questions, but there are 10 multiple choice questions per chapter so a lot of extra effort for little direct reward.

TurningPoint software (http://www.keepad-interactive.edu.au), which connects to Microsoft Powerpoint, was employed to run the clicker sessions. A screen shot of the appearance of a TurningPoint question in Powerpoint, is shown in Figure 1.

A practice clicker session was held in week 1, with students responding to questions such as "What is your gender?" and "Which degree are you enrolled in?" During the course of the semester, clickers were handed out at the start and collected at the end of each class that they were required. None were lost during the semester, and only one clicker stopped working during the semester. Class size was small (no more than 40 on any one day) so not a great deal of time was lost in this exercise. See Barnett (2006) for a discussion of the handling of clickers in larger classes.

The number of students who answered clicker questions varied from 10 to 24, with a mean of 19 and a standard deviation of 3. The results are as follows. The boxplot shows the spread of correct responses by category of question. Unlike the SCI, the clicker questions included four questions on design of surveys and experiments. There are only two probability questions, and the spread of percent correct is small for these two questions. For the other four categories, the percent correct shows similar spread. (see Figure 2)

An ANOVA was conducted to test whether question type has a significant effect on the mean percent correct. There was no significant difference ($F = 0.193$, $p = 0.939$, $df = 4$ and 20). This suggests that clickers can be used successfully across all parts of the introductory statistics curriculum.

Table 3 shows the questions categorised by type (5 categories) and percent correct (6 categories ranging from 30-39% up to 80-89%).

A graphical version of this table, produced using correspondence analysis, appears below.

Figure 1. Screen shot of a turning point multiple choice quiz question

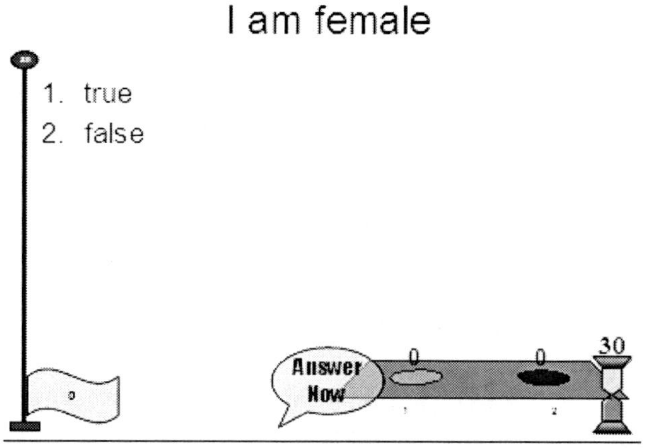

Figure 2. Percentage of clicker questions answered correctly by category of question (graph: graphical; desc: descriptive, design: design; prob: probability; inf: inferential)

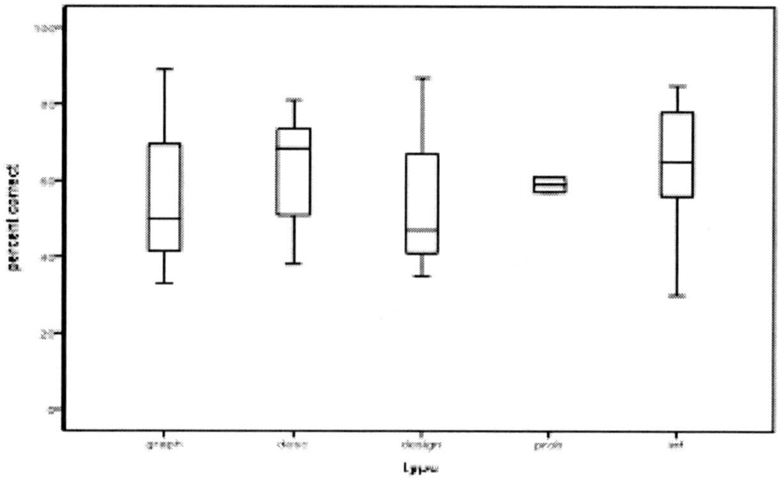

Table 3. Percentage of clicker questions answered correctly by category of question

Question category	Percentage answered correctly						
	30-39	40-49	50-59	60-69	70-79	80-89	Total
graphical	1	0	1	0	0	1	3
descriptive	1	1	1	1	3	1	8
design	1	1	0	0	0	1	3
probability	0	0	1	1	0	0	2
inferential	1	0	2	2	2	2	9
Total	4	2	5	4	5	5	25

Since the five question categories are widely spread around the diagram, there appears to be no strong relationships between them in terms of the percent correct. Descriptive statistics questions are loosely associated with approximately 70% correct responses, while design and graphical questions are loosely associated with approximately 30% correct responses. Probability and inference are loosely associated with approximately 50 – 60% correct responses. (see Figure 3)

Three main patterns of response can be identified across the 24 questions that had three alternative responses. One question had four alternative answers:

1. Clear majority, where the percentage choosing one response (not necessarily the right one) is more than the sum of the percentages choosing the other two responses. Twenty of the 24 questions were answered in this way. Recall that these questions were typically administered at the end of a lecture on a given topic, and so it is not surprising that students knew the answer. Only once did the majority get it wrong: is a question on the stems of stem-and-leaf plots, the majority chose 00, 10, 20, …, 90 when they were supposed to choose 0, 1, 2, …, 9. It is hardly surprising that a majority of students were led astray in their selection of a response when two of them are so similar and almost come down to a typesetting issue! We also note that the questions were taken directly from Moore (2010). The questions are "straightforward questions about basic facts from the chapter" and students are told that they "should expect all of your answers to be correct."

Figure 3. Correspondence analysis showing relationship between clicker question category and percentage answering correctly. (graph: graphical; desc: descriptive, design: design; prob: probability; inf: inferential)

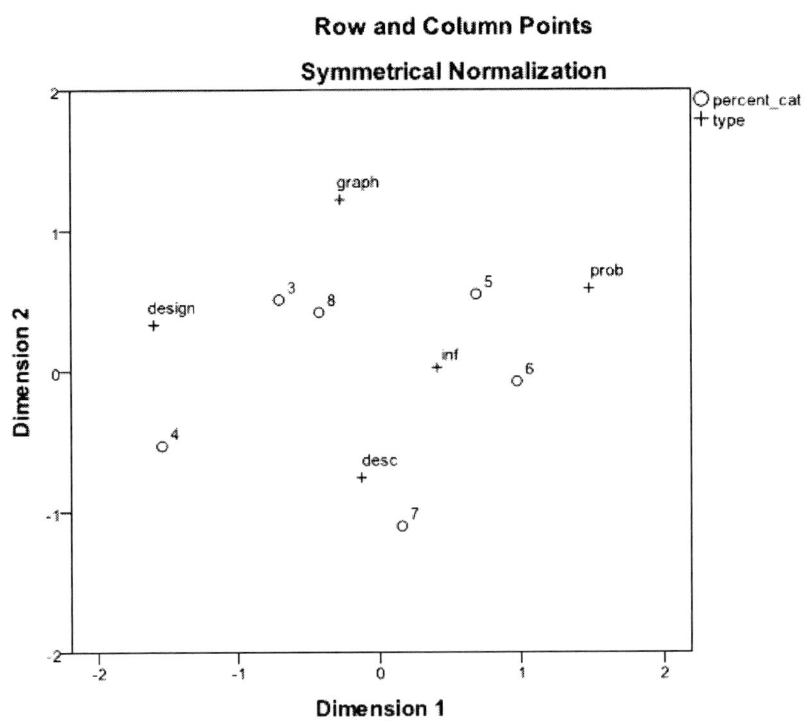

2. Split across two, either with percentages such as 50/45/5. Two questions went this way. One was a question about the scale invariance of the correlation, and the other involved students recognising that a lengthy description was of a completely randomised experiment.
3. Split across three e.g. 38/23/38. Three questions went this way. One involved calculating a five-number summary, one involved identifying a null hypothesis from a lengthy description and the other involved identifying an experiment that was not matched pairs from three lengthy descriptions.

From the very small number of questions that led to a split, it appears that problems can arise when a little bit of calculation is required, or when there is a fair amount of reading (70 - 80 words). Students also complained when questions were presented that required a page of output, followed by questions on the next slides. Our advice is to provide students with a handout with attached graphs and output to assist with timely comprehension of the background to the questions.

The responses to clicker questions cannot be tied to individual students, although some systems do allow for this to be done. Some authors note that anonymity of response is a selling point to students while others use the clickers as part of formal assessment and therefore need to be able to record individual students' responses.

Experiment Design: Response

Response variables recorded were as follows:

- SCI (end of semester): this is available for 2009 and 2010.
- Test and final examination marks: available every semester
- Survey of attitudes towards clickers (end of semester): 2010 only

Discriminatory Ability of SCI

We want to know if the SCI has captured student learning such as is measured by final grades well. We are also interested in whether SCI questions discriminate well between students overall, and whether the different categories behave differently in terms of their discriminatory ability.

In semester 1, 2009, the SCI was administered to a group of 27 Introduction to Statistics students at UC in the last two weeks of semester. This unit is an introductory statistics unit with 4 hours per week contact for 13 weeks. The questions were reordered to follow the order of topics in the course, and split into two sets of 11. This is because the SCI was being used primarily as a revision tool in the context of weekly online quizzes consisting of about 10 multiple choice questions. Furthermore, students were offered two chances to complete the test, the same conditions as their other weekly quizzes. The results from the first attempt are used here. There is a chance that some students would have answered without thinking, merely to access the feedback for each question.

First, we want to know if the SCI has captured student learning, such as is measured by final grades, well. The correlation between SCI score and final mark of the 27 Introduction to Statistics students was 0.49 ($p=0.005$), suggesting that there is a moderate but significant association between increased SCI score and increased final mark.

In semester 1, 2010, the 22 questions were administered to a group of 29 Introduction to Statistics students at UC in the last two weeks of semester. As before, the questions were re-ordered to follow the order of topics in the course, and split into two sets of 11. This time however, students were offered only one chance to complete the tests. Of the 29, 18 sat the finalexam and hence provided complete data. The correlation between SCI score and final mark of the 18 Introduction to Statistics students was 0.59 ($p = 0.01$), suggesting again that there is a moderate but significant

Figure 4. Discriminatory ability of SCI questions by category, 1/2009 and 1/2010. Number of significant tests shown in brackets. 2010 indicated by 0 e.g. prob0

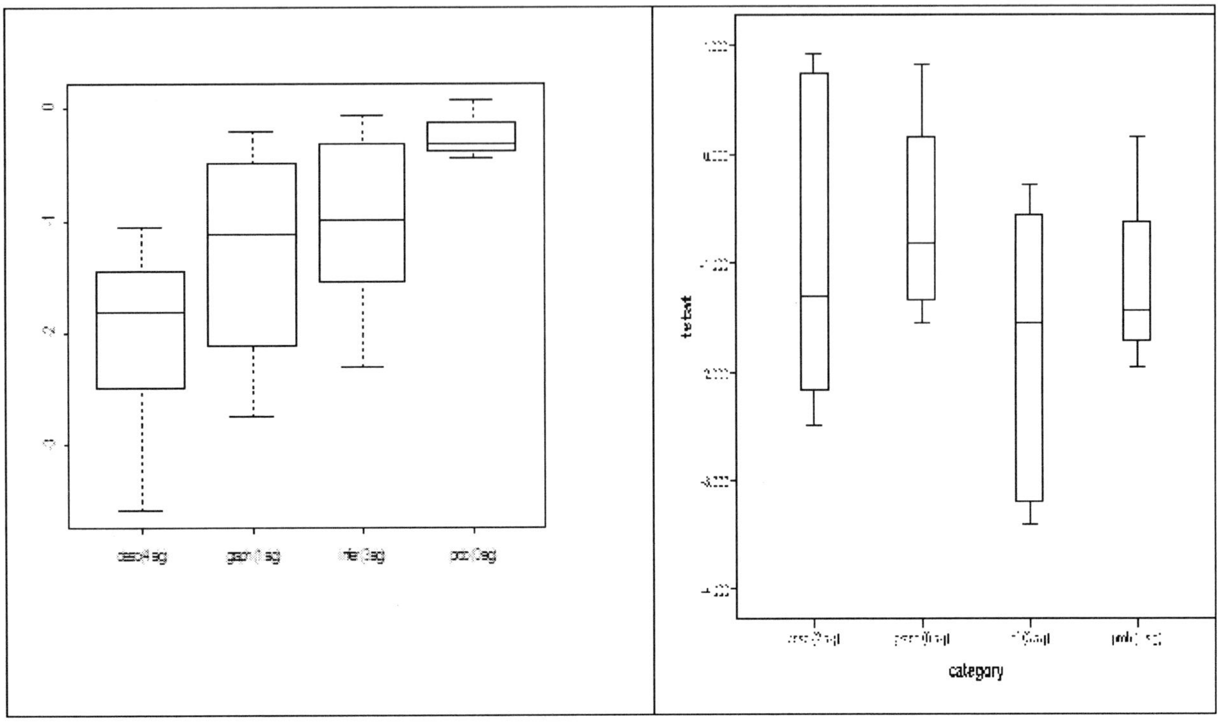

association between increased SCI score and increased final mark.

Second, we are interested in whether questions discriminate well between students overall. In particular, is there a significant difference in mean final mark between students who get a particular question right and students who get a particular question wrong? In semester 1, 2009, there were only 8 significant results (with $p < 0.05$) out of 22. In semester 1, 2010, there were only 6 significant results (with $p < 0.05$) out of 22. Even without a correction for multiple tests, it does not therefore appear that these questions discriminated well between students.

Third, do the different categories behave differently in terms of their discriminatory ability? Discriminatory ability is defined as a significant difference in mean exam mark for those who got a question right compared to those who got a question wrong. Figure 4 shows the t test statistics of the 22 SCI questions by category in semester 1, 2009 and semester 1, 2010. An analysis of variance produces $F = 3.01$ ($p = 0.058$) for 2009 and $F = 0.782$ ($p = 0.521$) for 2010. This suggests that with the quantity of data we have, it is unlikely that the different categories behave differently in terms of their discriminatory ability.

The number of significant differences per category is also shown on the horizontal axis. The largest two are 4 for descriptive statistics in 2009 and 3 for inferential statistics in 2010. These larger counts provide some evidence that these types of questions will discriminate better between students compared to graphical or probability questions.

If the discriminatory ability of SCI questions differs from category to category then the locations of the four sets of t statistics will be significantly different. This hypothesis is not supported by the analysis of variance conducted above.

Table 4. Number correct by nonparametric graduate students in different categories

Student	1	2	3
Inference (7 qns)	3	4	4
Probability (3 qns)	0	1	1
Descriptive (4 qns)	1	2	2
Graphical (8 qns)	5	6	8
Total (22 qns)	9	13	15

Fourth, we are interested in how well material is retained from first semester to subsequent semesters. There was one student from Introduction to Statistics in semester 1, 2009 who studied Nonparametric Statistics in semester 2. She scored 22/22 in semester 1, and 21/22 in semester 2. The incorrect one was an Inference question. Table 4 shows the results by category for the other 3 Nonparametric Statistics students. It appears that descriptive concepts are well retained by the students, along with inference. Graphical concepts and probability are not as well retained. It should be noted that these graduate students are not necessarily in their second statistics course at this point, but may be in their sixth or seventh.

A second way to study retention from semester to semester is through the SCI given to 17 Forensic students in 1/2010. These students are all supposed to have studied an introductory statistics course, generally in the previous semester. Students were given all 22 questions at once, and one attempt only. The mean mark was 43% and the standard deviation was 20%.

The same questions about the SCI that were asked at the end of Introduction to Statistics can be asked of the Forensic Statistics students, as we now have a group of students who have had to retain statistics concepts for at least the summer holidays. Their introductory statistics course was not necessarily Introduction to Statistics, and they did not necessarily study statistics in the semester immediately preceding Forensic Statistics, so

Figure 5. Comparison of SCI questions by category, forensic stats 2010

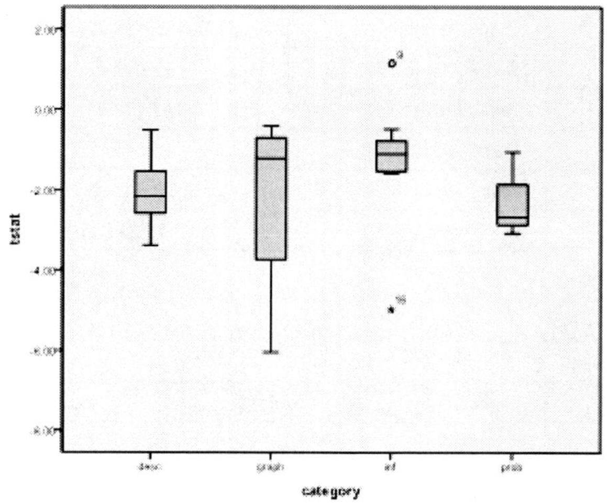

the retention period could be much longer than a couple of months.

First, the correlation between SCI score and final mark of the 18 Forensic Statistics students was 0.834 ($p < 0.0000$), suggesting that there is a moderately strong but significant association between increased SCI score and increased final mark.

Second, is there a significant difference in mean final mark between students who get a particular question right and students who get a particular question wrong? There were seven significant results (with $p < 0.05$) out of 22. It was noticeable that the seven questions were all from the descriptive and graphical categories. An analysis of variance of the t statistics produces $F = 0.376$ ($p = 0.771$) for Forensic Statistics. See Figure 5. The difference in patterns of significance from category to category is not statistically significant.

It is noticeable that some of the t statistics are quite large (well outside the range -2.0 to +2.0) but statistical significance is not achieved. This generally occurs when only two or three students have answered a question correctly or incorrectly. The small sample size in that group makes it difficult to achieve statistical significance with the underlying variability in final marks.

Table 5. Comparison of test marks for the three groups

	Control (CG)	Language (EG1)	Language + Clickers (EG2)
Assessment	2007 Mean (SD) (n=34)	2008 Mean (SD) (n=22)	2010 Mean (SD) (n=28)
Test 1 (out of 30)	19.4 (5.8)	21.8 (4.7)	21.0 (5.7)
Test 2 (out of 30)	17.8 (7.3)	20.5 (4.7)	20.0 (4.4)
Test 3 (out of 30)	14.1 (5.8)	19.1 (6.7)	20.1 (5.8)
Test 4 (out of 30)	22.4 (5.5)	20.5 (5.7)	22.7 (5.0)
Exam (out of 100)	61.8 (19.1)	61.5 (21.5)	64.0 (15.3)

Table 6. Examination grade distribution (%) for CG and EG1 and EG2

Group	Fail	Pass	Credit	Distinction	High Distinction
2007 (n=34)	17.6	32.4	26.5	17.6	5.9
2008 (n=20)	5.0	40.0	5.0	35.0	15.0
2010 (n=25)	12.0	16.0	20.0	24.0	28.0

Quantitative Results

Tests and Examination

The final piece of evidence demonstrating the efficacy of the new teaching model is provided by the four tests taken by the three groups of students. An analysis of variance of the data underlying the means in Table 5 shows that while the means of CG, EG1 and EG2 for test 1 and 2 are not significantly different, for test 3 the differences in means are significant at the 5% level. For test 4, the means are significantly different but EG1 scores lower than CG. This may be due to the fact that there were two significant errors in the EG1 paper, and furthermore different topics were tested in the two groups: two-sample hypothesis tests and regression in EG1 and regression and analysis of variance in CG.

Table 6 shows the distribution of grades in the final examination for 2007 - 2010, after eliminating students who passed the unit on in-term assessment and chose not to attempt to improve their grade by sitting the examination. A χ^2 test shows that there are no significant differences ($X^2 = 13.311$, $p = 0.102$, $df = 8$) between the grade distributions for the three groups. However the number of students who fail remains below the expected number in each of the experimental years. Data on admission scores and GPAs for these students was too sparse to be able to confirm the effect in 2008 where credit-level students were achieving at a Pass or a Distinction level in this unit.

These examination results can be interpreted as evidence that the new teaching model does promote learning in students, particularly in obtaining a Pass grade at a minimum. There does not appear to be a regression to the mean effect here.

Lucas (2009) reports a correlation of 0.57 between clicker scores and final exam marks. His sample size is reported to be 24, although the accompanying scatter plot appears to have about 40 observations. We are not able to comment on the correlation in this data, as it is not possible to link clicker scores and exam marks. It is however possible to link attitudes towards clickers and exam marks, as discussed in the next section.

Qualitative Evaluation of Clickers

Six questions asked on a five-point Likert scale:

- I used the clickers in class
- The clicker questions helped my understanding during class.
- The clicker questions helped my revision.
- I would recommend the use of clickers to other statistics classes in particular.
- I would recommend the use of clickers to other university classes in general.
- Please add any comments.

If the clicker questions are scored in the "desired" way, the maximum possible score is 25. The 22 scores followed an approximately Normal distribution with a mean of 19.95 and a standard deviation of 2.87. This suggests that the scores were positive, suggesting that only those who had enjoyed using the clickers may have bothered to respond.

So the students who responded to the clicker survey were uniformly positive about it, but did those students actually perform better than those who did? Independent-samples t tests comparing test and exam marks between the 22 students who did respond to the clicker survey and the 11 students who did not show no significant difference in mean scores ($p = 0.286, 0.226, 0.624, 0.831$ and 0.290 for tests 1 – 4 and the exam respectively.) So while they did not show a significant difference in their mean test results, we can take this to indicate that a student's attitude to the clickers may not influence their exam result, but their usage of them still might. Only one student who responded to the clicker survey had used them less than "three or four times."

Another way to look at whether good students use clickers is to study the relationship between the qualitative aspects of clicker use and exam marks. The relationship is weakly positive and not statistically significant ($r = 0.337$, $p = 0.202$, $n = 16$).

Students were also invited to comment on the clickers. Eleven students took up the invitation and their comments were uniformly positive. Some reinforced the positive aspects of clickers e.g. "It can help the students review the key points and find the weakness quickly"; "Clickers give us opportunity to learn from our mistakes."

Some recommended greater use e.g. "They were good, maybe the use of them some more would have been better, allow us to review more in class and allowing you to see where we needed extra help"; "should have clickers after every section"; "I think we might have attempted more clickers, if we had more time during our semester."

Some recommended variations in use. "I believe the clickers are excellent for on the spot feedback [to] help direct [the] lecture. The part that is hard to tell is when further explanation is given, how to tell that it is understood? Should there be a second similar question given after the further explanation from the first, this could see if the results increase to show that students understand!" It is worth reiterating that the questions selected were revision questions from Moore (2010) and the author himself emphasizes that by the end of a chapter, students should be able to answer all these questions correctly.

"I reckon clickers are a very good tool to understanding more in lectures, however you lose out when you cannot attend lectures and listen to them online." Lectures were recorded using the audio-only system Audacity. A system such as Echo360 which captures both voice and Powerpoint, should improve the capacity of students listening asynchronously to benefit from the clicker questions

Finally, there were several general comments such as "Thanks for making statistics easy by adding activities like clickers" and "Thanks, they've been a great help."

DISCUSSION

A distinction should be made between retention of knowledge from the first course to subsequent ones, and capstone courses. A capstone Statistics course, using a book such as Spurrier (2000) or Chatfield (1996), aims to integrate knowledge from many previous courses, not just one. Typically such a course is case study-based, involving a mixture of basic techniques such as exploratory data analysis with advanced techniques such as generalised linear and multilevel models.

We have described above a project that brought together experts from language teaching and statistics to produce a new teaching/learning model for beginning students. We evaluated this new approach and demonstrated that we have made a statistically significant difference to students' performance in tests and exams, improved their understanding of some key concepts and their ability to view statistics in relation to their professional practice. It is this multi-sensory approach to delivery that we would recommend to others. A caveat is that the groups of students involved in this project were fairly small (EG = 20). Consequently, we need to act with caution if we are to apply these new teaching strategies to a larger group of students, say, of size 200. Admittedly, the size of the student cohort might make group work more difficult but with attention paid to management issues related to a larger size group, it is still possible to implement such group work. For some group work management ideas, see Ebert-May, Brewer and Allred (1997).

Future work will naturally concentrate on collecting more data, separating the graduate and undergraduate data, and standardising the administration of the SCI. More data should permit adjustment of results for factors such as time between the first course and the course in which the SCI is administered, or the number of courses between the first course and the one in which the SCI is administered.

We would recognize there is a huge untapped resource in the recordings that were made of every lecture. Future work in this area could involve transcribing the tapes and capturing the best teachable moments in order to make them available for future students.

REFERENCES

Aliaga, M., Cobb, G., Cuff, C., Garfield, J., Gould, R., & Lock, R. … Witmer, J. (2005). *GAISE college report.* Retrieved from www.amstat.org/education/gaise/GAISECollege.pdf

Barnett, J. (2006). Implementation of personal response units in very large lecture classes: Student perceptions. *Australasian Journal of Educational Technology*, *22*(4), 474–494.

Berry, J. (2009). Technology support in nursing education: Clickers in the classroom. *Nursing Education Research*, *30*, 295–298.

Bowen, C. W., & Bunce, D. M. (1997). Testing for conceptual understanding in general chemistry. [New York, NY: Springer-Verlag.]. *Chemical Educator*, *2*(2), 1–17. doi:10.1007/s00897970118a

Chatfield, C. (1995). *Problem solving: A statistician's guide.* London, UK: Chapman and Hall.

Crossgrove, K., & Curran, K. L. (2008). Using clickers in nonmajors- and majors-level biology courses: Student opinion, learning and long-term retention of course material. *CBE Life Sciences Education*, *7*, 146–154. doi:10.1187/cbe.07-08-0060

Czekanski, A. J., & Roux, D.-M. P. (2008). The use of clicker technology to evaluate short- and long-term concept retention. *Proceedings of the 2008 ASEE Zone 1 Conference*, West Point, NY. Retrieved from www.loyola.edu/betasites/cas/midatlanticasee/publications/march2008/documents/ASEE12008_0035_paper.pdf

De Jong, T., Lane, J., Sharp, S., & Kershaw, P. (2009). Optimising personal audience response systems technology to enhance student learning in teacher education lectures. *Proceedings of the 32nd HERDSA Conference,* Darwin, Australia. Retrieved from www.herdsa.org.au/wp-content/uploads/conference/2009/papers/HERDSA2009_DeJong_T.pdf

Delmas, D., Garfield, J., Ooms, A., & Chance, B. (2006). *Assessing students' conceptual understanding after a first course in statistics*. Paper presented at the Annual Meetings of the American Educational Research Association, San Francisco, CA, April 9, 2006. Retrieved from https://app.gen.umn.edu/artist/articles/AERA_2006_CAOS.pdf

Ebert-May, D., Brewer, C., & Allred, S. (1997). Innovation in large lectures – Teaching for active learning. *Bioscience*, *47*(9), 601–607. doi:10.2307/1313166

Garfield, J. (1995). How students learn statistics. *International Statistical Review*, *63*(1), 25–34. doi:10.2307/1403775

Hestenes, D., Wells, M., & Swackhamer, G. (1992). Force concept inventory. *The Physics Teacher*, *30*, 141–166. doi:10.1119/1.2343497

Joiner, B. L. (1975). Living histograms. *International Statistical Review*, *3*, 339–340. doi:10.2307/1403117

Judson, E., & Sawada, D. (2002). Learning from past and present: Electronic response systems in college lecture halls. *Journal of Computers in Mathematics and Science Teaching*, *21*, 167–181.

Kaya, O. N. (2008). A student-centred approach: Assessing the changes in prospective science teachers' conceptual understanding by concept mapping in a general chemistry laboratory. *Research in Science Education*, *38*, 91–110. doi:10.1007/s11165-007-9048-7

Khodor, J., Halme, D. G., & Walker, G. C. (2004). A hierarchical biology concept framework: A tool for course design. *Cell Biology Education*, *3*, 111–121. doi:10.1187/cbe.03-10-0014

Krause, S., & Birk, J. R., Bauer, B., Jenkins, B., & Pavelich, M. J. (2004). *Development, testing, and application of a chemistry concept inventory*. Paper presented at the 34th ASEE/IEEE Frontiers in Education Conference, Savannah, GA, October 20 – 23.

Lidbury, B., & Zhang, F. (2008). Comprehension of scientific language as a strategy to enhance learning and engagement for molecular biology students. *Australian Biochemist*, *39*(3), 10–13.

Love, T. E. (1998). A project-driven second course. *Journal of Statistics Education*, *6*(1). Retrieved from www.amstat.org/publications/jse/v6n1/love.html.

Lowery, R. C. (2006). *Clickers in the classroom: A comparison of interactive student-response keypad systems*. National Technology and Social Science Conference, Las Vegas, 6 April 2006.

Lucas, A. (2009). Using peer instruction and I-clickers to enhance student participation in calculus. *PRIMUS (Terre Haute, Ind.)*, *19*, 219–231. doi:10.1080/10511970701643970

McClure, J. R., Sonak, B., & Suen, H. K. (1999). Concept map assessment of classroom learning: Reliability, validity, and logistical practicality. *Journal of Research in Science Teaching*, *36*(4), 475–492. doi:10.1002/(SICI)1098-2736(199904)36:4<475::AID-TEA5>3.0.CO;2-O

Mendenhall, W., & Sincich, T. (1996). *A second course in statistics: Regression analysis*. Upper Saddle River, NJ: Prentice Hall.

Michael, J. (2007). Conceptual assessment in the biological sciences: A National Science Foundation-sponsored workshop. *Advances in Physiology Education, 31*, 389–391. doi:10.1152/advan.00047.2007

Novak, J. D. (2001). *The theory underlying concept maps and how to construct them*. Retrieved June 8, 2008, from http://cmap.coginst.uwf.edu/info/

Patterson, B., Kilpatrick, J., & Woebkenberg, E. (2010). Evidence for teaching practice: The impact of clickers in a large classroom environment. *Nurse Education Today, 30*, 603–607. doi:10.1016/j.nedt.2009.12.008

Penuel, W. R., Boscardin, C. K., Masyn, K., & Crawford, V. M. (2007). Teaching with student response systems in elementary and secondary education settings: A survey study. *Educational Technology Research and Development, 55*, 315–346. doi:10.1007/s11423-006-9023-4

Richardson, A. M. (2009). Retention of knowledge between statistics courses: Results of a pilot study. *Proceedings of the 3rd ASEARC Research Conference*, Newcastle, December 7 – 8, 2009. Retrieved from www.uow.edu.au/content/groups/public/@web/@inf/@math/documents/doc/uow074264.pdf

Richardson, A. M., Lidbury, B. A., & Zhang, F. (2007). One potato, two potato, three potato, four: The use of Hot Potatoes software in science language comprehension. *Uniserve Science Symposium Proceedings, Science Teaching and Learning Research including Threshold Concepts*, Sydney 27 – 28 September, 2007 (pp. 83-88).

Richardson, A. M., & Zhang, F. (2008a). *Language support for statistics learners*. Paper presented at OZCOTS 08, Melbourne July 3 - 4, 2008.

Richardson, A. M., & Zhang, F. (2008b). Activating multiple senses in learning statistics. *Uniserve Science Symposium Proceedings, Visualisation and Concept Development*, Sydney 3 – 4 October, 2008 (pp. 98-103).

Schaeffer, R. L., Gnanadesikan, M., Watkins, A., & Witmer, J. A. (1996). *Activity-based statistics*. New York, NY: Springer.

Schau, C., Stevens, J., Dauphinee, T. L., & Del Vecchio, A. (1995). The development and validation of the survey of attitudes towards statistics. *Educational and Psychological Measurement, 55*, 868–875. doi:10.1177/0013164495055005022

Smith, M. K., Wood, B. W., & Knight, J. K. (2008). The genetics concept assessment: A new concept inventory for gauging student understanding of genetics. *CBE Life Sciences Education, 7*, 422–430. doi:10.1187/cbe.08-08-0045

Snavely, J. C., & Chan, K. C. (2009). Do clickers click in the classroom? *Journal of Financial Education, 35*, 42–65.

Spurrier, J. D. (2000). *The practice of statistics: Putting it all together*. Pacific Grove, CA: Duxbury.

Stone, A., Allen, K., Rhoads, T. R., Murphy, T. J., Shehab, R. L., & Saha, C. (2003). *The statistics concept inventory: Pilot study*. 3rd ASEE/IEEE Frontiers in Education Conference, November 5 – 8, Boulder CO.

Swern, A. S. (2010). A story of evidence-based medicine: hormone replacement therapy and coronary heart disease in post-menopausal women. *Chance, 23*, 52–56. doi:10.1007/s00144-010-0049-3

Zhang, F., & Lidbury, B. (2006). It's all foreign to me: Learning through the language of genetics and molecular biology. In *UniServe Science Symposium Proceedings, Assessment in Science Teaching and Learning* (pp. 153-159). Sydney, Australia: University of Sydney.

ADDITIONAL READING

Ancker, J. S. (2006). The language of conditional probability. *Journal of Statistics Education, 14*(2). Retrieved from www.amstat.org/publications/jse/v14n2/ancker.html.

Australian Bureau of Statistics. (2001). Australian standard classification of education (asced) Retrieved from www.ausstats.abs.gov.au/ausstats/free.nsf/0/2D08D604CC032093CA256AAF0080766F/$File/12720_2001.pdf

Bayer, H., Lau, B., & Glimcher, P. (2007). Statistics of dopamine neuron spike trains in the awake primate. *Journal of Neurophysiology, 98*, 1428–1439. doi:10.1152/jn.01140.2006

Dutton, J., & Dutton, M. (2005). Characteristics and performance of students in an online section of business statistics. Journal of Statistics Education, (Volume 13, no. 3). Retrieved from www.amstat.org/publications/jse/v13n3/dutton.html

Freeman, J. V., Collier, S., Staniforth, D., & Smith, K. J. (2008). Innovation in curriculum design: A multi-disciplinary approach to teaching statistics to undergraduate medical students. *BMC Medical Education, 8*(28).

Gal, I., & Ginsburg, L. (1994). The role of beliefs and attitudes in learning statistics: Towards an assessment framework. *Journal of Statistics Education, 2*(2). Retrieved from www.amstat.org/publications/jse/v2n2/gal.html.

Garfield, J. (1995). How students learn statistics. *International Statistical Review, 63*(1), 25–34. doi:10.2307/1403775

Hatch, E., & Lazaraton, A. (1991). *The research manual: Design and statistics for applied linguistics*. New York: Newbury House.

Johnson, H. D., & Dasgupta, A. N. (2005). Traditional versus non-traditional teaching: Perspectives of students in introductory statistics classes Journal of Statistics Education 11(2). Retrieved from www.amstat.org/publications/jse/v13n2/johnson.html

Lovett, M. C., & Greenhouse, J. B. (2000). Applying cognitive theory to statistics instruction . *The American Statistician, 54*, 196–217. doi:10.2307/2685590

Malone, C. J., & Bilder, A. C. R. (2001). Statistics course web sites: Beyond syllabus.Html. *Journal of Statistics Education, 9*(2). Retrieved from www.amstat.org/publications/jse/v9n2/malone.html.

Moore, D. S. (2007). *The basic practice of statistics* (4th ed.). New York: W. H. Freeman and Company.

Nijhuis, J., Segers, M., & Gijselaers, W. (2007). The interplay of perceptions of the learning environment, personality and learning strategies: A study amongst international business studies students. *Studies in Higher Education, 32*(1), 59–77. doi:10.1080/03075070601099457

Nooriafshar, M., & Marasent, T. N. (2005). A comparison of learning preferences and perceptions of students for statistics concepts and techniques. Proceedings of the 4th Annual International Conference on Statistics, Hawaii. Retrieved from eprints.usq.edu.au/849/1/Maraseni_Nooriafshar_2005_Hawaii.pdf

Richardson, A., & Zhang, F. (2008). Language support for statistics learners. Paper presented at the OZCOTS 2008 - 6th Australian Conference on Teaching Statistics, Melbourne. Retrieved from sky.scitech.qut.edu.au/~macgilli/ozcots2008/

Richardson, A., Zhang, F., & Lidbury, B. A. (2008). Activating multiple senses in learning statistics. Paper presented at the Assessment in Science Teaching and Learning Symposium, The University of Sydney, Sydney. Retrieved from sydney.edu.au/

Roiter, K., & Petocz, P. (1996). Introductory statistics courses-a new way of thinking. *Journal of Statistics Education*, *4*(2). Retrieved from www.amstat.org/publications/jse/v4n2/roiter.html.

Utts, J., Sommer, B., Acredolo, C., Maher, M. W., & Matthews, A. H. M. (2003). A study comparing traditional and hybrid internet-based instruction in introductory statistics classes Journal of Statistics Education, 11(3). Retrieved from www.amstat.org/publications/jse/v11n3/utts.html

Warner, B. A., Pendergraft, D., & Webb, T. (1998). That was venn, this is now. *Journal of Statistics Education*, *6*(1). Retrieved from www.amstat.org/publications/jse/v6n1/warner.html.

Yilmaz, M. R. (1996). The challenge of teaching statistics to non-specialists. *Journal of Statistics Education*, *4*(1). Retrieved from www.amstat.org/publications/jse/v4n1/yilmaz.html.

KEY TERMS AND DEFINITIONS

Clickers: Wireless devices that transmit student responses to multiple choice questions to a central receiver. Answers are collated and displayed in real time.

Correspondence Analysis: A statistical technique to display the counts in a two-way table in a two-dimensional scatter plot. Row and column categories that are closely related appear close to each other in the scatterplot.

Statistics Concept Inventory: A 25 question multiple choice quiz designed to assess students' grasp of important statistical concepts from an introductory statistics course.

APPENDIX A: STATISTICS CONCEPT INVENTORY (STONE ET AL., 2003)

- **Question 1.** A certain diet plan claims that subjects lose an average of 10 kg in 6 months on their plan. A dietician wishes to test this claim and recruits 15 people to participate in an experiment. Their weight is measured before and after the 6-month period. Which is the appropriate test statistic to test the diet company's claim?
  a. independent samples t test
  b. paired samples t test
  c. independent samples z test
  d. paired samples z test
- **Question 2.** Which location would be more likely to have 70% boys born on a given day: a small rural hospital or a large urban hospital?
  a. both equally likely
  b. rural
  c. both are extremely unlikely
  d. urban
- **Question 3.** Three of the following are graphical presentations of the same set of data. Which graph is of a different data set?

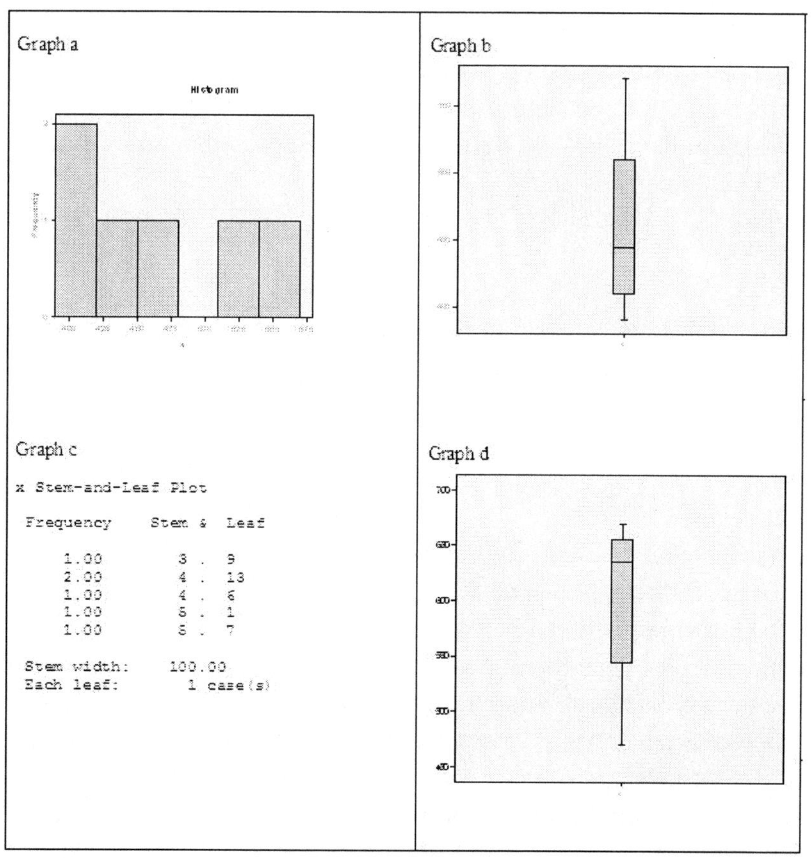

a. a
b. b
c. c
d. d

- **Question 4.** A student scores in the 75th percentile in his Chemistry class. Which of the following statements is always true?
  a. the student's grade is at least as high as 75% of their classmates
  b. none of these is always true
  c. the student earned at least 75% of the total possible marks
  d. The student's grade will be DI
- **Question 5.** The following are temperatures for a week in Adelaide in December: 33, 33, 34, 36, 37, 37, 38. By how much could the highest temperature increase without changing the median?
  a. increase by 2 degrees
  b. it can increase by any amount
  c. it cannot increase without changing the median
  d. increase by 5 degrees
- **Question 6.** A bottling company believes a machine is under-filling 375 ml bottles. What will be the alternative hypothesis to test this belief?
  a. on average, the bottles are not being filled to 375 ml
  b. on average, the bottles are being filled with less than 375 ml
  c. on average, the bottles are being filled to 375 ml
  d. on average, the bottles are being filled with more than 375 ml
- **Question 7.** Which of the following statistics is least impacted upon by extreme outliers?
  a. standard deviation
  b. 3rd quartile
  c. range
  d. mean
- **Question 8.** For the following set of numbers, which measure will most sensibly reflect the location? 3, 4, 5, 5, 6, 8, 10, 12, 19, 36, 83
  a. mode
  b. mean
  c. median
  d. standard deviation
- **Question 9.** A researcher conducts an experiment and reports a 95% confidence interval for the mean. Which of the following must be true?
  a. 95% of the measurements can be considered valid
  b. 95% of the time, the experiment produces an interval that contains the population mean
  c. 5% of the measurements should be considered outliers
  d. 95% of the measurements will be between the upper and lower limits of the confidence interval
- **Question 10.** The mean height of male Australian university students is 175 cm with a s.d. of 7.5 cm. The mean height of female Australian university students is 167.5 cm with a s.d. of 10 cm. You conduct an experiment at the University of Canberra measuring the height of 100 Australian males and 100 Australian females. Which result would most surprise you?

a. the mean height of UC females is 170 cm
b. one female with height 185 cm
c. the mean height of UC males is 182.5 cm
d. one male with height 197.5 cm

- **Question 11.** You perform the same two significance tests on large samples from the same population. The two samples have the same mean and the same standard deviation. The first test results in a p-value of 0.01; the second, a p-value of 0.02. The sample mean is equal for the 2 tests. Which test has a larger sample size?
  a. sample sizes are not equal but there is not enough information to determine which sample is larger
  b. sample sizes equal
  c. second test
  d. first test
- **Question 12.** Which of the following variables would you expect to have a Normal distribution? I: height of Australian women II: shoe size of Australian men III: age in years of University of Canberra first year students
  a. I and III
  b. I and II
  c. II and III
  d. all three
- **Question 13.** Estimate the correlation for the two variables X and Y from the following scatter plot.

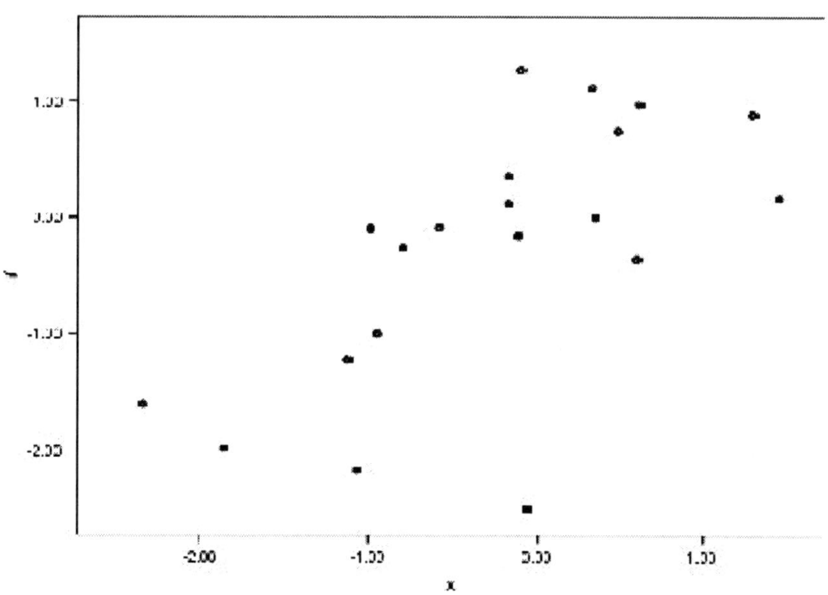

a. -0.6
b. 0.6
c. 0
d. 0.9
e. 1.6

- **Question 14.** Consider the following sample distribution. This sample was most likely taken from what kind of population distribution?

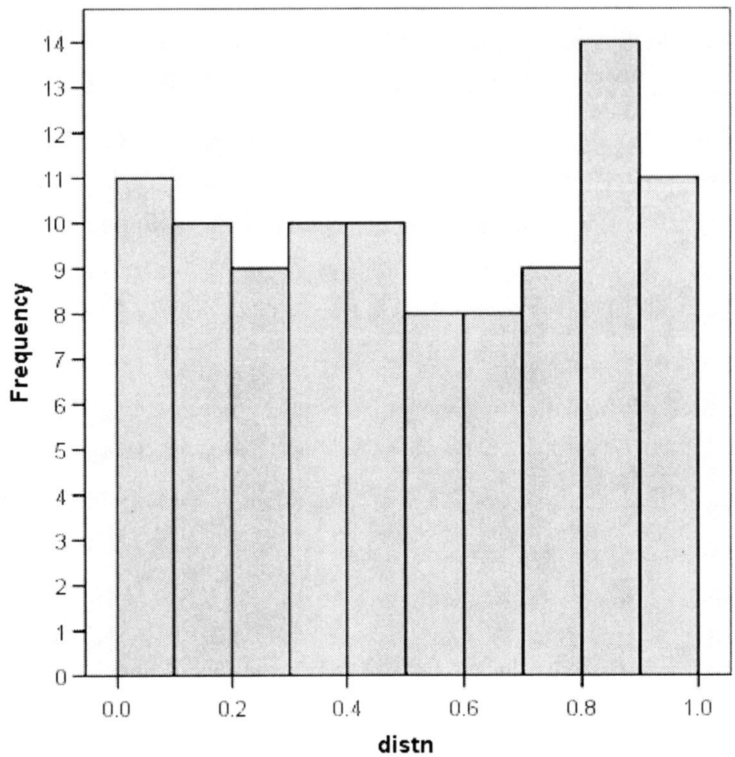

  a. skewed
  b. Normal
  c. Uniform
  d. bimodal
- **Question 15.** You have a set of 30 numbers. The standard deviation from these numbers is reported as 0. You can be certain that:
  a. all of the numbers in the set are 0
  b. all of the numbers in the set are equal
  c. half the numbers are above the mean
  d. the numbers are evenly spaced on both sides of the mean
- **Question 16.** In order to determine the mean height of Australian unversity students, which sampling method would NOT introduce bias?
  a. use a random number table to select students based on their student ID
  b. randomly select from the university basketball team
  c. none of the methods would introduce bias
  d. roll a pair of dice to select from among your friends
- **Question 17.** A scientist takes a set of 50 measurements. The standard deviation is reported as -2.30. Which of the following must be true?
  a. the standard deviation was calculated incorrectly
  b. all of the measurements were less than the mean

c. all of the measurements were negative

d. most of the measurements were negative

- **Question 18.** The following are histograms of quiz scores for four different classes. Which distribution shows the most variability?

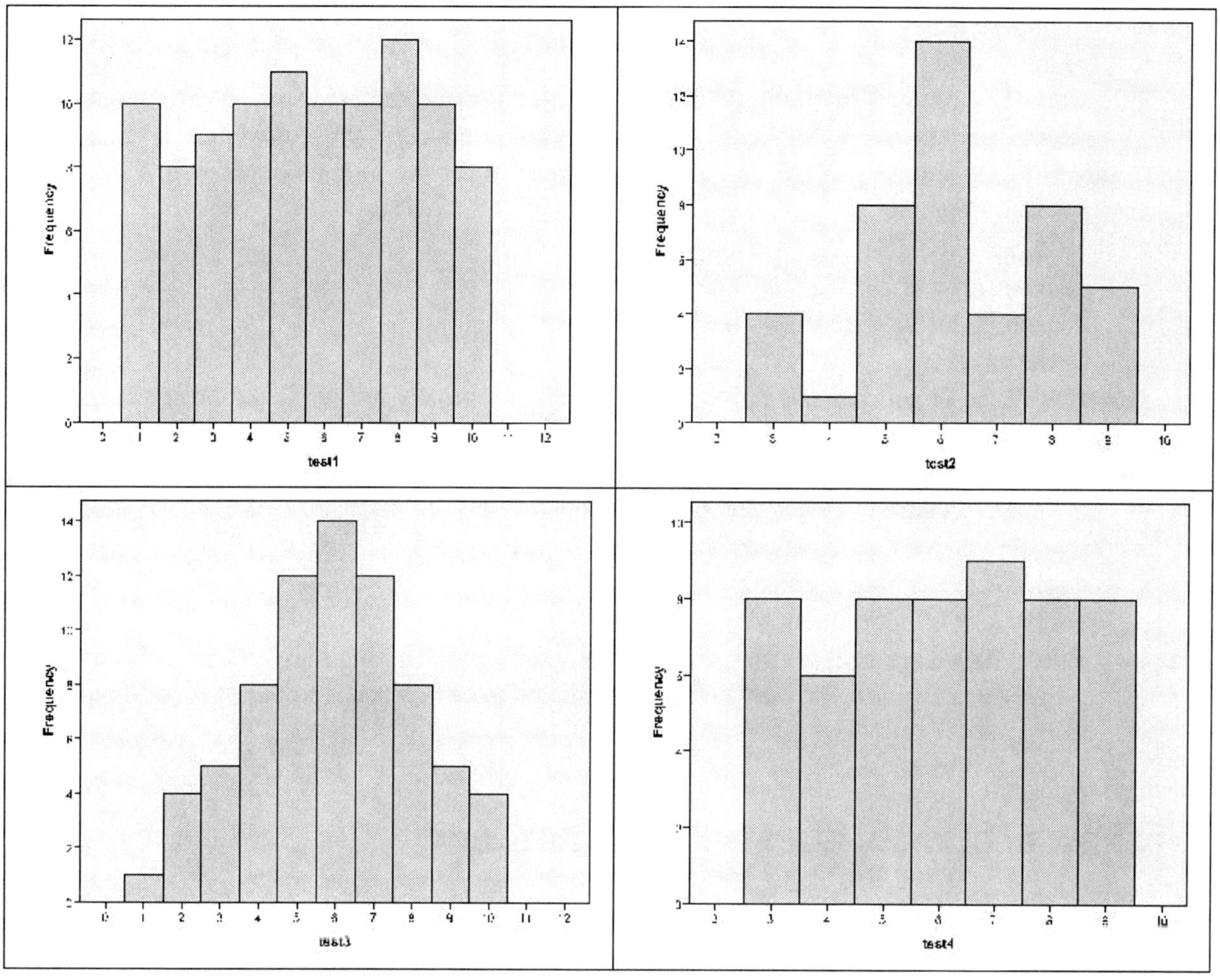

a. a

b. b

c. c

d. d

- **Question 19.** For the past 100 years, the mean maximum temperature on March 1 is 25 degrees with a s.d. of 2 degrees. What is the probability that the maximum temperature on March 1 next year will be between 23 and 27 degrees?

a. 0.95

b. 0.68

c. 0.997

d. 1.00

- **Question 20.** You are rolling dice. You roll 2 dice and compute the mean of the numbers rolled, then 6 dice and compute the mean, then 10 dice and compute the mean. Under which scenario would you be most surprised to find a mean of at least 4.5?
    a. rolling 2 dice
    b. there is no way this can happen
    c. rolling 10 dice
    d. rolling 6 dice
- **Question 21.** Two confidence intervals are claculated for two samples from a given population. Assume the samples have the same standard deviation and that the confidence level is fixed. Compard to the smaller sample, the confidence interval for the larger sample will: Choose one answer.
    a. be wider
    b. be the same width
    c. depend on the confidence level
    d. be narrower
- **Question 22.** Information about different car models is routinely printed in public sources such as consumer magazines and new car buying guides. Data was obtained from these sources for 2003 models of cars. For each car, engine size in cc was compared to the number of engine revolutions per km. The correlation between the two was found to be -0.824. Which of the following statements would you most agree with? Choose one answer.
    a. engine size is a poor predictor of revolutions
    b. engine size is independent of revolutions
    c. a car with a large engine size would be predicted to have a high number of revolutions
    d. a car with a large engine size would be predicted to have a low number of revolutions

APPENDIX B: CLICKER QUESTIONS (MOORE, 2010)

- **Question 1.** To display the distribution of grades (HD, DI, CR, P, N) in a unit, it would be correct to use
    a. a pie chart but not a bar graph
    b. a bar graph but not a pie chart
    c. either a pie chart or a bar graph
    d. neither a pie chart nor a bar graph
- **Question 2.** Here are the amounts of money (cents) in coins carried by 10 students in a statistics class: 50, 35, 0, 95, 75, 0, 0, 85, 20, 65. To make a stemplot of the data, you would use stems
    a. 0, 1, 2, 3, 4, 5, 6, 7, 8, 9
    b. 0, 2, 3, 5, 6, 7, 8, 9
    c. 00, 10, 20, 30, 40 50, 60, 70, 80, 90
- **Question 3.** The five number summary of the data below is 50, 35, 0, 95, 75, 0, 0, 85, 20, 65
    a. 0, 0, 42.5, 75, 95
    b. 50, 37.5, 85, 65
    c. 0, 10, 42.5, 70, 95

- **Question 4.** Which if the following is least affected if an extreme outlier is added to your data?
  a. the median
  b. the mean
  c. the standard deviation
- **Question 5.** The scores of adults on an IQ test are approximately Normal with mean 100 and standard deviation 15. Corinne scores 118 on such a test. Her z-score is about
  a. 1.2
  b. 7.87
  c. 18
- **Question 6.** The scores of adults on an IQ test are approximately Normal with mean 100 and standard deviation 15. Corinne scores 118 on such a test. She scores higher than what percent of all adults?
  a. about 12%
  b. about 88%
  c. about 98%
- **Question 7.** In a scatterplot of the average price of a barrel of oil and the average retail price of a liter of petrol, you expect to see
  a. a positive association
  b. a negative association
  c. very little association
- **Question 8.** For a biology project, you measure the weight in grams and the tail length in mm of a group of 10 mice. The correlation is r = 0.7. If you had measured tail length in cm instead of mm, what would be the correlation?
  a. 0.7/10 = 0.07
  b. 0.7
  c. (0.7)(10) = 7
- **Question 9.** Measurements on young children in Mumbai, India found this least-squares line for predicting height, y, from armspan, x: $\hat{y} = 6.4 + 0.93x$. All measurements are in cm. How much on average does height increase for each additional cm of armspan?
  a. 0.93 cm
  b. 6.4 cm
  c. 7.33 cm
- **Question 10.** According to the regression line $\hat{y} = 6.4 + 0.93x$, the predicted height of a child with armspan 100 cm is about
  a. 106.4 cm
  b. 99.4 cm
  c. 93 cm
- **Question 11.** A committee on community relations in a university town plans to survey local businesses about the importance of students as customers. From telephone book listings, the committee chooses 150 businesses at random. Of these, 73 return the questionnaire mailed by the committee. The population for this study is
  a. all businesses in the university town
  b. the 150 businesses chosen
  c. the 73 businesses that returned the questionnaire

- **Question 12.** A sample of households in a community is selected at random from the telephone directory. In this community 4% of households have no telephone, 10% have only mobile phones and another 25% have unlisted phone numbers. The sample will certainly suffer from
  a. nonresponse
  b. undercoverage
  c. false responses
- **Question 13.** Can changing diet reduce high blood pressure? Vegetarian diets and low-salt diets are both promising. Men with high blood pressure are assigned at random to four diets [details omitted]. The 240 subjects are label 001 to 240. Software assigns an SRS of 60 subjects to diet 1, an SRS of 60 of the remaining 180 to diet 2, and an SRS of 60 of the remaining 120 to diet 3. The 60 who are left get diet 4. This is a
  a. completely randomised design
  b. block design, with four blocks
  c. matched pairs design
- **Question 14.** The Community Intervention Trial for Smoking Cessation asked whether a community-wide advertising campaign would reduce smoking. The researchers located 11 pairs of communities, each pair similar in location, size, economic status, and so on. One community in each pair participated in the advertising campaign and the other did not. This is
  a. an observational study
  b. a matched pairs experiment
  c. a completely randomised experiment
- **Question 15.** You read in a book on poker that the probability of being dealt three of a kind in a five-card poker hand is 1/50. This means that
  a. if you deal thousands of poker hands, the fraction of them that contain three of a kind will be very close to 1/50.
  b. if you deal 50 poker hands, exactly 1 f them will contain three of a kind
  c. if you deal 10,000 poker hands, exactly 200 of them will contain three of a kind
- **Question 16.** Choose an American household at random and let the random variable X be the number of cars (including SUVs and light trucks) they own. Here ID the probability model if we ignore the few households that own more than 5 cars.

Number of cars X	0	1	2	3	4	5
Probability	0.09	0.36	0.35	0.13	0.05	0.02

  A housing company builds houses with two-car garages. What percent of households have more cars than the garage can hold?
  a. 20%
  b. 45%
  c. 55%
- **Question 17.** The Bureau of Labour Statistics announces that last month it interviewed all members of the labour force in a sample of 60,000 households, **4.9%** of the people interviewed were unemployed. The boldface number is a
  a. sampling distribution
  b. parameter
  c. statistic

- **Question 18.** The number of hours a light bulb burns before failing varies from bulb to bulb. The distribution of burnout times is strongly skewed to the right. The central limit theorem says that
  a. as we look at more and more bulbs, their average burnout time gets ever closer to the mean μ for all bulbs of this type
  b. the average burnout time for a large number of bulbs has a distribution of the same shape (strongly skewed) as the distribution for individual bulbs.
  c. the average burnout time of a large number of bulbs has a distribution that is close to Normal
- **Question 19.** To give a 96% confidence interval for a population mean μ, you would use the critical value
  a. $z^* = 1.96$
  b. $z^* = 2.054$
  c. $z^* = 2.326$
- **Question 20.** Experiments on learning in animals sometimes measure how long it takes mice to find their way through a maze. The mean time is 18 seconds for one particular maze. A researcher thinks that a loud noise will cause the mice to complete the maze faster. She measures how long each of 10 mice take with a noise as stimulus. The sample mean is $\overline{x} = 16.5$ seconds. The null hypothesis for the significance test is
  a. $H_0: \mu = 18$
  b. $H_0: \mu = 16.5$
  c. $H_0: \mu < 18$
- **Question 21.** The most important condition for sound conclusions from statistical inference is usually
  a. that the data can be thought of as a random sample from the population of interest
  b. that the population distribution is exactly Normal
  c. that the data contain no outliers
- **Question 22.** Vigorous exercise helps people live several years longer, on average. Whether mild activities like slow walking extend life is not clear. Suppose that the added life expectancy from regular slow walking is just 2 months. A statistical test is more likely to find a significant increase in mean life if
  a. it is based on a very large random sample
  b. it is based on a very small random sample
  c. the size of the sample doesn't have any effect on the significance of the test
- **Question 23.** We prefer the t procedures to the z procedures for inference about a population mean because
  a. z can only be used for large samples
  b. z requires that you know the population standard deviation σ
  c. z requires that you can regard your data as an SRS from the population
- **Question 24.** Which of these settings does not allow the use of a matched pairs procedure?
  a. you interview both the husband and the wife in 64 married couples and ask each about their ideal number of children
  b. you interview a sample of 64 unmarried male students and another sample of 64 unmarried female students and ask each about their ideal number of children

c. you interview 64 female students in their freshman year and again in their senior year and ask each about their ideal number of children

- **Question 25.** In the 2005 NEAP sample of Texas eighth-graders, the mean mathematics scores were 279 for female students and 283 for male students. To see if there is a significant difference between the mean scores of all female and male students in Texas, you would use
  a. the one-sample t test
  b. the matched pairs t test
  c. the two-sample t test
- **Question 26.** There are two common methods for measuring the concentration of a pollutant in fish tissue. Do the two methods differ on the average? You apply both methods to a sample of 18 carp and use
  a. the one-sample t test
  b. the matched pairs t test
  c. the two-sample t test
- **Question 27.** Florida reappraises real estate every year, so the county appraiser's Web site lists the current "fair market value" of each piece of property. Property usually sells for somewhat more than the appraised market value. Here are the appraised market values and actual selling prices (in $000) of condominium units sold in a beachfront building over a 19-month period.

Selling price	appraised value	selling price	appraised value
850	758.0	790	605.9
900	812.7	700	483.3
625	504.0	715	585.8
1075	956.7	825	707.6
890	747.9	675	493.9
810	717.7	1050	802.6
650	576.6	1325	1031.8
845	648.3	845	586.7

Here is part of the SPSS output for regressing selling price on appraised value.

Model Summary

Model	R	R Square	Adjusted R Square	Std. Error of the Estimate
1	.928[a]	.861	.851	69.730

a. Predictors: (Constant), appraise

Coefficients[a]

Model	Unstandardized Coefficients		Standardized Coefficients		
	B	Std. Error	Beta	t	Sig.
1 (Constant)	127.270	79.489		1.601	.132
appraise	1.047	.113	.928	9.295	.000

a. Dependent Variable: sell

The equation of the least-squares regression line for predicting selling price from appraised value is:

a. predicted price = 79.489 + 0.113 × appraised value

b. predicted price = 127.270 + 1.047 × appraised value
c. predicted price =1.047 + 127.270 × appraised value

- **Question 28.** Is there significant evidence that selling price increases as appraised value increases? The p-value for this test is
  a. 0.132
  b. less than 0.001
  c. 0.861
- **Question 29.** The air in poultry processing plants often contains fungus spores. Large spore concentrations affect the health of the workers. To measure the presence of spores, air sample are pumped to an agar plate and colony forming unis (CFUs) are counted after an incubation period. Here are data from the "kill room" of a plant that slaughters 37.000 turkeys per day, taken at four seasons of the year. Each observation was made on a different day. The units are CFUs per cubic metre of air.

Autumn	winter	spring	Summer
1231	384	2105	3175
1254	104	701	2526
752	251	2947	1763
1088	97	842	1090

	N	Mean	Std. Deviation	Std. Error	95% Confidence Interval for Mean	
					Lower Bound	Upper Bound
autumn	4	1081.25	231.458	115.729	712.95	1449.55
winter	4	209.00	136.575	68.287	-8.32	426.32
spring	4	1648.75	1071.244	535.622	-55.84	3353.34
summer	4	2138.50	906.429	453.215	696.17	3580.83
Total	16	1269.38	978.454	244.613	747.99	1790.76

Here is SPSS output for ANOVA to compare mean CFUs in the four seasons.

ANOVA					
spores					
	Sum of Squares	df	Mean Square	F	Sig.
Between Groups	8236359.250	3	2745453.083	5.380	.014
Within Groups	6124210.500	12	510350.875		
Total	1.436E7	15			

The most striking conclusion from the numerical summaries for the plant is that
  a. there appears to be little difference between the seasons
  b. on the average, CFUs are much lower in winter than in other seasons
  c. the air in the plant is clearly unhealthy
- **Question 30.** The conclusion of the ANOVA test is that
  a. there is quite strong evidence (p = 0.014) that the mean CFUs are not the same in all four seasons

b. there is quite strong evidence ($p = 0.014$) that the mean CFUs are much lower in winter than in any other season
c. the data give no evidence ($p = 0.014$) to suggest that mean CFUs differ from season to season

APPENDIX C: LANGUAGE QUESTIONS

The questions below are a selection of those devised by the author to reinforce statistical language in Introduction to Statistics in 2008 and 2010.

- **Question 1.** Fill in the gaps with the appropriate word, selected from the drop-down list.
  a. An ... observation is a long way from the bulk of the observations. [outlying]
  b. An observation which is a long way from the bulk of the observations is called an .... [outlier]
  c. The term ... is not used in Statistics. [outliner]
- **Question 2.** Twenty percent of American households own three or more cars. You choose 12 households at random. Let X be the number of households among the 12 that own three or more cars. Match the terms associated with this scenario to their definitions.
  a. X [name of random variable]
  b. Binomial [name of distribution]
  c. p [probability of success, 0.2]
  d. n [sample size, 12]
- **Question 3.** What does the statement "X ~ B(n, p") mean? Match the parts of the response to the symbols.
  a. ~ [follows a]
  b. B [Binomial distribution]
  c. p [probability of success]
  d. n [smaple size]
  e. X [X]
- **Question 4.** What does the statement "X ~ B(n, p)" mean?
  a. X follows a Binomial distribution with mean n and standard deviation p
  b. X follows a Normal distribution with sample size n and probability of success p
  c. X folows a Normal distribution with mean n and standard deviation p
  d. X follows a Binomial distribution with sample size n and probability of success p
- **Question 5.** Match the definitions to the terms.
  a. to divide a population into groups [stratify]
  b. several subgroups in a population [strata]
  c. the process of dividing a population into groups is called this [stratification]
  d. one subgroup of a population [stratum]
  e. a population that has been divided into groups is called this [stratified]
- **Question 6.** In a simple linear regression of y on x, we found that the coefficient of determination, Rsquared, was 0.035. Select the best answer from the list below to explain what this value means.
  a. 3.5% of the variation in y is explained by the variation in x.
  b. The correlation between y and x is 0.035.

c. 3.5% of the variation in y is explained by the variation in x.
d. 3.5% of the variation in x is explained by the variation in y.

- **Question 7.** Suppose the correlation, r, between two variables x and y, was 0.188. Select the best answer from the list below to explain what this value means.
  a. The correlation between x and y is 0.188
  b. 3.5% of the variation in x is explained by the variation in y
  c. 18.8% of the variation in y is explained by the variation in x.
  d. There is a weak positive linear relationship between x and y
- **Question 8.** Match the characteristics of scatter plots to the selected descriptors.

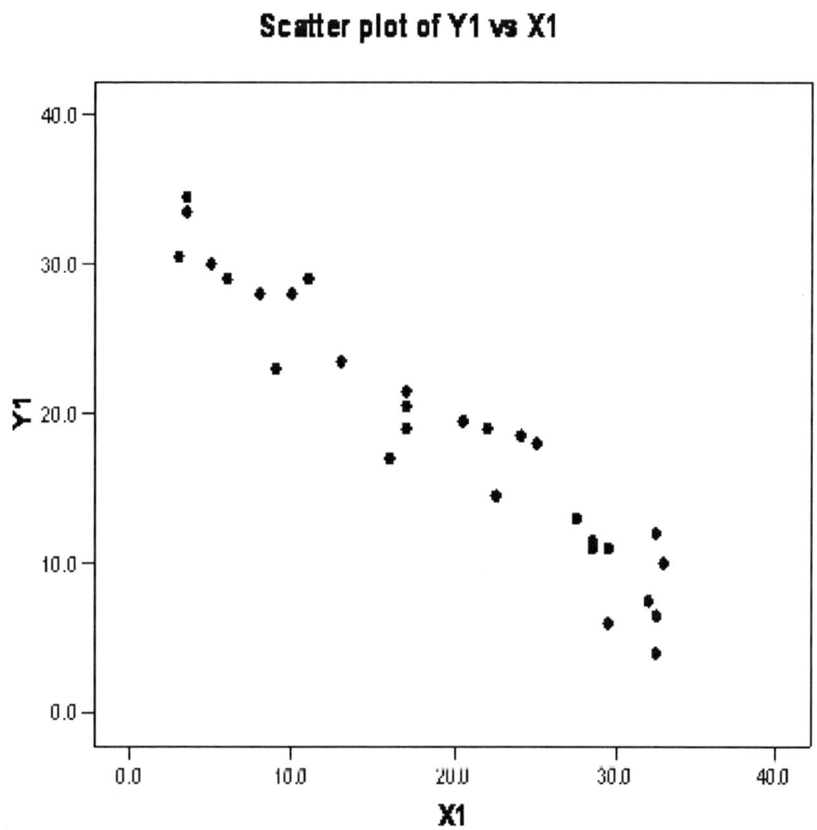

  a. strength [weak / moderate /strong]
  b. direction [positive / negative]
  c. form [linear / non-linear]
- **Question 9.** Suppose the equation of the regression line is $\hat{y} = 4.202 + 0.405 * x$. Select the best interpretation of the intercept from the list below.
  a. When the x increases by 1 unit, the predicted y increases by 0.405.
  b. When the x = 0, the predicted value of y is 0.405.
  c. When x = 0, the predicted value of y is 4.202.
  d. When the x increases by 1 unit the predicted value of y increases by 4.202.
- **Question 10.** Suppose the equation of the regression line is $\hat{y} = 4.202 + 0.405 * x$. Select the best interpretation of the slope from the list below.
  a. When x increases by 1 unit, the predicted value of y increases by 4.202

b. When x increases by 1 unit, the predicted value of y increases by 0.405
c. When x = 0, the predicted value of y is 0.405.
d. When x = 0, the predicted value of y is 4.202.

- **Question 11.** Use the following residual plot to fill in the gaps by selecting the appropriate option given just before the gap.

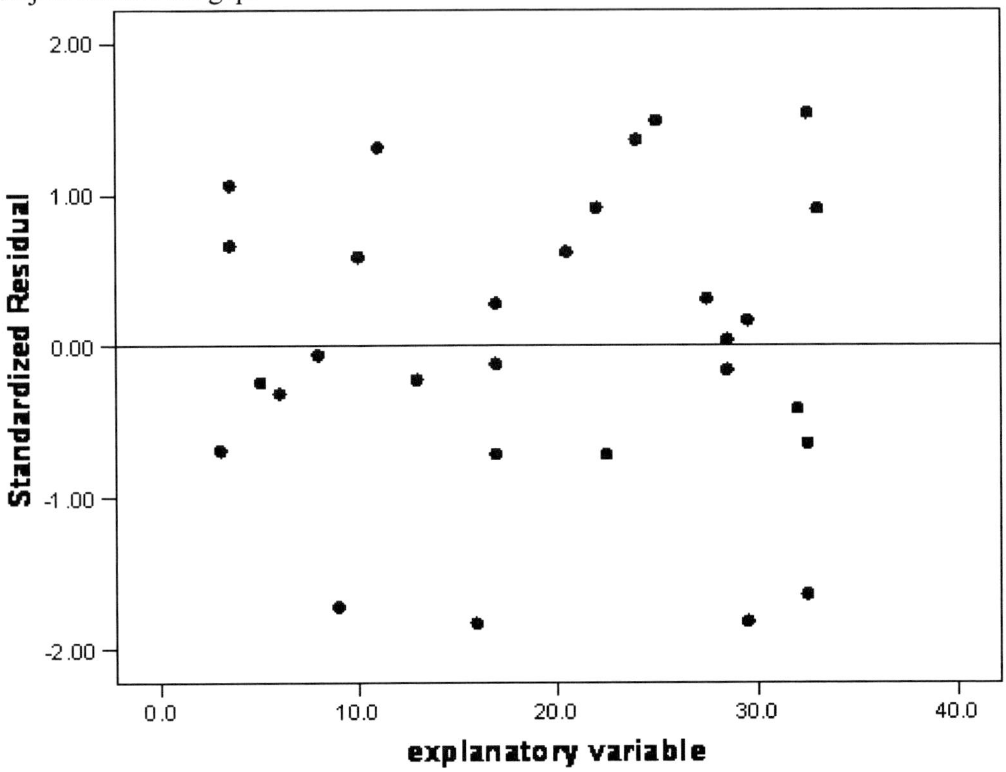

The scatter of points appears to be [*random / not random*] and so the assumption of [*independent observations / homoscedasticity / no outliers*] appear to be [*met / not met*]. The scatter of points shows [*the same spread / increasing spread / decreasing spread*] and so the assumption of [*independent observations / homoscedasticity / no outliers*] appears to be [*met / not met*]. There [*are / are not*] many residuals beyond 2 and so the assumption of appears to be [*met / not met*].

- **Question 12.** How should we interpret a 95% confidence interval for the population mean?
  a. 95% of the sample values lie within the interval
  b. We are 95% confident that the population mean falls in the interval
  c. We are 95% confident that an interval like the one we calculated contains the population mean
- **Question 13.** What three factors determine the margin of error?
  a. the location, spread and shape
  b. The level of confidence, the population standard deviation and the sample size
  c. The mean, the standard deviation and the sample size
  d. The intercept, the slope and the standard deviation

- **Question 14.** [Moore (2010)] Healthy women aged 18 to 40 participated in a study of eating habits. Subjects were given bags of potato chips and bottled water and invited to snack freely. Interviews recorded that some women were trying to restrict their diet out of concern about their weight. Here are the data on grams of potato chips consumed.

Group	n	mean	s.d.
Unrestricted	9	59	21
Restricted	11	32	33

Which statistical test would you use to find out how much effect did these intentions have on their eating habits?

a. independent samples t test
b. one sample t test
c. paired samples t test
d. paired samples z test
e. independent samples z test
f. one sample z test

- **Question 15.** The p-value for the test described above turns out to be 0.0335. The researchers are using $\alpha = 0.05$. Fill in the gaps by selecting the appropriate option given just before the gap. There [is / is not] sufficient evidence to conclude at the [1 /5] percent level that the population mean of the differences is [different from / greater than / less than] zero.

Chapter 9
A Successful Collaboration Between Language Arts and Science

Felicia Zhang
University of Canberra, Australia

ABSTRACT

This chapter contains findings from the preceding chapters in r the project addressing language difficulties in science education in large classes.

INTRODUCTION

It is difficult to merge the worlds of language ("arts") and science largely due to the persistent insouciant stance of scientists and science educators on the primacy of language comprehension for scientific learning (Yore & Treagust, 2006). Through genuine cross-collaboration with a colleague from the humanities, science academics have bridged this traditional gap and have effectively embedded language into tertiary science education. While not a panacea to the challenges of teaching science, for motivated students the benefits are clear as they look to enhance their deep and meaningful understanding of complex scientific concepts and principles. To ensure an impact across the broad range of student attitudes and ability, instructional methods like those described herein may now also need to examine how to augment motivation to study with enhanced comprehension and practice via the multiple languages of science (Lemke, 1998; Pickersgill & Lock, 1991).

In this casebook, we have described a project that brought together experts from language teaching and various fields of science to produce a new

DOI: 10.4018/978-1-61350-062-0.ch009

teaching/learning model for first year science students. We evaluated this new approach and demonstrated that we have made a statistically significant difference to students' performance in tests and exams (in terms of better performance) and improved their understanding of concepts in science.

THIS PROJECT SPECIFICALLY AIMED TO

- target the issue of language in science and suggest ways of solving some of the language issues by importing techniques and strategies frequently used in the teaching of foreign languages;
- create innovative online teaching modules which directly address the language difficulties in the targeted disciplines in science
- expand and reshape the current teaching approach to include a language focus in the teaching of science in the face to face mode;
- increase student awareness of the language used in the targeted disciplines by presenting student and staff insights of the particular types of language used in that context;
- rigorously evaluate the implementation of these learning strategies on student learning to enable their transportability to other teaching contexts in higher education in Australia.

A number of positive outcomes were obtained. These are:

- Students demonstrated better achievement scores at every university and in almost all disciplines;
- The failure rates for almost all subjects in the disciplines decreased;
- Passing, credit, distinction and high distinction rates in each discipline increased dramatically;
- Students' perceptions of lecturers' abilities to teach were dramatically improved.

For example, in the discipline of physics, even within the limit of a 3 hour lecture plus a bi-weekly workshop model, integrating language learning practices in first year physics has shown to have some merit especially where there was a fairly large proportion of students coming from a non-English speaking background. Students' grades improved for students who participated in using the language strategies. In addition, their perception of staff's ability to teach also dramatically improved. This trend is similarly observed in other disciplines as well. In the physics program at the University of Technology, Sydney, analyses of the overall performance of students in the group that used clickers was an average 15% better than their peers who did not use clickers as formative assessment.

In the Chemistry discipline, strategies developed within the context of large classes and limited resources are fully adaptable to other topics beyond first chemistry and to other sciences. They are also necessarily designed to be scalable to large or small classes and to be sustainable. Online resources are useful for helping students become familiar with chemical language and symbolism and to provide them with a means of practicing their use. Online quizzes are an invaluable means of students self-assessing their progress and of providing meaningful assessment of their level of mastery. In-class activities involving student response systems and student-centred, inquiry based approaches are built around active learning and on-going formative assessment. To develop language skills, social interaction via peer instruction and group discussions can be utilized to ensure that classes are vibrant and engaging.

Australian (AUSSE: Australian Survey of Student Engagement (2008); K-L. Krause, Hartley,

James, & McInnis, 2005) and international studies (Hounsell, McCune, Hounsell, & Litjens, 2008; Rust, 2002; Yorke & Longdon, 2008) points to low student satisfaction with the amount, quality and promptness of feedback that they receive. These studies report that this area often scores lowest on student evaluations. In contrast to satisfaction in other areas, satisfaction with feedback provision is actually in decline.

The project provides evidence that quality and prompt feedback can be provided, even in large first year classes, through a variety of means with great success. The provision of feedback through a combination of formative and summative assessment items is the way forward. Furthermore, formative feedback such as those provided through the use of audience response systems, or a show of hands in lectures, or through small group tutorial discussions is significantly more timely and effective for learning and therefore should not be ignored. As pointed out by Bridgeman (in this volume) "The ability to gauge the level of comprehension of a topic whilst the students are still present is a privilege".

While most of the achievement mechanisms used at the various participating universities were lecturer-constructed exams and tests, steps were also taken to test the reliability and validity of some of the published achievement tests. For instance, in Chapter 5, the reliability and discriminatory power of the UC GenCI (which was based on the Genetics Concept Inventory (Elrod, 2007)) was evaluated by four statistical tests, two of them focus on individual items and two of which focus on the test as a whole. The results indicated that the UC GenCI is a reliable test with adequate discriminatory power to assess student learning in the unit Genetics. Similar work is currently being undertaken in testing the veracity of the Statistical Concept Inventory. These benchmarking efforts are important as they offer researchers a means of comparing the achievement of their Australian cohorts of students against achievement of student cohorts in countries such as the United States of America. Since most of these concept inventories were constructed and produced based on data collected in the US, implementation of the same concept inventories in Australian universities would necessarily require adaptation. Hopefully, chapter 5 of this casebook would offer a model on how to verify the fitness of concept inventories (which were based on US data) into Australian universities.

Another key achievement of this project was that it succeeded in changing the teaching practices of lecturers, not by imposing a set of 'best practices' onto them, but by directly involving them in designing, testing and implementing practices that they would use in their own teaching. This volume contains chapters that detail precisely how this could be done thus offering future researchers a point of departure for future research.

Finally, this book contains a wealth of ideas, practical exercise sheets and technological advice targeted at a number of specific topics in the Physics, Chemistry and Biology curricular. In some disciplines such as Chemistry, materials concerning all the usual topics have been included. Though coverage in other disciplinary areas has been isolated to a few topics, the model of creating language oriented exercises, creating exercises that integrate the use of technology can be readily transferable to other areas of science and other disciplines such as the discipline of Business and Health. What distinguishes the materials in this book from other curriculum advice is that our learning materials are not like those in any science textbooks; are produced by the disciplinary science academics themselves rather than by the educationalist; can be easily implemented with or without technology and most importantly have proven to produce significant improvement in first year science students' achievement. As such, it is also a proven and sustainable method for academic staff development.

Finally, the diverse and comprehensive coverage of the interaction between scientific content, learning technology and language learning strategies in this casebook will contribute to our understanding of how successful integrations of content, learning technology and strategy use can be carried out. The detailed coverage of the three science disciplines will be useful for educators who seek inspiration and new ideas to implement in the first year science disciplines. The extensive coverage of the language exercises provided within the book will make it an indispensable resource should educators wish to address the issue of scientific language in their curricula. This project also provides a sustainable model for professional development for academic staff. This model should be of interest to academic developers as an alternative way of changing staff.

REFERENCES

AUSSE. Australian Survey of Student Engagement. (2008). *Engaging students for success*. Retrieved March 22, 2010, from http:// ausse.acer.edu.au/ images/ docs/ AUSSE_2008_Australasian_Student_Engagement_Report.pdf

Hounsell, D., McCune, V., Hounsell, J., & Litjens, J. (2008). The quality of guidance and feedback to students. *Higher Education Research & Development*, *27*, 55–67. doi:10.1080/07294360701658765

Lemke, J. L. (1998). *Teaching all the languages of science: Words, symbols, images, and actions*. Paper presented at the International Conference on Ideas for a Scientific Culture. Barcelona. Retrieved from http:// www-personal.umich.edu/ ~jaylemke/ sci-ed.htm

Pickersgill, S., & Lock, R. (1991). Student understanding of selected non-technical words in science. *Research in Science & Technological Education*, *9*(1), 71–79. .doi:10.1080/0263514910090107

Rust, C. (2002). The impact of assessment on student learning: How can the research literature practically help to inform the development of departmental assessment strategies and learner-centred assessment practices? *Active Learning in Higher Education*, *3*, 145–158. doi:10.1177/1469787402003002004

Yore, L. D., & Treagust, D. F. (2006). Current realities and future possibilities: Language and science literacy-empowering research and informing instruction. *International Journal of Science Education*, *28*(2-3), 291–314. .doi:10.1080/09500690500336973

Yorke, M., & Longdon, B. (2008). *The first-year experience of higher education in the UK: Final report of a project funded by the Higher Education Academy*. York, UK: Higher Education Academy. Retrieved March 22, 2010, from http:// www.heacademy.ac.uk/ news/ detail/ fye_final_report

Compilation of References

Abraham, M. (2005). Inquiry and the learning cycle approach. In Pienta, N. J., Cooper, M. M., & Greenbowe, T. J. (Eds.), *Chemists' guide to effective teaching*. Upper Saddle River, NJ: Prentice Hall.

Active Learning in University Science. (2010). *Website*. Retrieved 11th November 2010, from http:// www.alius.edu.au/

Aliaga, M., Cobb, G., Cuff, C., Garfield, J., Gould, R., & Lock, R. … Witmer, J. (2005). *GAISE college report*. Retrieved from www.amstat.org/education/gaise/GAISECollege.pdf

Alonzo, A. C., & Steedle, J. T. (2008). Developing and assessing a force and motion learning progression. *Science Education, 93*(3), 389-421. doi: 10.1002/ sce.2030310.1002/ (SICI)1098-237X(199907)83:4<397::AID-SCE1>3.0.CO;2-#

Angell, C., Guttersrud, O., Henriksen, E. K., & Isnes, A. (2004). Physics: Frightful, but fun: Pupils' and teachers' views of physics and physics teaching. *Science Education, 88*, 683–706. doi:10.1002/sce.10141

Angell, C., Guttersrud, O., Henriksen, E. K., & Isnes, A. (2004). Physics: Frightful, but fun- Pupils' and teachers' views of physics and physics teaching. *Science Education, 88*, 683–706. doi:10.1002/sce.10141

Arons, A. B. (1990). *A guide to introductory physics teaching*. New York, NY: Wiley.

Asirvatham, M. (2009). *Clickers in action: Increasing student participation in general chemistry*. New York, NY: W.W. Norton & Co.

AUSSE. Australian Survey of Student Engagement. (2008). *Engaging students for success*. Retrieved March 22, 2010, from http:// ausse.acer.edu.au/ images/ docs/ AUSSE_2008_Australasian_Student_Engagement_Report.pdf

Baldwin, J. A., Ebert-May, D., & Burns, D. J. (1999). The development of a college biology self-efficacy instrument for nonmajors. Science Education, 83(4), 397-408. doi: 10.1002/(SICI)1098-237X(199907)83:4<397::AID-SCE1>3.0.CO;2-#

Barnett, J. (2006). Implementation of personal response units in very large lecture classes: Student perceptions. *Australasian Journal of Educational Technology, 22*(4), 474–494.

Barrie, S. C., Buntine, M. A., Jamie, I. M., & Kable, S. H. (2001). *APCELL: Developing better ways of teaching in the laboratory*. Paper presented at the Research and Development into University Science Teaching and Learning Workshop.

Basler, C. F., Mikulasova, A., Martinez-Sobrido, L., Paragas, J., Mühlberger, E., & Bray, M. (2003). The Ebola virus VP35 protein inhibits activation of interferon regulatory factor 3. *Journal of Virology, 77*, 7945–7956. doi:10.1128/JVI.77.14.7945-7956.2003

Beall, H. (1994). Probing student misconceptions in thermodynamics with in-class writing. *Journal of Chemical Education, 71*(12), 1056–1057. doi:10.1021/ed071p1056

Bedgood, D. R. Jr, Gardiner, M. B. Y., Bridgeman, A. J., Morris, G., Buntine, M., & Mocerino, M. (2010). The development of teaching skills to support active learning in university science (ALIUS). *Journal of Learning Design, 1*(3), 10–19.

Bedgood, D. R. Jr, Yates, B., Buntine, M., Pyke, S., Lim, K., & Mocerino, M. (2008). Why are we still teaching the way we were taught in the 1980s? *Chemistry in Australia*, *11*, 22–23.

Bedgood, D. R. Jr, Yates, B., Buntine, M., Pyke, S., Lim, K., & Mocerino, M. (2010). Leading change in Australian science teaching. *Chemistry in Australia*, *7*(5), 18–19.

Bedgood, D. R. Jr, Gardiner, M. G., Yates, B. F., Bridgeman, A. J., Morris, G., & Buntine, M. (2010). The development of teaching skills to support active learning in university science (ALIUS). *Journal of Learning Design*, *1*(3), 10–19.

Bello, Y. M., Falabella, A. F., & Schachner, L. A. (2003). Management of epidermolysis bullosa in infants and children. *Clinics in Dermatology*, *21*, 278–282. doi:10.1016/S0738-081X(03)00050-6

Ben-Zvi, R., Eylon, B., & Silberstein, J. (1988). Theories, principles and laws. *Education in Chemistry*, *5*, 89–92.

Berry, J. (2009). Technology support in nursing education: Clickers in the classroom. *Nursing Education Research*, *30*, 295–298.

Birge, K. (1995). Nutrition management of patients with epidermolysis bullosa. *Journal of the American Dietetic Association*, *95*, 575–579. doi:10.1016/S0002-8223(95)00157-3

Boud, D. (1993). Experience as the base for learning. *Higher Education Research & Development*, *12*(1), 33–44. doi:10.1080/0729436930120104

Bowen, C. W., & Bunce, D. M. (1997). Testing for conceptual understanding in general chemistry. [New York, NY: Springer-Verlag.]. *Chemical Educator*, *2*(2), 1–17. doi:10.1007/s00897970118a

Breckler, J., Joun, D., & Ngo, H. (2009). Learning styles of physiology students interested in health professions. *Advances in Physiology Education*, *33*, 30–36. doi:10.1152/advan.90118.2008

Bridgeman, A. J. (2005). Spectroscopy, mechanisms and calculations online. *Chemical Reviews*, *15*(3).

Bridgeman, A. J. (2010). *Chemical games*. Retrieved 11th November, 2010, from http:// firstyear.chem.usyd.edu.au/ games/ index.shtml

Bridgeman, A. J., & Schmid, S. (2010). *Collaborative laboratory for quantitative data analysis*. Paper presented at the Uniserve Science.

Brown, D. E. (1992). Using examples and analogies to remediate misconceptions in physics: Factors influencing conceptual change. *Journal of Research in Science Teaching*, *29*(1), 17–34. .doi:10.1002/tea.3660290104

Bruner, J. S. (1986). *Actual minds, possible worlds*. Cambridge, MA: Harvard University Press.

Bunce, D. M., Flens, E. A., & Neiles, K. Y. (2010). How long can students pay attention in class? A study of student attention decline using clickers. *Journal of Chemical Education*, *87*(12), 1438–1443. doi:10.1021/ed100409p

Bunce, D. M., Vandenplas, J. R., & Havanki, L. L. (2006). Comparing the effectiveness on student achievement of a student response system versus online WebCT quizzes. *Journal of Chemical Education*, *83*(3), 488–493. doi:10.1021/ed083p488

Burton, L. J., & Dowling, D. (2005). *In search of the key factors that influence student success at university*. Paper presented at the HERDSA 2005 International Conference: Higher Education in a Changing World, Sydney, Australia. Retrieved from http:// eprints.usq.edu.au/

Cary, W. R. (1984). State of the art in the high school curriculum. *Journal of Chemical Education*, *61*(10), 856–857. doi:10.1021/ed061p856

Cavallo, A. M. L., & Schafer, L. E. (1994). Relationships between students' meaningful learning orientation and their understanding of genetics topics. *Journal of Research in Science Teaching*, *31*, 393–418. doi:10.1002/tea.3660310408

Chatfield, C. (1995). *Problem solving: A statistician's guide*. London, UK: Chapman and Hall.

Chin, C., & Brown, D. E. (2002). Student-generated questions: A meaningful aspect of learning in science. *International Journal of Science Education*, *24*(5), 521–549. .doi:10.1080/09500690110095249

Crossgrove, K., & Curran, K. L. (2008). Using clickers in nonmajors- and majors-level biology courses: Student opinion, learning and long-term retention of course material. *CBE Life Sciences Education*, *7*, 146–154. doi:10.1187/cbe.07-08-0060

Crouch, C. H., & Mazur, E. (2001). Peer instruction: Ten years of experience and results. *American Journal of Physics*, *69*(9), 970–977. .doi:10.1119/1.1374249

Cummings, K., Marx, J., Thornton, R., & Kuhl, D. (1999). Evaluating innovation in studio physics. *American Journal of Physics*, *67*(S1), S38–S44. .doi:10.1119/1.19078

Czekanski, A. J., & Roux, D.-M. P. (2008). The use of clicker technology to evaluate short- and long-term concept retention. *Proceedings of the 2008 ASEE Zone 1 Conference*, West Point, NY. Retrieved from www.loyola.edu/betasites/cas/midatlanticasee/publications/march2008/documents/ASEE12008_0035_paper.pdf

De Jong, T., Lane, J., Sharp, S., & Kershaw, P. (2009). Optimising personal audience response systems technology to enhance student learning in teacher education lectures. *Proceedings of the 32nd HERDSA Conference,* Darwin, Australia. Retrieved from www.herdsa.org.au/wp-content/uploads/conference/2009/papers/HERDSA2009_DeJong_T.pdf

Delmas, D., Garfield, J., Ooms, A., & Chance, B. (2006). *Assessing students' conceptual understanding after a first course in statistics*. Paper presented at the Annual Meetings of the American Educational Research Association, San Francisco, CA, April 9, 2006. Retrieved from https://app.gen.umn.edu/artist/articles/AERA_2006_CAOS.pdf

Ding, L., Chabay, R., Sherwood, B., & Beichner, R. (2006). Evaluating an electricity and magnetism assessment tool: Brief electricity and magnetism assessment. *Physics Education Research, 2*(1).

Dobson, J. L. (2009). Learning style preferences and course performance in an undergraduate physiology class. *Advances in Physiology Education*, *33*, 308–314. doi:10.1152/advan.00048.2009

Doran, D. (1980). *Basic measurement and evaluation of science instruction*. Washington, DC: NSTA.

Dreyfus, A., Jungwirth, E., & Eliovitch, R. (1990). Applying the cognitive conflict strategy for conceptual change - Some implications, difficulties, and problems. *Science Education*, *74*(5), 555–569. .doi:10.1002/sce.3730740506

Dupin, J. J., & Johsua, S. (1989). Analogies and "modeling analogies" in teaching: Some examples in basic electricity. *Science Education*, *73*(2), 207–224. .doi:10.1002/sce.3730730207

Ebert-May, D., Brewer, C., & Allred, S. (1997). Innovation in large lectures – Teaching for active learning. *Bioscience*, *47*(9), 601–607. doi:10.2307/1313166

Eilks, I., & Byers, B. (2010). The need for innovative methods of teaching and learning chemistry in higher education – Reflections from a project of the European Chemistry Thematic Network. *Chemistry Education Research and Practice*, *11*, 233–240. doi:10.1039/c0rp90004d

Ellem, G. K., & Mclaughlin, E. A. (2005). *Tales from the coalface: From tragedy to triumph in a blended learning approach to the teaching of 1st year biology*. Paper presented at the 2005 National UniServe Science Symposium: Blended Learning in Science Teaching and Learning. Retrieved from http:// sydney.edu.au/

Elrod, S. (2007). *Genetics concepts inventory*. Retrieved from http:// www.bioliteracy.net/ Readings/ papersSubmittedPDF/ Elrod.pdf

Fagen, A. P. (2003). *Assessing and enhancing the introductory science course in physics and biology: Peer instruction, classroom demonstrations, and genetic vocabulary*. Cambridge, Massachusetts: PhD, Harvard University.

Flavell, J. H. (1976). Metacognitive aspects of problem solving. In Resnick, L. B. (Ed.), *The nature of intelligence* (pp. 231–236). Hillsdale, NJ & New York, NY: Lawrence Erlbaum Associates.

Fuchs, E. (1996). The cytoskeleton and disease: Genetic disorders of intermediate filaments. *Annual Review of Genetics*, *30*, 197–231. doi:10.1146/annurev.genet.30.1.197

Gao, W. J., Peng, X. M., Xie, D. Y., Xie, Q. F., Gao, Z. L., & Yao, J. L. (2004). Construction of exogenous multiple epitopes of helper T lymphocytes and DNA immunization of its chimeric plasmid with HBV pre-S2/S gene. *World Journal of Gastroenterology*, *10*, 2979–2983.

Gardner, P. L. (1977). Logical connectives in science: A summary of the findings. *Research in Science Education*, *7*(1), 9–24. .doi:10.1007/BF02643108

Gardner, P. L., & Australian Science Education Project. (1972). *Words in science: An investigation of non-technical vocabulary difficulties amongst form I, II, III & IV science students in Victoria, a report*. Melbourne, Australia: Australian Science Education Project.

Gardner, P. L., & Australian Science Education Project. (1972). *Words in science: An investigation of non-technical vocabulary difficulties amongst form I, II, III, & IV science students in Victoria, a report*. Melbourne, Australia: Australian Science Education Project.

Garfield, J. (1995). How students learn statistics. *International Statistical Review*, *63*(1), 25–34. doi:10.2307/1403775

Garvin-Doxas, K., & Klymkowsky, M. W. (2008). Understanding randomness and it impact on student learning: Lessons learning from building the biology concept inventory (BCI). *CBE Life Sciences Education*, *7*(2), 227–233. doi:10.1187/cbe.07-08-0063

Gijbels, D., Dochy, F., Van Den Bossche, P., & Segers, M. (2005). Effects of PBL: A meta-analysis from the angle of assessment. *Review of Educational Research*, *75*(1), 27–61. doi:10.3102/00346543075001027

Glynn, S. M., Taasoobshirazi, G., & Brickman, P. (2009). Science motivation questionnaire: Construct validation with nonscience majors. *Journal of Research in Science Teaching*, *46*, 127–146. doi:10.1002/tea.20267

Granville, M. F. (1985). Student misconceptions in thermodynamics. *Journal of Chemical Education*, *62*(10), 847–848. doi:10.1021/ed062p847

Gunstone, R. F. (1994). The importance of specific science content in the enhancement of metacognition. In Fensham, P. J., Gunstone, R. F., & White, R. T. (Eds.), *The content of science: A constructivist approach to its teaching and learning*. London, UK: Routledge.

Hake, R. (2007). Six lessons from the physics education reform effort. *Latin-American Journal of Physics Education, 1*(1), 24-31. Retrieved from http://www.journal.lapen.org.mx/

Halliday, D., Resnick, R., & Walker, J. (2008). *Fundamentals of physics* (8th ed.). Hoboken, NJ: John Wiley & Sons, Inc.

Halloun, I. A., & Hestenes, D. (1985). The initial knowledge state of college physics students. *American Journal of Physics*, *53*(11), 1043–1055. .doi:10.1119/1.14030

Halloun, I. (1998). Schematic concepts for schematic models of the real world: The Newtonian concept of force. *Science Education, 82*(2), 239-63. doi: 10.1002/(SICI)1098-237X(199804)82:2<239::AID-SCE7>3.0.CO;2-F

Harris, K.-L., Krause, K., Gleeson, D., Peat, M., Taylor, C., & Garnett, R. (2007). Enhancing assessment in the biological sciences: Ideas and resources for university educators. Retrieved 1st April, 2007, from http:// www.bioassess.edu.au

Hestenes, D., Wells, M., & Swackhamer, G. (1992). Force concept inventory. *The Physics Teacher*, *30*(3), 141–158. doi:10.1119/1.2343497

Hestenes, D., & Wells, M. (1992). A mechanics baseline test. *The Physics Teacher*, *30*(3), 159–166. doi:10.1119/1.2343498

Hewson, W. P., Robert Tabachnick, B., Zeichner, K. M., Blomker, K. B., Meyer, H., & Lemberger, J. (1999). Educating prospective teachers of biology: Introduction and research methods. *Science Education*, *83*(3), 247–273. doi:10.1002/(SICI)1098-237X(199905)83:3<247::AID-SCE1>3.0.CO;2-C

Hounsell, D., McCune, V., Hounsell, J., & Litjens, J. (2008). The quality of guidance and feedback to students. *Higher Education Research & Development*, *27*, 55–67. doi:10.1080/07294360701658765

Itza-Ortiz, S. F., Rebello, N. S., Zollman, D. A., & Rodriguez-Achach, M. (2003). The vocabulary of introductory physics and its implications for learning physics. *The Physics Teacher*, *41*(6), 330–336. .doi:10.1119/1.1607802

Jacobs, G. (1989). Word usage misconceptions among first-year university physics students. *International Journal of Science Education*, *11*(4), 395–399. .doi:10.1080/0950069890110404

Johnstone, A. H. (2000). Teaching of chemistry - logical or psychological? *Chemistry Education Research and Practice in Europe, 1*(1), 9-15. Retrieved from http:// www.uoi.gr/ cerp/

Joiner, B. L. (1975). Living histograms. *International Statistical Review*, *3*, 339–340. doi:10.2307/1403117

Judson, E., & Sawada, D. (2002). Learning from past and present: Electronic response systems in college lecture halls. *Journal of Computers in Mathematics and Science Teaching, 21*, 167–181.

Kaya, O. N. (2008). A student-centred approach: Assessing the changes in prospective science teachers' conceptual understanding by concept mapping in a general chemistry laboratory. *Research in Science Education, 38*, 91–110. doi:10.1007/s11165-007-9048-7

Kempa, R. F., & Ayob, A. (1995). Learning from group work in science. *International Journal of Science Education, 17*(6), 743–754. doi:10.1080/0950069950170606

Khan, T., Atherton, D., & Denyer, J. (2002). Epidermolysis bullosa and the podiatrist's role. *British Society of Podiatry Conference Abstracts, 2002*, 121.

Khodor, J., Halme, D. G., & Walker, G. C. (2004). A hierarchical biology concept framework: A tool for course design. *Cell Biology Education, 3*(2), 111–121. doi:10.1187/cbe.03-10-0014

Kline, P. (1986). *A handbook of test construction: Introduction to psychometric design*. London, UK: Methuen.

Klymkowsky, M. W., & Garvin-Doxas, K. (2008). Recognizing student misconceptions through ed's tools and the biology concept inventory. *PLoS Biology, 6*(1), e3. .doi:10.1371/journal.pbio.0060003

Knight, J. K., & Smith, M. K. (2010). Different but equal? How nonmajors and majors approach and learn genetics. *CBE Life Sciences Education, 9*, 34–44. doi:10.1187/cbe.09-07-0047

Koch, A. (2001). Training in metacognition and comprehension of physics texts. *Science Education, 85*(6), 758–768. .doi:10.1002/sce.1037

Kolb, D. A. (1984). *Experiential learning: Experience as the source of learning and development*. Englewood Cliffs, NJ: Prentice-Hall.

Krause, K.-L., Hartley, R., James, R., McInnis, C., & Centre for the Study of Higher Education. University of Melbourne. (2005). *The first year experience in Australian universities: Findings from a decade of national studies*. Retrieved from http:// www.dest.gov.au/ sectors/ higher_education/ publications_resources/

Krause, K.-L., Hartley, R., James, R., Mcinnis, C., & Centre for the Study of Higher Education, University of Melbourne (2005). *The first year experience in Australian universities: Findings from a decade of national studies.* Retrieved from http://www.dest.gov.au/

Krause, S., & Birk, J. R., Bauer, B., Jenkins, B., & Pavelich, M. J. (2004). *Development, testing, and application of a chemistry concept inventory*. Paper presented at the 34th ASEE/IEEE Frontiers in Education Conference, Savannah, GA, October 20 – 23.

Krause, S., Birk, J., Bauer, R., Jenkins, B., & Pavelich, M. J. (2004). *Development, testing, and application of a chemistry concept inventory*. Paper presented at the 34th ASEE/IEEE Frontiers in Education Conference. from http://fie-conference.org/fie2004/

Ladyshewsky, R. K. (2006). Peer coaching: A constructivist methodology for enhancing critical thinking in postgraduate business education. *Higher Education Research & Development, 25*(1), 67–84. .doi:10.1080/13600800500453196

Lave, J., & Wenger, E. (1991). *Situated learning: Legitimate peripheral participation*. Cambridge, England & New York, NY: Cambridge University Press.

Lawson, A., E. (2002). Sound and faulty arguments generated by preservice biology teachers when testing hypotheses involving unobservable entities (Vol. 39, pp. 237-52).

Lemke, J. L. (1998). *Teaching all the languages of science: Words, symbols, images, and actions.*

Lidbury, B., & Zhang, F. (2008). Comprehension of scientific language as a strategy to enhance learning and engagement for molecular biology students. *Australian Biochemist, 39*(3), 10–13.

Love, T. E. (1998). A project-driven second course. *Journal of Statistics Education, 6*(1). Retrieved from www.amstat.org/publications/jse/v6n1/love.html.

Lowery, R. C. (2006). *Clickers in the classroom: A comparison of interactive student-response keypad systems*. National Technology and Social Science Conference, Las Vegas, 6 April 2006.

Lucas, A. (2009). Using peer instruction and I-clickers to enhance student participation in calculus. *PRIMUS (Terre Haute, Ind.)*, *19*, 219–231. doi:10.1080/10511970701643970

Maloney, D. P., O'Kuma, T. L., Hieggelke, C. J., & Van Heuvelen, A. (2001). Surveying students' conceptual knowledge of electricity and magnetism. *American Journal of Physics*, *69*(S1), S12–S23. .doi:10.1119/1.1371296

Marais, P., & Jordaan, F. (2000). Are we taking symbolic language for granted? *Journal of Chemical Education*, *77*(10), 1355. .doi:10.1021/ed077p1355

Markow, P. G. (1988). Teaching chemistry like the foreign language it is. *Journal of Chemical Education*, *65*(1), 57–58. doi:10.1021/ed065p57

Martini, F. H., & Nath, J. L. (2009). *Fundamentals of anatomy & physiology* (8th ed.). New York, NY: Pearson.

Marton, F., Hounsell, D., & Entwistle, N. J. (Eds.). (1984). *The experience of learning*. Edinburgh, UK: Scottish Academic Press.

Martyn, M. (2007). Clickers in the classroom: An active learning approach. *EDUCAUSE Quarterly*, *30*(2). Retrieved from http:// www.educause.edu/.

Maskill, R., & Cachapuz, A. F. C. (1989). Learning about the chemistry topic of equilibrium: The use of word association tests to detect developing conceptualizations. *International Journal of Science Education*, *11*, 57–69. doi:10.1080/0950069890110106

Mason, D. S. (2006). Persistent misconceptions. *Journal of Chemical Education*, *83*(5), 681. doi:10.1021/ed083p681

Mayer, K. (2011). Addressing students misconceptions about gases, mass, and composition. *Journal of Chemical Education*, *88*(1), 111–115. doi:10.1021/ed1005085

Mazur, E. (1997). *Peer instruction: A user's manual*. Upper Saddle River, NJ: Prentice-Hall.

McCleary, V. L., Aasen, G., & Slotnick, H. B. (1999). Predictors of success in undergraduate human physiology. *Advances in Physiology Education*, *22*, 119–126.

McClure, J. R., Sonak, B., & Suen, H. K. (1999). Concept map assessment of classroom learning: Reliability, validity, and logistical practicality. *Journal of Research in Science Teaching*, *36*(4), 475–492. doi:10.1002/(SICI)1098-2736(199904)36:4<475::AID-TEA5>3.0.CO;2-O

McDaniel, C. N., Lister, B. C., Hanna, M. H., & Roy, H. (2007). Increased learning observed in redesigned introductory biology course that employed Web-enhanced, interactive pedagogy. *CBE Life Sciences Education*, *6*(3), 243–249. .doi:10.1187/cbe.07-01-0003

McDermott, L. C. (1991). Millikan lecture 1990: What we teach and what is learned-Closing the gap. *American Journal of Physics*, *59*, 301–315. doi:10.1119/1.16539

Mckeachie, W. J. (2002). *McKeachie's teaching tips* (11th ed.). New York, NY: Houghton Mifflin.

Mehmet, S. (2009). Exploring university students' expectations and beliefs about physics and physics learning in problem based learning context. *Eurasia Journal of Mathematics. Science & Technology Education*, *5*(4), 321–333.

Mendenhall, W., & Sincich, T. (1996). *A second course in statistics: Regression analysis*. Upper Saddle River, NJ: Prentice Hall.

Michael, J. (2007). Conceptual assessment in the biological sciences: A national science foundation-sponsored workshop. *Advances in Physiology Education*, *31*(4), 389–391. .doi:10.1152/advan.00047.2007

Michael, J., McFarland, J., & Wright, A. (2008). The second conceptual assessment in the biological sciences workshop. *Advances in Physiology Education*, *32*(3), 248–251. .doi:10.1152/advan.90122.2008

Michael, J. (2007). Conceptual assessment in the biological sciences: A National Science Foundation-sponsored workshop. *Advances in Physiology Education*, *31*, 389–391. doi:10.1152/advan.00047.2007

Millar, R. (1991). Why is science hard to learn? *Journal of Computer Assisted Learning*, *7*(2), 66–74. doi:10.1111/j.1365-2729.1991.tb00229.x

Miller, V., Gregg, G., & Kelly, B. (2000). *Peer Assisted Study Sessions (PASS) in a first year biology teaching program-strategies developed by undergraduate PASS leaders*. Paper presented at the Effective Teaching and Learning at University Conference. Retrieved from http:// www.tedi.uq.edu.au/ conferences/

Moog, R. S. (2008). *Process Oriented Guided Inquiry Learning (POGIL) (Vol. 994)*. American Chemical Society.

Nakhleh, N. B. (1992). Why some students don't learn chemistry: Chemical misconceptions. *Journal of Chemical Education*, *69*(3), 191–196. doi:10.1021/ed069p191

Newport, E. (1991). Contrasting conceptions of the critical period for language. In Carey, S., & Gelman, R. (Eds.), *The epigenesis of mind: Essays on biology and cognition*. Hillsdale, N.J.: Erlbaum.

Nieswandt, M. (2001). Problems and possibilities for learning in an introductory chemistry course from a conceptual change perspective. *Science Education, 85*(2), 158-79. doi: 10.1002/1098-237X(200103)85:2<158::AID-SCE40>3.0.CO;2-3

Novak, J. D. (2001). *The theory underlying concept maps and how to construct them*. Retrieved June 8, 2008, from http://cmap.coginst.uwf.edu/info/

Nussbaum, J., & Novick, S. (1983). Alternative frameworks, conceptual conflict and accommodation: Toward a principled teaching strategy. *Instructional Science*, *11*(2), 183–200.

O'Byrne, J. W., Britton, S., George, A., Franklin, S., & Frey, A. (2009). Using academic predictors to identify first year science students at risk of failing. *CAL-laborate International, 17*(1), 15-25. Retrieved from http:// sydney.edu.au/

Odom, A. L., & Kelly, P. V. (2001). Integrating concept mapping and the learning cycle to teach diffusion and osmosis concepts to high school biology students. *Science Education*, *85*(6), 615–635. .doi:10.1002/sce.1029

Odom, A. L., & Barrow, L. H. (1995). Development and application of a two-tier diagnostic test measuring college biology students' understanding of diffusion and osmosis after a course of instruction. *Journal of Research in Science Teaching*, *32*(1), 45–61. doi:10.1002/tea.3660320106

Oliver, J. S., & Simpson, R. D. (1988). Influences of attitude toward science, achievement motivation, and science self concept on achievement in science: A longitudinal study. *Science Education*, *72*(2), 143–155. doi:10.1002/sce.3730720204

Patterson, B., Kilpatrick, J., & Woebkenberg, E. (2010). Evidence for teaching practice: The impact of clickers in a large classroom environment. *Nurse Education Today*, *30*, 603–607. doi:10.1016/j.nedt.2009.12.008

Pearsall, N. R., Skipper, J. E. J., & Mintzes, J. J. (1997). Knowledge restructuring in the life sciences: A longitudinal study of conceptual change in biology. Science Education, 81(2), 193-215.

Penuel, W. R., Boscardin, C. K., Masyn, K., & Crawford, V. M. (2007). Teaching with student response systems in elementary and secondary education settings: A survey study. *Educational Technology Research and Development*, *55*, 315–346. doi:10.1007/s11423-006-9023-4

Perry, M. D., & Wight, R. D. (2008). Using an ACS general chemistry exam to compare traditional and POGIL instruction. [American Chemical Society.]. *Process Oriented Guided Inquiry Learning*, *994*, 240–247. doi:10.1021/bk-2008-0994.ch020

Perry, W. G. (1988). Different worlds in the same classroom. In Ramsden, P. (Ed.), *Improving learning: New perspectives*. London, UK: Kogan Page.

Piattelli-Palmarini, M. (1989). Evolution, selection and cognition: From "learning" to parameter setting in biology and in the study of language. *Cognition*, *31*(1), 1–44. doi:10.1016/0010-0277(89)90016-4

Pickersgill, S., & Lock, R. (1991). Student understanding of selected non-technical words in science. *Research in Science & Technological Education*, *9*(1), 71–79. doi:10.1080/0263514910090107

Pipes, R. B., & Wilson, J. M. (1996). A multimedia model for undergraduate education. *Technology in Society*, *18*(3), 387–401. doi:10.1016/0160-791X(96)00020-6

Pollock, S. J. (2008). *Comparing student learning with multiple research-based conceptual surveys: CSEM and BEMA*. Paper presented at the Physics Education Research Conference 2008. Retrieved from http:// www.colorado.edu/

Redish, E. F. (1994). The implications of cognitive studies for teaching physics. *American Association of Physics Teachers*, *62*(6), 796–803.

Richardson, A. M. (2009). Retention of knowledge between statistics courses: Results of a pilot study. *Proceedings of the 3rd ASEARC Research Conference*, Newcastle, December 7 – 8, 2009. Retrieved from www.uow.edu.au/content/groups/public/@web/@inf/@math/documents/doc/uow074264.pdf

Richardson, A. M., Lidbury, B. A., & Zhang, F. (2007). One potato, two potato, three potato, four: The use of Hot Potatoes software in science language comprehension. *Uniserve Science Symposium Proceedings, Science Teaching and Learning Research including Threshold Concepts,* Sydney 27 – 28 September, 2007 (pp. 83-88).

Richardson, A., & Zhang, F. (2008). *Language support for statistics learners*. Paper presented at the OZCOTS 2008 - 6th Australian Conference on Teaching Statistics, Melbourne. Retrieved from http:// sky.scitech.qut.edu.au/ ~macgilli/ ozcots2008/

Richardson, A., Zhang, F., & Lidbury, B. A. (2008). *Activating multiple senses in learning statistics*. Paper presented at the Assessment in Science Teaching and Learning Symposium. Retrieved from http:// sydney.edu.au/

Ritchie, S. M. (2001). Actions and discourses for transformative understanding in a middle school science class. *International Journal of Science Education*, *23*(3), 283–299. .doi:10.1080/09500690117500665 29

Roth, W. M., & Roychoudhury, A. (1994). Science discourse through collaborative concept mapping: New perspectives for the teacher. *International Journal of Science Education*, *16*(4), 437–455. .doi:10.1080/0950069940160405

Roth, W. M., Tobin, K., Zimmermann, A., Bryant, N., & Davis, C. (2002). Lessons on and from the dihybrid cross: An activity-theoretical study of learning in coteaching. *Journal of Research in Science Teaching*, *39*(3), 253–282. .doi:10.1002/tea.10018

Roth, W.-M. (1998). Science teaching as knowledgability: A case study of knowing and learning during coteaching. *Science Education, 82*(3), 357-77. doi: 10.1002/(SICI)1098-237X(199806)82:3<357::AID-SCE4>3.0.CO;2-B

Roy, H. (2003). Studio vs interactive lecture demonstration–effects on student learning. Bioscene. *Journal of College Biology Teaching*, *29*(1), 3–6. Retrieved from http:// acube.org/ pdf/bioscene/ v29-1.pdf.

Rust, C. (2002). The impact of assessment on student learning: How can the research literature practically help to inform the development of departmental assessment strategies and learner-centred assessment practices? *Active Learning in Higher Education*, *3*, 145–158. doi:10.1177/1469787402003002004

Sanger, M. J., & Greenbowe, T. J. (1999). An analysis of college chemistry textbooks as sources of misconceptions and errors in electrochemistry. *Journal of Chemical Education*, *76*(6), 853–860. doi:10.1021/ed076p853

Savinainen, A., Scott, P., & Viiri, J. (2005). Using a bridging representation and social interactions to foster conceptual change: Designing and evaluating an instructional sequence for Newton's third law. *Science Education*, *89*(2), 175–195. .doi:10.1002/sce.20037

Schaeffer, R. L., Gnanadesikan, M., Watkins, A., & Witmer, J. A. (1996). *Activity-based statistics*. New York, NY: Springer.

Schau, C., Stevens, J., Dauphinee, T. L., & Del Vecchio, A. (1995). The development and validation of the survey of attitudes towards statistics. *Educational and Psychological Measurement*, *55*, 868–875. doi:10.1177/0013164495055005022

Schon, D. A. (1987). *Educating the reflective practitioner: Toward a new design for teaching and learning in the professions*. San Francisco, CA: Jossey-Bass.

Schulte, J. (2006). *Delivering first year physics assignments with limited resources - An Australian three-centre study*. Paper presented at the 2006 National UniServe Conference, Symposium of Science Teaching and Learning Research, The University of Sydney, Sydney. Retrieved from http:// sydney.edu.au/

Shepherd, G. M. (1994). *Neurobiology* (3rd ed.). Oxford: Oxford University Press.

Simpson, O. (2006). Predicting student success in open and distance learning. *Open Learning: The Journal of Open and Distance Learning*, *21*(2), 125–138.

Smith, M. K., Wood, W. B., Adams, W. K., Wieman, C., Knight, J. K., Guild, N., & Su, T. T. (2009). Why peer discussion improves student performance on in-class concept questions. *Science*, *323*(5910), 122–124. doi:10.1126/science.1165919

Smith, M. K., Wood, B. W., & Knight, J. K. (2008). The genetics concept assessment: A new concept inventory for gauging student understanding of genetics. *CBE Life Sciences Education*, *7*, 422–430. doi:10.1187/cbe.08-08-0045

Snavely, J. C., & Chan, K. C. (2009). Do clickers click in the classroom? *Journal of Financial Education*, *35*, 42–65.

Spurrier, J. D. (2000). *The practice of statistics: Putting it all together*. Pacific Grove, CA: Duxbury.

Stone, A., Allen, K., Rhoads, T. R., Murphy, T. J., Shehab, R. L., & Saha, C. (2003). *The statistics concept inventory: Pilot study*. 3rd ASEE/IEEE Frontiers in Education Conference, November 5 – 8, Boulder CO.

Svinicki, M. D. (2004). *Learning and motivation in the postsecondary classroom*. Boston, MA: John Wiley and Sons.

Swern, A. S. (2010). A story of evidence-based medicine: hormone replacement therapy and coronary heart disease in post-menopausal women. *Chance*, *23*, 52–56. doi:10.1007/s00144-010-0049-3

Tinto, V., & Pusser, B. (2006). *Moving from theory to action: Building a model of institutional action for student success*. Paper presented at the National Symposium on Postsecondary Student Success. Retrieved from http://nces.ed.gov/npec/papers.asp

Touger, J. S. (1991). When words fail us. *The Physics Teacher*, *29*(2), 90–95. doi:10.1119/1.2343227

Towns, M. (2010). Crossing the chasm with classroom response systems. *Journal of Chemical Education*, *87*(12), 1317–1319. doi:10.1021/ed9000624

Walker, R. (2010). Motivating science undergraduates: Ideas and interventions. *International Journal of Innovation in Science and Mathematics Education*, *18*, 1–13.

Wallace, J., & Mintzes, J. (1990). The concept map as a research tool: Explaining conceptual change in biology. *Journal of Research in Science Teaching*, *27*(10), 1033–1052. doi:10.1002/tea.3660271010

Wandersee, J. H. (1988). The terminology problem in biology education: A reconnaissance. *The American Biology Teacher*, *50*(2), 97–100. Retrieved from http://www.jstor.org/.

Wellington, J. J., & Osborne, J. (2001). *Language and literacy in science education*. Buckingham, UK & Philadelphia, PA: Open University.

Williams, H. T. (1999). Semantics in teaching introductory physics. *American Journal of Physics*, *67*(8), 670–680. doi:10.1119/1.19351

Wu, H.-K., Krajcik, J. S., & Soloway, E. (2001). Promoting understanding of chemical representations: Students' use of a visualization tool in the classroom. *Journal of Research in Science Teaching*, *38*(7), 821–842. .doi:10.1002/tea.1033

Yager, R. E. (1983). The importance of terminology in teaching K-12 science. *Journal of Research in Science Teaching*, *20*, 577–588. doi:10.1002/tea.3660200610

Yeo, S., & Zadnik, M. (2001). Introductory thermal concept evaluation: Assessing students' understanding. *The Physics Teacher*, *39*(8), 496–504. doi:10.1119/1.1424603

Yore, L. D., Florence, M. K., Pearson, T. W., & Weaver, A. J. (2006). Written discourse in scientific communities: A conversation with two scientists about their views of science, use of language, role of writing in doing science, and compatibility between their epistemic views and language. *International Journal of Science Education*, *28*(2-3), 109–141. .doi:10.1080/09500690500336601

Yore, L. D., Hand, B. M., & Prain, V. (2002). Scientists as writers. *Science Education*, *86*(5), 672–692. .doi:10.1002/sce.10042

Yore, L. D., & Treagust, D. F. (2006). Current realities and future possibilities: Language and science literacy-empowering research and informing instruction. *International Journal of Science Education*, *28*(2 - 3), 291–314. .doi:10.1080/09500690500336973

Yore, L. D., & Treagust, D. F. (2006). Current realities and future possibilities: Language and science literacy-empowering research and informing instruction. *International Journal of Science Education*, *28*(2-3), 291–314. doi:10.1080/09500690500336973

Yorke, M., & Longdon, B. (2008). *The first-year experience of higher education in the UK: Final report of a project funded by the Higher Education Academy*. York, UK: Higher Education Academy. Retrieved March 22 2010 from http:// www.heacademy.ac.uk/ news/ detail/ fye_final_report

Zeegers, P. (2001). Approaches to learning in science: A longitudinal study. *The British Journal of Educational Psychology*, *71*, 115–132. doi:10.1348/000709901158424

Zhang, F., & Barber, B. (Eds.). (2008). *Handbook of research on computer-enhanced language acquisition and learning*. Hershey, PA: Information Science Reference. doi:10.4018/978-1-59904-895-6

Zhang, F. Z. (2006). *The teaching of Mandarin prosody: A somatically-enhanced approach for second language learners*. Unpublished PhD thesis, University of Canberra, Canberra.

Zhang, F., & Lidbury, B. A. (2006). *It's all foreign to me: Learning through the language of genetics and molecular biology*. Paper presented at the 2006 National UniServe Science Symposium: Assessment in Science Teaching and Learning Research, The University of Sydney, Sydney. Retrieved from http:// sydney.edu.au/

Zhang, F., & Lidbury, B. A. (2006). *It's all foreign to me: Learning through the language of genetics and molecular biology*. Paper presented at the 2006 National UniServe Science Symposium: Assessment in Science Teaching and Learning. from http:// sydney.edu.au/

Zhang, F., Lidbury, B., Schulte, J., Bridgeman, A., Yates, B., & Rodger, J. (2008). *Language difficulties in first year science - An interim report*. Paper presented at the 2008 National UniServe Science Symposium: Assessment in Science Teaching and Learning. Retrieved from http:// sydney.edu.au/

Zhang, F., Lidbury, B., Schulte, J., Bridgeman, A., Yates, B., & Rodger, J. (2009). *A cross-disciplinary approach to language support for first year students in the science disciplines*. Paper presented at the HERDSA 2009.

Zhang, F., Lidbury, B. A., Schulte, J., & Bridgeman, A. J. B., Y., & Rodger, J. (2010). *Catering for the language needs of diverse first year science students*. Paper presented at the First Year in Higher Education.

Zoller, U. (1990). Students' misunderstandings and misconceptions in college freshman chemistry (general and organic). *Journal of Research in Science Teaching*, *27*(10), 1053–1065. doi:10.1002/tea.3660271011

About the Contributors

Felicia Zhang possesses a Master of Arts degree in Applied Linguistics from the University of Melbourne, Australia; Holder of a Teaching English as a Foreign Language Certificate (TEFLA) issued by the Royal Society of Arts, United Kingdom; and a Doctorate in Education from the University of Canberra. She is currently a senior lecturer in Applied Linguistics and Chinese at the University of Canberra, Australia. Her research interests include the use of active learning techniques in foreign language teaching, the use of technology in language teaching and acquisition, Ee-learning, integrating computer technology in curriculum design in Educationeducation. She has just completed an Australian Learning and Teaching Council grant on science education which also won her and her team at the University of Canberra, Australia, a University of Canberra Teaching Award for Programs that Enhanced Learning. She published a the "Handbook of research on computer-enhanced Language Acquisition and Learning" in 2008. She is also the 2003 winner of Australian Awards for University Teaching.

Adam Bridgeman is the Director of First Year Studies and a Senior Lecturer in the School of Chemistry, The University of Sydney. He has published 60 peer-reviewed articles. He was awarded the 2005 RSC Higher Education Teaching Award in the UK for his work on using the internet to enhance and support student learning. As a recent import to Australia, he is focusing on developing the effectiveness of service teaching and electronic resources in the School of Chemistry.

Michael Gardiner is a Senior Lecturer in the School of Chemistry at the University of Tasmania, Hobart. Prior to commencing this current role, he obtained B.Sc. (Hons) and Ph.D. degrees from Griffith University. He then undertook postdoctoral research at Monash University, the Technical University of Munich, and the University of Sydney. His research is in the area of synthetic and catalytic applications of organometallic chemistry. He teaches into a range of undergraduate and postgraduate units covering a number of chemical sub-disciplines. He has been the first year chemistry coordinator since 2006 and has led the program through a number of administrative and teaching based changes in that time.

Brett Lidbury is an Associate Professor in Biomedical Sciences, The John Curtin School of Medical Research, The Australian National University. Post-graduate studies encouraged a research path into immunology, cell biology, and the molecular biology of virus infection and disease. A/Professor Lidbury lectured and taught *Molecular Biology* and *Genetics* for 10 years, and recognized the barrier of molecular "foreign language" to student learning. With Dr. Felicia Zhang, funding was obtained to study language pedagogy in the science context, with the results of this study forming the foundation for a successful *Australian Learning and Teaching Council* (ALTC) Project grant in 2007. In addition to the role of language, other empirical observations on student learning at a tertiary level are reported, particularly in terms of motivation.

Karen Mate is a part-time lecturer in the School of Biomedical Sciences and Pharmacy (SBSP) at the University of Newcastle. Over the past three years she has taught and coordinated several first year undergraduate subjects in the areas of human physiology, biochemistry, and biomolecular analysis, and is also coordinator of the Peer Assisted Study program for SBSP. Her research interests are in the area of reproductive physiology and more recently, the diagnosis and management of dementia.

Alice Richardson is an Assistant Professor in Statistics, in the Faculty of Information Sciences & Engineering at the University of Canberra. Her PhD research, carried out at the Australian National University, was focused on robust estimation and mixed linear models. Her main research area is now statistics education. She has published several papers on active learning in the statistics classroom, informed by her experiences teaching semester courses to undergraduates, postgraduates, and short courses to government employees. She has also collaborated with researchers on quantitative projects ranging from pattern recognition in pathology databases to change management in small businesses to drivers of learning outcomes in universities.

John Rodger is the Director of the Tom Farrell Institute for the Environment at the University of Newcastle. His area of research interest for nearly 40 years has been marsupial reproduction and the development of innovative biotechnological 'tools' for manipulation of marsupial fertility. This work has lead to over 80 peer reviewed publications. From 1995-2003 he was Director of the Australian Government funded Cooperative Research Centre for Conservation and Management of Marsupials, an Australian-New Zealand outcomes-focused collaborative joint venture. Professor Rodger has been teaching biology at first year university level and is interested in seeking new ways to effectively bridge the transition from school to university. The second major teaching interest is in preparing later stage students for the transition to the workplace, in particular, skills to meet the challenges of the application of students' science and technology knowledge and expertise in practical outcomes–focused situations.

Jurgen Schulte is Lecturer in Physics and Advanced Materials at the University of Technology, Sydney. He has published 37 peer-reviewed articles, 11 peer-reviewed proceedings, edited 7 books, holds 4 industrial patents, and presented at over 30 international conferences. He is the Reviews Editor of CAL-laborate, a peer-reviewed journal focusing on information Information technology Technology in tertiary teaching and learning for the sciences. He is currently undertaking a study on the efficacy of on-line assignment delivery in large service teaching classes with respect to student performance.

Brian Yates, Following following a PhD at the Australian National University, Brian Yates completed postdoctoral work at the University of California, Berkeley and the University of Georgia (USA) before taking up an academic position at the University of Tasmania in 1989. Following his appointment as an academic in 1989, he has built up a strong reputation for teaching excellence. He has taught across a range of units in years one to four of undergraduate chemistry; involved in a number of competitively funded teaching development projects at the national (CAUT/CUTSD, ALTC) and state (UTAS) levels and has been rewarded with the 2006 Australian University Teaching Council Award for University Teaching Excellence in Physical Sciences. From 2006 to 2010, Brian was Head of the School of Chemistry at UTAS. In 2010 Brian was appointed as Discipline Scholar in Science for the Australian Learning and Teaching Council.

Index

A

achievement mechanisms 228
active learning 1-2, 14, 18, 23, 26, 31, 35-36, 39-40, 63, 77, 117, 132, 181, 207, 227, 229-232, 235, 237, 240-241
Active Learning in University Science (ALIUS) 31, 33, 35-36, 63, 77, 230-231
active learning skills 1
active role 31, 166
activitymania 7, 164
allied health programs 129
aminopeptidase 100
analysis of variance (ANOVA) 198, 202-204, 221
animal function 130
animal kingdom 130
assessment anxiety 113, 135
atomic symbols 26-27
audience response devices 10-11, 166
audience response systems 196, 207, 228, 232
Australian Learning and Teaching Council (ALTC) 1, 31, 60, 63, 77, 163, 179, 186, 240-242
Australian Science Education Project 6, 13, 37, 61, 77, 117, 126, 233
Australian Survey of Student Engagement (AUSSE) 28, 36, 227, 229-230

B

BamHI 102
biology 1-2, 5-6, 9, 12-16, 21, 37-42, 76, 98-102, 106-108, 113-117, 123-124, 126-130, 137-144, 150, 164, 169, 180-183, 196-197, 206-207, 209, 217, 228, 230-241
Biology Concept Inventory (BCI) 9, 13, 37, 126, 143, 164, 180, 233-234
Biology Self-Efficacy questionnaire 137
Blackboard LMS 3, 8, 131, 165
Brief Electricity Magnetism Assessment (BEMA) 9, 14, 40, 164, 181, 184, 236

C

calcium homeostasis and osteoporosis 131
California Polytechnic State University 119
chemical kinetics 61, 64
chemical songs 58, 67-68
chemical Sudoku puzzles 58, 72-74
chemical symbolism 24
Chemistry Concept Inventory (CCI) 9, 13, 38, 43, 78, 164, 181, 207, 234
chemistry discipline 1-4, 6, 9-14, 18-26, 28, 30-40, 42-43, 50, 52, 58-65, 67-73, 75-79, 85, 88, 98-99, 107-109, 116, 164, 169, 181, 196, 206-207, 212, 227-228, 230-237, 239-242
Chi-square ($\chi2$) statistical analysis 104, 118, 133, 136, 144-145
chromatids 124
clicker student interaction tool 10-12, 14, 17, 27-28, 30, 33, 36, 39, 50, 63, 138, 145, 150, 166-167, 174, 179, 181, 195-201, 204-208, 210, 216, 227, 230-232, 234-236, 238
Cloze (Gap-Fill) Exercises 102
cognitive conflicts 9, 17
common sense theory 9, 164
comprehension 2, 13, 19-21, 25-26, 29, 38, 61, 69, 98-100, 103, 107-109, 111-113, 115, 117, 126, 132, 143, 161, 170, 180, 184, 201, 207-208, 226, 228, 234, 237
Comprehensive Assessment of Outcomes in Statistics (CAOS) 196, 207, 232
concentration 2, 4, 19, 21, 61, 73, 107-108, 140, 146, 148-149, 152, 220
concept inventories 9, 164, 228
concept mapping exercises 132
concept mapping tools 10, 166
Concept Tests 69-71, 76-77
Conceptual Survey of Electricity and Magnetism (CSEM) 9, 14, 40, 164, 181, 184, 236
concept words 6

conjugate base 24, 32
correspondence analysis 198, 200, 210
Credit (CR) 104-106, 108, 110-113, 118, 133-134, 216, 227
cross-collaboration, science and humanities 113, 226
crossword activities 24, 72, 103, 118, 132
cystic fibrosis 131

D

delusion index 3, 5, 22
diagnostic instruments 9, 164
dimerization 100
discriminatory ability 201-202
Discussion Board 136
Disease of the Week modules 130-132, 134-136, 142
Distinction (DI) 23, 65, 104-106, 108, 111, 113, 118, 132-134, 155, 204, 206, 212, 216, 227
DNA cloning 99, 102
drosophila 100-101

E

EcoRI 99, 102
electrophile 19
elicit-confront-resolve teaching model 166
engagement 9-10, 17, 20, 26, 28, 33, 36, 38, 73-74, 76, 99, 101, 104, 107, 109-112, 115-116, 126, 129-130, 142-143, 165, 167, 184, 207, 227, 229-230, 234
Epidermolysis bullosa (EB) 131, 142, 231, 234
equation 2, 20-21, 61, 107-108, 140, 147-149, 220, 223
equilibrium 2, 19, 21, 39, 61, 107-108, 140, 146, 148-149, 170, 186, 235
experiment 2, 21, 61, 107-108, 117-118, 140, 147-149, 187, 192, 198, 201, 211-212, 218

F

Fill in the Blanks exercises 25, 132, 155, 157
Force Concept Inventory (FCI) 9-10, 13, 38, 164-165, 180, 207, 233
foreign languages 99, 196, 227
formative feedback 2, 24, 27, 228
Full Time Student Equivalents (FTSE) 131

G

genetic language 99-100
genetics 1, 5, 9, 13, 16, 37, 41-42, 76, 98-121, 123-128, 136, 142, 144, 164, 180, 196, 208-209, 228, 231-232, 234, 238-240
Genetics Concept Inventory (GenCI) 9, 13, 37, 111, 116, 119-121, 124-125, 126, 164, 180, 228
Gibbs free energy 61
Grade Point Average (GPA) 106, 109, 118
graph 2, 21, 61, 63, 107-108, 140, 147-149, 167, 170, 187, 199-200, 211, 216

H

High Distinction (HD) 104-106, 108, 132-134, 216, 227
home language 7, 163
homologous chromosomes 124
Hot Potatoes 99, 101-103, 198, 208, 237
humanities 113, 226
Human Physiology 129-130, 132-136, 142, 235, 241

I

Immunology 102, 240
in-class activities 18, 35, 58, 62-63, 227
in-class audience surveys 58
in-class student surveys 63
inquiry-based practices 8, 164
instructional language 7, 163
Interactive Engagement Strategies (IE) 9-10, 17, 165, 184
interactive mode of teaching 11, 166

J

jumbled sentence 25, 71-72
Jumbled sentence questions 71

L

language-centred 98-99, 104-105, 118
language comprehension 98-100, 103, 107-109, 111-113, 132, 208, 226, 237
language focused 5, 99, 160
language of physics 160, 170, 177, 179
language of science 6, 11, 117, 129, 161
Language Strategy/Approach 115
language teaching 7, 23, 117, 163, 196, 206, 226, 240
Learning Management System (LMS) 2-3, 8, 10, 21, 102-104, 108-109, 111-112, 131-132, 136, 165, 170, 198

learning modality 135
learning modality, auditory 135
learning modality, kinaesthetic 135
learning modality, reading-writing 135-136
learning modality, visual 7, 24, 135, 163
learning modules 131-132, 134-135, 137, 140-141, 227
learning strategies 2, 11, 18, 58, 63, 98, 112, 129, 132, 197, 209, 227
lecture warm ups 64-65
living boxplot 196
living histogram 196
low group 122-123

M

Materials Safety Data Sheets (MSDS) 26
mathematical modeling 161
MCQ Quizzes 137
Mechanical baseline Test 9, 164
memory aids 68
metacognition 8, 13, 37-38, 164, 180, 184, 233-234
Mix and Match exercises 24, 99, 101-102, 104, 109, 132, 140
mnemonic sentences 68
molecular biology 16, 38, 42, 76, 98-102, 106, 114-117, 126, 128, 130, 143-144, 207, 209, 234, 239-240
Moodle 102, 104, 108-109, 198
multiple choice quizzes 27, 195
multiple sclerosis 131
MyLO assignment 70

N

native English speakers 160, 162, 179
neurotoxins 131
New South Wales (NSW) 131
nomenclature 26-27, 64-65
non-native English speakers 160, 162, 170, 177-179, 227
Nonparametric Statistics 203
nuclear translocation 100

O

Online quizzes 18, 25, 27, 29, 35, 85, 168, 198, 201, 227

P

Pass (P) 6-8, 10, 13-15, 36-40, 76-78, 85, 100, 104-108, 110-115, 118, 121-122, 126-128, 132-135, 140-144, 161, 163-164, 171-172, 177, 180-181, 183-184, 188, 198, 201-207, 209-210, 216, 221-222, 230-237, 239
pathobiology 119
Peer Assisted Study (PASS) 10, 14, 39, 76, 104-106, 108, 112-113, 127, 132-134, 142, 144, 177, 204, 236, 241
Peer Instruction (PI) 10, 12, 17-18, 35, 39, 85, 115, 126, 143, 165, 183-184, 207, 227, 232, 235
phosphorylation 100
Physical Modeling 4, 11, 161-162, 166, 170
physics 1-2, 4-7, 9-16, 21, 36-41, 98-99, 107-108, 115, 126, 143, 160-162, 164-172, 174, 176-184, 186, 188-189, 196, 207, 227-228, 230-238, 241
physics lingo 161
physiology 43, 78, 127, 129-130, 132-137, 142-143, 208, 231-232, 235, 241
Plasmid 99-100, 102, 114, 232
Polymerase Chain Reaction (PCR) 99-100
post-lecture activities 62, 69
power 2-3, 6, 21-22, 107-108, 112, 122-125, 128, 161, 167, 170, 186, 228
pre-lecture strategies 64
Process Oriented Guided Inquiry Learning (POGIL) 31, 39-40, 43, 63, 78, 236

R

rate constant 61
reflection 10, 27, 30, 34, 117, 138, 165
regression 109-112, 204, 207, 217, 220, 222-223, 235
representation 7, 15, 19-20, 30, 42, 78, 147, 161, 163, 183, 237
research 2-10, 12-16, 21-24, 26, 28, 35-38, 40-42, 60-62, 77-78, 100, 107, 114, 117, 126-127, 131, 140, 142-144, 147-149, 161, 163-165, 170-171, 178, 180-184, 186, 195, 206-209, 228-241
revision survey 64

S

Sapir-Whorf hypothesis 185
science language 7, 101, 112, 163, 208, 237
science language learners 7, 163

scientific doing 7, 163
scientific learning 68, 113, 226
scientific literacy 7, 163
scientific terms 6-7, 19, 22, 61, 130, 161, 163
scientific vocabulary 6, 117, 161
sense-making 7, 163-164
service teaching 11, 59-60, 166, 178-179, 240-241
small group work 11
Spearman Method 107
specialist terminology 1, 6
specific heat capacity 66, 72, 162
spontaneous process 61
Statistics Concept Inventory (SCI) 195-198, 201-203, 206, 208, 210-211, 238
statistics course 195-196, 203, 206, 209-210
stem-and-leaf plots 200
stoichiometry 19
student-centred pedagogies 28
Student Evaluation of Teaching and Learning (SETL) surveys 76
student feedback 2, 11, 24-30, 32, 38, 58, 63-64, 67, 69-73, 76-77, 85, 106, 118, 130, 132, 140, 167, 177-178, 201, 205, 228-229, 233
Student Feedback Results (SFR) 177
student learning 14-15, 20, 37, 40, 104, 125-127, 135, 140, 143-144, 180-182, 184, 198, 201, 207, 227-229, 232-233, 236-237, 240-241
student response system 28, 37, 195, 231
studio reaching 17
subcohort 142
summative assessment 228
survey 9, 11, 21-23, 28, 36, 61-65, 73, 76, 107, 117, 119, 137-138, 164, 168-172, 183, 186, 188, 201, 205, 208, 217, 227, 229-230, 236-237
symbolic language 6, 14, 19-20, 35, 38, 235
symbolic notation 19
system 2-3, 10, 12, 17, 20-22, 30, 32, 37, 61, 63, 68, 71-72, 107-108, 131-132, 138-140, 145-146, 148-149, 151-152, 170, 195, 198, 205, 231

T

teaching modules 227
tertiary education 1, 59, 98
The Item Difficulty Index (P) 6-8, 13-15, 36-40, 77-78, 85, 100, 104-108, 110-112, 114-115, 118, 121-122, 126-128, 133-135, 140-144, 161, 163-164, 171-172, 180-181, 183-184, 188, 198, 201-207, 209-210, 216, 221-222, 230-237, 239
The Item Discrimination Index (D) 2, 12-16, 24, 27-30, 35-42, 71, 73, 77-78, 85-87, 114-115, 122-123, 126, 128, 133, 137, 142-143, 146-149, 169, 172, 180-184, 186-194, 206-216, 222-225, 229-238, 240
The Point Biserial Coefficient 124-125, 128
Thermodynamics 36-37, 61, 65-66, 71, 91, 169, 172, 230, 233
traditional 9-11, 20, 23, 28, 31, 33-34, 40, 59, 63, 99, 104-106, 113, 115, 117-118, 132, 138, 162, 165-166, 183, 209-210, 226, 236
transmissive 10-11, 99, 104-106, 115, 117-118, 166
transmissive mode 10-11, 117-118, 166
Turning Point Software 199

U

University of Newcastle (Australia) 1, 5, 10, 12, 129, 131, 134-138, 141, 146, 150, 241
University of Newcastle, Callaghan campus 131-134, 136
University of Newcastle, Ourimbah campus 131-134, 136
University of Tasmania, Hobart 58, 240
University of Technology, Sydney 1, 4-5, 10-11, 160-161, 166, 169, 186, 188, 227, 241
urvoting.com 11-12, 17, 63, 138, 145

V

VotApedia 11-12, 17, 30, 63-65, 76-77, 79, 137-139, 142, 145, 150
VotApedia tool 12, 17

W

WebCT 3, 37, 102, 104, 108-109, 231
WileyPLUS website 168
word matching exercises 58

CPSIA information can be obtained at www.ICGtesting.com
Printed in the USA
269236BV00002B/2/P

9 781613 500620